Preserving Nature in the National Parks

Preserving Nature in the National Parks

A History

RICHARD WEST SELLARS

Yale University Press New Haven & London

This book is published partially through a generous grant from Eastern National Park and Monument Association.

All royalties from this book go to the Albright-Wirth Employee Development Fund to advance the professional skills of National Park Service employees.

Earlier versions of portions of this book have appeared in the *Washington Post*, *Wilderness*, *Journal of Forestry*, *Montana The Magazine of Western History*, and *The George Wright Forum*.

Designed by James J. Johnson and set in Caledonia types by Keystone Typesetting, Inc. Printed in the United States of America by Edwards Brothers, Inc., Ann Arbor, Michigan.

Library of Congress Cataloging-in-Publication Data

Sellars, Richard West, 1935–
Preserving nature in the national parks: a history/Richard West Sellars.
p. cm.
Includes bibliographical references and index.
ISBN 0-300-06931-6 (cloth)
0-300-07578-2 (pbk.)

1. United States. National Park Service—History. 2. National parks and reserves—United States—Management—History. 3. Nature conservation—United States—History. 4. Natural resources—United States—Management—History. I. Title
SB482.A4 S44 1977
333.7′0973—dc21 97-16154

A catalogue record for this book is available from the British Library.
The paper in this book meets the guidelines for permanence and durability of the Committee on Production Guidelines for Book Longevity of the Council on Library Resources.

10 9 8 7 6 5 4 3 2

For John E. Cook,
of the National Park Service,
and for my wife,
Judith Stevenson Sellars

But our national heritage is richer than just scenic features; the realization is coming that perhaps our greatest national heritage is nature itself, with all its complexity and its abundance of life, which, when combined with great scenic beauty as it is in the national parks, becomes of unlimited value. This is what we would attain in the national parks.
—GEORGE M. WRIGHT, JOSEPH S. DIXON, and BEN H. THOMPSON, *Fauna of the National Parks of the United States*, 1933

A national park should represent a vignette of primitive America. . . . Yet if the goal cannot be fully achieved it can be approached. A reasonable illusion of primitive America could be recreated, using the utmost in skill, judgment and ecologic sensitivity. This in our opinion should be the objective of every national park and monument.
—A. STARKER LEOPOLD et al., "Wildlife Management in the National Parks" (The Leopold Report), March 1963

I have always thought of our Service as an institution, more than any other bureau, engaged in a field essentially of morality—the aim of man to rise above himself, and to choose the option of quality rather than material superfluity.
—FREEMAN TILDEN to GEORGE B. HARTZOG, JR., ca. 1971

Many of our problems are historical, but history can't be wiped out.
—JOHN A. CARVER, JR., Assistant Secretary of the Interior, to the National Park Superintendents "Conference of Challenges," Yosemite National Park, October 1963

Contents

Preface

The national park system contains some of the most recognizable natural features on this continent. Such sublime scenery as the Grand Canyon, the Yosemite Valley and Half Dome, Old Faithful, and the Teton Mountains are familiar to millions. These and other landscape icons of the system symbolize the romantic nationalism that has always sustained public support of national parks. The celebrated geography of high mountains and vast open spaces has helped perpetuate a kind of "From the New World" fantasy—the parks as virgin land—which has long enhanced America's national park movement.

In part because of their great symbolic beauty, the national parks have been easy to write about with enthusiasm and effusion. Early studies, and many works published by the National Park Service itself, have tended to glorify the founding fathers of the Park Service and extol the expansion of the system. Although the founders deserve much credit, and expansion has certainly been important, the appeal of this zealous approach has diminished. Recent scholars have written not so much about how the parks came to be created and who promoted them, but about how they were treated after their establishment. As a study of the management of nature in the parks, this book belongs in the latter category.

Nature preservation—especially that requiring a thorough scientific understanding of the resources intended for preservation—is an aspect of park operations in which the Service has advanced in a reluctant, vacillating way. The analysis that follows is at times critical of the Park Service. Indeed, writing National Park Service history from within runs some risks—but it also enjoys certain advantages. As a historian with the Park Service for more than two decades, I have had the opportunity to observe the Service closely and to refine my understanding of its culture and corporate psyche. I have had ready access to the files and to the thoughts of fellow employees and retirees. Each individual held strong opinions about what the Service has been and should be, and discussed national park management with a high degree of candor and openness.

It is my hope that this book will inform future efforts of the Park Service, the public, and the Congress to address national park issues. To prepare for the future, it is important first to analyze the past with as much clarity and impartiality as can be mustered.

RICHARD WEST SELLARS
Santa Fe, New Mexico

Introduction

There was a time, through the middle of the twentieth century, when the national parks reigned indisputably as America's grandest summertime pleasuring grounds. Managed by the National Park Service after 1916, the spectacular mountains, canyons, forests, and meadows set aside to provide for the public's enjoyment appealed tremendously to a public increasingly mobile and enamored of sightseeing and automobile touring. To make the parks accessible to millions of vacationers, graceful winding roads were constructed, with romantic names like Going to the Sun Highway or Trail Ridge Road. Huge rustic hotels built of log and stone, such as Yellowstone's Old Faithful Inn and Grand Canyon's El Tovar, welcomed overnight visitors to the parks. In hotel lobbies or in nearby museums, courteous park rangers stood ready to take eager visitors on nature walks—out into the crisp, pine-scented mountain air to enjoy the wonders of trailside forests and streams. In parks such as Sequoia and Yellowstone, visitors fed bears along roadsides or gathered in specially constructed bleachers to watch rangers feed bears; and at dusk each summer a firefall of burning embers cascaded from the heights of Yosemite's Glacier Point.

Enjoying immense popularity, the national park system grew to include areas in the East and Midwest while continuing to expand in the West, where it had begun and where the majority of the older and more famous parks are located. Preserving remnants of the wild landscapes of the frontier, the parks were from the beginning a part of frontier history and romantic western lore. Most national parks were truly isolated, and the nearby lands were little developed and sparsely populated. For many park rangers, working in the vast, majestic parks seemed a kind of lingering frontier experience: long assignments in remote backcountry areas; horse patrols along park boundaries; and primitive, wood-heated log cabins to house the family.

In recent decades the situation has changed. Today many national parks, although still beautiful, are marred by teeming, noisy crowds in campgrounds, visitor centers, grocery stores, and restaurants, and by traffic jams on roads and even on trails. The push and shove of hordes of tourists and the concomitant law-enforcement problems eclipse the unalloyed pleasure that earlier generations surely experienced. Bland, unattractive modern structures have replaced many of the rustic park administrative buildings and tourist facilities of the past. Housing for rangers and other employees frequently is comparable at best to urban tract homes. Spending fewer hours in the backcountry, rangers more and more find themselves encumbered by office work. In addition, the National Park Service has experienced a decline in its discretionary authority, as it must confront powerful, competing special-interest groups that watch every move. With their natural conditions degraded by air and water pollution, accelerated development of adjacent lands, extensive public use, and inappropriate actions taken by the Park Service itself, the national parks have become the focus of angry battles over environmental issues that often result in litigation by batteries of lawyers.

Set within the context of this broad array of national park operations and issues, the environmental and ecological aspects of national park management—principally the treatment of natural resources—form the central theme of this volume. This study traces over many decades the interaction of bureaucratic management with the flora, fauna, and other natural elements in parks of scenic grandeur that are intended to be visited and enjoyed by large numbers of people yet in some fashion to be preserved. The book begins in the late nineteenth century, when the earliest parks were established and when management principles were first set in place. It extends almost to the present day, when the recency of issues—many yet unresolved—flattens the perspective from historical to journalistic.

The first chapter, based mostly on secondary sources, summarizes the period before the National Park Service was founded in 1916. Subsequent chapters, drawing extensively on primary documents such as internal memoranda and reports (most of them never before researched), include an analysis of the legislative history of the act creating the Service and the intent of that act. Next is a detailed account of national park management over time—in effect, how the act was implemented: the growth and development of the park system during the 1920s, the rise of biological science within the Park Service, and the bureau's triumphs in recreational planning and development during the New Deal. The story continues with the World War II–era retrenchment and declining interest in biological science, the Park Service's reinvigoration during the tourism explosion of the 1950s, and the Service's clash with the environmental movement of the 1960s and 1970s even as it began to revitalize its biology programs.

Rather than presenting a broad study of conservation history, this book focuses chiefly on internal Park Service concerns—on how a bureau created to administer the national parks arrived at management policies for natural resources, put them into practice, and in time changed many of them. Especially since its wildlife biology programs gained strength in the 1930s, the Park Service has not been of one mind about how to care for the parks' natural resources; philosophical and political disagreements have been persistent.

Indeed, present-day management of nature in the parks differs substantially from that in the early decades of national park history—the most fundamental difference being the degree to which science now informs the Service's natural resource practices. And in an age of ecological science, the extent to which the Service manages parks in a scientifically informed way may be seen as a measure of its true commitment to ecological principles. It may also be a measure of its commitment to the ethical purposes always implicit in the national park concept, but more recognized today—principally, that within these specially designated areas native species will be protected and preserved.

It might be assumed that management of national parks with the intent of preserving natural conditions would necessarily require scientific knowledge adequate to understand populations and distributions of native species and their relation to their environment, and that without such information the parks' natural history is fraught with too many questions, too many unknowns. At least from the early 1930s, this argument was voiced within the Park Service's own ranks. Yet it has not been the view of park management throughout most of the Service's history.

Because National Park Service decisionmaking most often has not been scientifically informed, the question arises as to what kind of management has been taking place, and why. Thus, in this study the management of nature in the parks is placed in the larger context of overall park operations and bureaucratic behavior—in ecological terms, it is placed within its "bureaucratic habitat."

The analysis is also expanded to embrace the corporate culture of the National Park Service. Of special interest is the extensive development of the parks for what might be called recreational tourism—pleasure travel focusing on appreciation of nature and enjoyment of the out-of-doors. This overriding emphasis on tourism development fostered the ascendancy of certain professions such as landscape architecture and engineering, and largely determined the Service's organizational power structure and its perception of what is right and proper for the parks.

Implementing its 1916 congressional mandate as it deemed proper, the Park Service engaged in two basic types of nature management: development for tourism, and what was later termed natural resource management. Both affected natural conditions in the parks. Although not generally perceived as such, tourism development amounted to a kind of de facto management of nature. It often resulted in extensive alterations to natural conditions, especially along road and trail corridors, and in pockets of intensive use (for example, along the south rim of the Grand Canyon or throughout the Yosemite Valley). By contrast, natural resource management involved direct, purposeful manipulation of natural elements—including the nurturing of favored species, such as bison, bears, and game fish; or the reduction of populations of so-called problem species, such as certain predators or tree-killing insects. These two basic types of nature management, factors in park management from the earliest decades, affected plants and animals throughout the parks, to the point of eliminating some species. This alteration of natural conditions created perplexing situations for later generations of managers and scientists.

The central dilemma of national park management has long been the question of exactly what in a park should be preserved. Is it the scenery—the resplendent landscapes of forests, streams, wildflowers, and majestic mammals? Or is it the integrity of each park's entire natural system, including not just the biological and scenic superstars, but also the vast array of less compelling species, such as grasses, lichens, and mice? The incredible beauty of the national parks has always given the impression that scenery alone is what makes them worthwhile and deserving of protection. Scenery has provided the primary inspiration for national parks and, through tourism, their primary justification. Thus, a kind of "facade" management became the accepted practice in parks: protecting and enhancing the scenic facade of nature for the public's enjoyment, but with scant scientific knowledge and little concern for biological consequences.

Criticism of this approach began in the 1930s, increased during the environmental era of the 1960s and 1970s, and is commonly voiced today. Nevertheless, facade management based largely on aesthetic considerations remains quite acceptable to many. Far easier to undertake, and aimed at ensuring public enjoyment of the parks, facade management has long held more appeal for the public, for Congress, and for the National Park Service than has the concept of exacting scientific management.

Yet aesthetics and ecological awareness are not unrelated. Whatever benefit and enjoyment the national parks have contributed to American life, they have undoubtedly intensified the aesthetic response of millions of people to the beauty and the natural history of this continent—a response that could then be pleasurably honed in more ordinary surroundings closer to home. Beyond the sheer enjoyment of scenery, a heightened aesthetic sensibility may have inspired in many a deeper understanding of, and concern for, the natural environment. This benefit defies quantification, but surely it has had consequences of immense value, both for individuals and for the nation.

The persistent tension between national park management for aesthetic purposes and management for ecological purposes underlies much of the following narrative.

CHAPTER 1

Creating Tradition: The Roots of National Park Management

> It is important to do something speedily [about the Yellowstone park proposal], or squatters and claimants will go in there, and we can probably deal much better with the government in any improvements we may desire to make for the benefit of our pleasure travel than with individuals.—JAY COOKE, October 30, 1871

On March 1, 1872, Congress established Yellowstone Park—the world's first "national park," more than two million acres located mostly in the northwest corner of present-day Wyoming—to be preserved and managed by the federal government for the enjoyment and benefit of the people. In the midst of the Gilded Age's rampant exploitation of public lands, the concept of federally managed parks protected from the extractive uses typical of the late-nineteenth-century American West abruptly gained congressional sanction. Yellowstone's awesome natural phenomena had inspired a political phenomenon.

Despite its eventual worldwide implications, the Yellowstone Park Act attracted minimal public attention; Congress only briefly debated the bill, giving little indication of what it intended for the park. The act came during an era when the federal government was aggressively divesting itself of the public domain through huge railroad land grants and, among others, homestead, mining, and timber acts. Although a few Americans were voicing concern about the preservation of nature and decrying the exploitation of natural resources, no broad, cohesive conservation movement existed in 1872. Yet the proposal to save the wonders of Yellowstone (principally the great falls of the Yellowstone River and the spectacular geysers) triggered legislation creating what was until very recently the largest national park in the contiguous forty-eight states.

The origin of the national park idea—who conceived it, and whether it was inspired by altruism or by profit motives—has been disputed. One account became a revered part of national park folklore and tradition: that the idea originated in September 1870 during a discussion around a campfire near the Madison Junction, where the Firehole and Gibbon rivers join to form the Madison River in present-day Yellowstone National Park. Nearing the conclusion of their exploration of the Yellowstone country, members of the Washburn-Doane Expedition (a largely amateur party organized to investigate tales of scenic wonders in the area) had encamped at Madison Junction on the evening of September 19. As they relaxed and mused around their wilderness campfire, the explorers recalled the spectacular sights they had seen. Then, after considering the possible uses of the area and the profits they might make from tourism, they rejected the idea of private exploitation. Instead, in a moment of high altruism, the explorers agreed that Yellowstone's awe-inspiring geysers, waterfalls, and canyons should be preserved as a public park.[1] This proposal was soon relayed to high political circles, and within a year and a half Congress established Yellowstone Park.

Through the decades, as the national park concept gained strength and other nations followed the American example, the Madison Junction campfire emerged as the legendary birthplace not just of Yellowstone but of all the world's national parks. Although the Yosemite Valley had been established as a California state park from federally donated lands in 1864 and the term "national park" had been occasionally used in the past, the belief that the national park idea truly began around a wilderness campfire at the Madison Junction evolved into a kind of creation myth: that from a gathering of explorers on a late summer evening in the northern Rocky Mountains came the inspiration for Yellowstone National Park, the prototype for hundreds of similar parks and reserves around the world. In the wilderness setting and with a backdrop of the vast, dramatic landscape of the western frontier, the origin of the national park idea seemed fitting and noble. Surely the national park concept deserved a "virgin birth"—under a night sky in the pristine American West, on a riverbank, and around a flaming campfire, as if an evergreen cone had fallen near the fire, then heated and expanded and dropped its seeds to spread around the planet.[2]

The campfire story may be seen in another light, however. Romantic imagery aside, the element of monopolistic business enterprise is notably absent from the traditional campfire story—the profit motive obscured by the altruistic proposal for a public park. In fact, corporate involvement with America's national parks has its roots in that same 1870 Washburn-Doane

Expedition and campfire discussion. Amid the great rush to settle the West after the Civil War, the Northern Pacific Railroad Company was by 1870 planning to extend its tracks from the Dakota Territory across the Montana Territory. With easiest access to Yellowstone being from the north, through Montana, the company believed that once it extended its tracks west it could monopolize tourist traffic into the area.

Alert to this potential, Northern Pacific financier Jay Cooke took special interest in the scenic Yellowstone country. In June 1870 he met in Philadelphia with Nathaniel P. Langford, politician and entrepreneur, who subsequently proceeded to Montana and, with Northern Pacific backing, successfully promoted the Washburn-Doane Expedition. This exploration of Yellowstone began in August, with Langford as a participant. Still supported by the Northern Pacific, Langford followed up the expedition with lectures to audiences in Montana and in East Coast cities, extolling the wonders of Yellowstone, while local boosters in Montana began promoting the park idea. The following year, the railroad company subsidized artist Thomas Moran's participation in the expedition into Yellowstone led by geologist Ferdinand V. Hayden. Moran's sketches from the Hayden Expedition (his impressive paintings were not yet completed) were displayed in the Capitol in Washington as part of the campaign to enact the Yellowstone legislation.[3]

Ever advancing Northern Pacific interests, Jay Cooke sought to ensure that the Yellowstone country did not fall into private hands, but rather remained a federally controlled area. He observed in October 1871, just before the legislation to create a park was introduced, that a government "reservation" (or park) would prevent "squatters and claimants" from gaining control of the area's most scenic features. Government control would be easier to deal with; thus, it was "important to do something speedily" through legislation.[4]

Subsequent to the Hayden Expedition, the Northern Pacific lobbied for the park with swift success: the Yellowstone bill was introduced on December 18, 1871, and enacted the following March. Like most future national parks, Yellowstone remained under the jurisdiction of the Department of the Interior, which managed the public lands of the West. The park's immense size came not because of an effort to preserve vast tracts of undisturbed wilderness, but largely as a result of recommendations by Ferdinand Hayden, who sought to include the lands most likely to contain spectacular thermal features.

From the first, then, the national parks served corporate profit motives, the Northern Pacific having imposed continuous influence on the Yellowstone park proposal, beginning even before the 1870 expedition that gave birth to the campfire tradition.[5] With their land grants stretching across the continent, American railroads were already seeking to establish monopolistic trade corridors. By preventing private land claims and limiting competition for tourism in Yellowstone, the federal reservation of the area served, in effect, as a huge appendage to the Northern Pacific's anticipated monopoly across southern Montana Territory.

Indeed, in historical perspective, the 1872 Yellowstone legislation stands as a resounding declaration that tourism was to be important in the economy of the American West. A matter of considerable consequence in the Yellowstone story, the collaboration between private business and the federal government fostered a new kind of public land use in the drive to open the West. A portion of the public domain was reserved for largely non-consumptive use, with unrestricted free enterprise and exploitation of natural resources prohibited. With magnificent scenery as the principal fount of profit, tourism was emerging in the nineteenth century as an economic land use attractive to business investment. The success of such investment depended in part on the preservation of scenery through prevention of haphazard tourism development and other invasive commercial uses such as mining and lumbering. The possibility of federal cooperation to manage vast scenic areas in the West and control development appealed to the Northern Pacific—and soon to other tourism interests.

Over time, accommodation for tourism in the national parks would become truly extensive and have enormous consequences for the parks. It is a significant, underlying fact of national park history that once Yellowstone and subsequent park legislation codified the commitment to public use and enjoyment, managers of the parks would inevitably become involved in design, construction, and long-range maintenance of park roads, trails, buildings, and other facilities. Allowing tourists to stay overnight in the parks meant that hotels, restaurants, campgrounds, garbage dumps, electrical plants, and water and sewage systems would sooner or later be seen as indispensable. The practical necessities for accommodating thousands, then millions, of tourists (the primary constituents of the national parks and a key source of political support) would increasingly demand park management's attention and seriously affect allocation of funds and staffing.

Moreover, such developmental concerns would foster a capitalistic, business-oriented approach to national parks, emphasizing the number of miles of roads and trails constructed, the number of hotel rooms and campsites available, the number of visitors each year, and the need for continued tourism development. Principally in an effort to ensure public enjoyment, nature itself would be manipulated in the national parks; to a large extent, natural resource management would serve tourism purposes.

Growth of the National Park Concept

After Yellowstone there was no rush to create additional national parks. Yellowstone came into existence during the Indian wars on the northern plains and in advance of extensive white settlement of the West—not an auspicious time and place for tourism. Created nearly two decades before the 1890 census announced the closing of the frontier, Yellowstone came close to becoming a historical anomaly rather than a trendsetter in public land policy.

In 1875 Congress established Mackinac National Park—the second such park, but one that occupied only about a thousand acres of Mackinac Island, located at the westernmost point of Lake Huron and the site of Fort Mackinac, a small U.S. Army post. Already a federal presence on the island, the army managed the national park until 1895, just after the fort was deactivated. With the army's departure, the State of Michigan was persuaded to operate Mackinac as a state park; thus the park lost its "national" designation.[6] Mackinac seems not to have advanced the national park concept. The park was created in part because the army was conveniently available to manage the area, and it was redesignated after the army departed.

In fact, after Yellowstone nearly two decades passed before the national park idea spread to any significant degree. In 1890 Congress established two large parks in California: Sequoia and Yosemite. (The latter comprised the High Sierra country surrounding the 1864 Yosemite grant to the State of California; the grant remained under state control until 1906, when it was added to the national park.) Also in 1890 came establishment of the relatively small General Grant National Park, four square miles of giant sequoia forest (incorporated into Kings Canyon National Park in 1940).

Following the flurry of new parks in 1890, Congress waited nine years before creating another large natural park—Mount Rainier, in 1899. Thus, by the turn of the century—nearly three decades after Yellowstone—there were in existence no more than four large parks, plus General Grant National Park. (In Arkansas, the "Hot Springs Reservation," established in 1832 as a small, approximately four-square-mile preserve containing thermal springs of medicinal value, was also managed by the Department of the Interior; not until 1921 would this preserve be designated a national park.)

In the early twentieth century, prior to establishment of the National Park Service in 1916, the number of parks began to grow steadily: Crater Lake (1902), Wind Cave (1903), Sully's Hill (1904), Mesa Verde (1906), Platt (1906), Glacier (1910), Rocky Mountain (1915), Hawaii (1916), and Lassen Volcanic (1916).

Led by the Northern Pacific, Southern Pacific, and Great Northern railroad companies and influenced by the rising concern for conservation, tourism interests exerted a powerful influence in creating new parks. Like Yellowstone, parks such as Sequoia, Yosemite, Mount Rainier, and Glacier were to a large degree the result of the railroads' political pressure.[7] In addition to the economic potential of tourism in the national parks, other profit-oriented motives arose. For instance, the Northern Pacific promoted the Mount Rainier legislation, which enabled the company to swap its lands in the park for more valuable timberlands elsewhere. And owners of nearby agricultural lands (including railroad companies) urged establishment of Sequoia and Yosemite, in part to protect watersheds through high-country forest conservation, which would benefit their investments in the valleys below. This factor was evidenced in the enabling legislation for each park, which referred to the parks as "reserved forest lands."[8]

Beginning with Ferdinand Hayden's proposal to include all of Yellowstone's major thermal features, the early national parks helped establish the important precedent that immense tracts of land could be put to use as public parks. Both the concern for watershed protection and an emerging interest in preserving wilderness (a consideration in the 1890 Yosemite legislation) seem to have influenced Congress to include in Sequoia and Yosemite much more land than necessary for the protection of key scenic features. Mount Rainier National Park, by comparison, was made sufficiently large to encompass a huge scenic feature—a splendid glacier-capped volcanic mountain—in addition to wilderness and watershed concerns, heroic scenery fostered the creation of some exceedingly large parks. Given the size of many of the parks, the extensive tourism development that would take place would still leave thousands of acres of undeveloped park "backcountry"—a factor that would become increasingly important in national park preservation concerns.

Vast and spectacularly beautiful, Yellowstone provided not only the first but also the most enduring image of a national park: a romantic landscape of mountains, canyons, abundant wildlife, and fantastic natural phenomena. Surely the park's great size and the fame and popularity it achieved by the early twentieth century helped fix the fledgling national park idea in the American mind. Moreover, the spacious, majestic scenery being preserved in such parks as Yellowstone, Sequoia, and Yosemite aroused a strong sense of patriotism and a romanticized pride in America's most dramatic landscapes, helping stimulate national tourism and the park movement.[9]

Yet Congress did not define national parks as being solely large natural areas. In addition to General Grant, other small parks were created. Platt National Park, about eight hundred acres of a mineral springs area in south-central Oklahoma, and Sully's Hill National Park, a few hundred acres of low, wooded hills in eastern North Dakota, had more in common with the defunct, diminutive Mackinac National Park—and all three varied substantially from the standards of size and scenery set by Yellowstone and the other large parks.[10]

In another deviation from the large natural park standard, Mesa Verde National Park was created to preserve impressive archeological sites. Moreover, in June 1906, within a few days of Mesa Verde's establishment, Congress passed the Antiquities Act, providing for creation of "national monuments"—a different kind of federal land reservation, which would in time be added to the national park system. The monuments were to include areas of importance in history, prehistory, or science, and be no larger than necessary to protect the specific cultural or scientific values of concern. The result of political pressure brought mainly by anthropologists seeking to prevent vandalism to the nation's prehistoric treasures, the act authorized the President to establish national monuments by proclamation (the same means by which national forest reserves were then created).

During President Theodore Roosevelt's administration, and as the conservation movement gathered steam, this means of establishing federal reserves without further congressional authorization promptly brought about the creation of numerous monuments, among them Devils Tower (1906), Chaco Canyon (1907), Muir Woods (1908), Mount Olympus (1908), and Grand Canyon (1908). Placed under the administration of the Interior, Agriculture, or War departments, depending on where the monuments were located, almost all of the national monuments would eventually be made part of the national park system and would come under the same management policies, with public use as the principal focus.

The Antiquities Act made illegal the unauthorized taking of antiquities from federal lands and legislated penalties for punishment of violators. It also authorized a permit system, allowing excavation of antiquities within the monuments only for professional research purposes.[11] Other than these stipulations, the act gave no directions for day-to-day management of the monuments. Although the act was passed because of concern for preserving prehistoric sites, it was also used to set aside especially scenic lands, such as the Grand Canyon and Mount Olympus. These two monuments established another significant precedent—that the Antiquities Act could be used to preserve very large tracts of public land, far larger than its supporters (or opponents) had envisioned.[12]

The Antiquities Act was conceived with much less concern for tourism and public use than were the national parks, and many monuments remained neglected and inaccessible for years by other than archeologists (the most striking exception being Grand Canyon National Monument, managed by the U. S. Forest Service until 1919). However, this neglect did not reflect a permanent policy of limited use and strict preservation of the monuments. In time, and under favorable funding and staffing circumstances, they would be targeted for extensive recreational tourism development, similar to that in the national parks. But with majestic scenery that could attract swarms of tourists, and with specific mandates for nature preservation, the national parks themselves—rather than the national monuments—would dominate the formulation of natural resource management policy in the growing park system.

Characteristically, the national parks featured outstanding natural phenomena: Yellowstone's geysers, Sequoia's and General Grant's gigantic trees, and Hot Springs' thermal waters. Such features greatly enhanced the potential of the parks as pleasuring grounds that would attract an increasingly mobile American public interested in the outdoors. Writing about Yellowstone in 1905, more than three decades after its establishment as a park, President Theodore Roosevelt observed that the preservation of nature was "essentially a democratic movement," benefiting rich and poor alike.[13] Even with the prospect of monopolistic control of tourist facilities, the national park idea was a remarkably democratic concept. The parks would be open to all—undivided, majestic landscapes to be shared and enjoyed by the American people.

Moreover, in preventing exploitation of scenic areas in the rapacious manner typical for western lands in the late nineteenth century, the Yellowstone Park Act marked a truly historic step in nature preservation. The act forbade "wanton destruction of the fish and game" within the park, and provided for the

> preservation, from injury or spoilation, of all timber, mineral deposits, natural curiosities, or wonders within said park, and their *retention in their natural condition* (emphasis added).[14]

Natural resources in Yellowstone and subsequent national parks were to be protected—by implication, the sharing would extend beyond the human species to the flora and fauna of the area. Indeed, this broad sharing of unique segments of the American landscape came to form the vital core of the national park idea, endowing it with high idealism and moral purpose as it spread to other areas of the country and ultimately around the world.

Toward the end of the nineteenth century, an emerging interest in protecting wilderness was apparent in national park affairs. In the mid-1880s, the congressional defeat of proposals by railroad and mining interests to build a railroad through northern Yellowstone and reduce the park in size underscored the importance of both the park's wildlife and its wild lands—thus moving beyond the original, limited concern for specific scenic wonders of Yellowstone. Interest in more general preservation within the parks also was evident with the creation of Yosemite National Park in 1890, which included extensive and largely remote lands surrounding the Yosemite Valley. John Muir, a leading spokesman for wilderness, sought to preserve the High Sierra in as natural a state as possible and was especially active in promoting the Yosemite legislation. For the new park, Muir envisioned accommodating tourism in the Merced River drainage (which encompasses the Yosemite Valley), while leaving the Tuolumne River drainage to the north (including the Hetch Hetchy Valley) as wilderness, largely inaccessible except on foot or by horseback.[15]

With the early national park movement so heavily influenced by corporate tourism interests such as the railroad companies, Muir's thinking regarding Yosemite and other parks stands out as the most prominent juncture between the park movement and intellectual concerns for nature's intrinsic values and meanings, as typified by the writings of Ralph Waldo Emerson and Henry David Thoreau. Moreover, except perhaps for Muir's efforts to understand the natural history of California's High Sierra, the advances in ecological knowledge taking place by the late nineteenth century had little to do with the national park movement. Busy with development, the parks played no role in leading scientific efforts such as the studies of plant succession by Frederic Clements in Nebraska's grasslands, or by Henry C. Cowles along Indiana's Lake Michigan shoreline.[16] Once national parks became more numerous and more accessible, an ever-increasing number of scientists would conduct research in them. But within national park management circles, awareness of ecological matters lay in the distant future, and genuine concern in the far-distant future.

In many ways, the national park movement pitted one utilitarian urge—tourism and public recreation—against another—the consumptive use of natural resources, such as logging, mining, and reservoir development. In the early decades of national park history, the most notable illustration of this conflict came with the controversy over the proposed dam and reservoir on the Tuolumne River in Yosemite's Hetch Hetchy Valley. The vulnerability of this national park backcountry, which John Muir wanted preserved in its wild condition, was made clear when Congress voted in December 1913 to dam the Tuolumne in order to supply water to San Francisco. Even though located in a national park, the Hetch Hetchy Valley was vulnerable to such a proposal in part because it was indeed wilderness, undeveloped for public use and enjoyment. The absence of significant utilitarian recreational use exposed the valley to reservoir development, a far more destructive utilitarian use.

This relationship Muir recognized; he had already come to accept tourism and limited development as necessary, and far preferable to uses such as dams and reservoirs. Yet the extensive, unregulated use of the state-controlled Yosemite Valley alerted Muir and his friends in the newly formed Sierra Club to the dangers of too much tourism development (and provided impetus for adding the valley to the surrounding national park in 1906).[17] Still, the national park idea survived and ultimately flourished because it was fundamentally utilitarian. From Yellowstone on, tourism and public enjoyment provided a politically viable rationale for the national park movement; concurrently, development for public use was intended from the very first. Becoming more evident over time, the concept that development for public use and enjoyment could foster nature preservation on large tracts of public lands would form an enduring, paradoxical theme in national park history.

Resorts, Spas, and Early National Parks

From the very beginning, the Northern Pacific Railroad Company's interest in the Yellowstone legislation anticipated the direction that national park management would take. The legendary 1870 campfire discussion itself foretold that the public would want to see Yellowstone—that "tourists and pleasure seekers" would visit the area. Certainly during the more than four decades between Yellowstone's establishment in 1872 and the creation of the National Park Service in 1916, management of the parks for public use and enjoyment was the overriding concern. The enthusiastic promotion of recreational tourism in the parks generated a tradition that the Park Service would eagerly embrace.[18] Given the extraordinary dominance of this concern, surely it reflected the chief intent behind the national park concept.

By the time of Yellowstone's establishment in 1872 as a "public park or pleasuring-ground," tourism activity in other parts of the country had established important precedents for development in the national parks. To accommodate tourism in scenic areas or around health-giving thermal springs, entrepreneurs, often backed by railroad companies, had built resort facilities, some of them fancy, others primitive. Although early national park management seems not to have looked collectively to such resorts for guidance, a pattern nevertheless evolved as, more than anything else, park development simulated resort development. Areas selected for intensive public use in the national parks took on the appearance of resorts, and effectively served that purpose.[19]

Emerging soon after the era of canal building, railroads played a major role in boosting tourism in the United States. Completed in 1825, the Erie Canal had made Niagara Falls more accessible to East Coast populations; and the coming of the railroad to western New York soon secured for Niagara its position as the nation's premier resort. More comfortable and faster than stagecoaches and canal boats, railroads enabled tourists to reach scenic attractions at increasing distances from the principal population centers. The growth of urban middle and upper classes after the Civil War, the desire to escape the summer heat of cities, and feverish postwar railroad construction accelerated interest in traveling for pleasure. In addition to Niagara, resorts and spas were developed in the Catskills and the Adirondacks, and at Lake George, Saratoga, White Sulphur Springs, and other scenic areas. Hotels and cabins were clustered near thermal springs, or situated with views of spectacular scenery. Relatively primitive at first, facilities improved as the popularity and prosperity of resorts increased; in some resorts, accommodations evolved into imposing, luxurious hotels. Yet also present at many scenic spots were ramshackle souvenir shops or cabins—the very type of small-time entrepreneurial activity that Jay Cooke sought to exclude from Yellowstone through establishment of a government "reservation."

At midcentury, railroads began to penetrate the upper Midwest, making this area accessible to travelers and extending farther west the phenomenon of popular tourist resorts.[20] Beyond the Mississippi River, early resort development (much of it in California and Colorado) included two places that would become important in national park history: the thermal springs of Hot Springs, Arkansas, and the Yosemite Valley of California.

These sites—one a spa and the other a dramatically scenic valley—formed the nuclei of the only present-day national parks that were in some way set aside before Yellowstone. Both places experienced intensive resort development.

In the decades after the 1832 establishment of the Hot Springs Reservation, primitive bathhouses were clustered around the springs, but the Civil War stalled development. Yet by 1873 the city of Hot Springs had six bathhouses and two dozen boarding houses and hotels. The first luxury accommodations appeared when the Arlington Hotel opened in 1875, about the time the first railroad line reached the city. By the late nineteenth century, the reservation's "Bathhouse Row" would begin to undergo extensive renovation, including a landscaping program of formal gardens and promenade and the replacement of older structures with imposing new bathhouses. The new Bathhouse Row became a national attraction and launched the heyday of therapeutic bathing at Hot Springs.[21]

Meanwhile, the Yosemite Valley also was experiencing extensive development. The 1864 federal grant to the State of California required that the valley and the nearby Mariposa Grove of big trees be managed as a park for the public's "use, resort, and recreation." Surrounded by a dramatic, vertical landscape of granite cliffs and majestic waterfalls, Yosemite's rather flat valley floor served as a kind of viewing platform from which to enjoy the scenery. And despite the cautionary recommendations of Frederick Law Olmsted's 1865 report on the new state park, much of the valley floor was developed to satisfy the whims of the tourist industry. Under lax state management, the Yosemite Valley emerged as a crazy quilt of roads, hotels, and cabins, and pastures and pens for cattle, hogs, mules, and horses. Tilled lands supplied food for residents and visitors, and feed for livestock; irrigation dams and ditches supported agriculture; and timber operations supplied wood for construction, fencing, and heating. Amid the clutter of development stood one "luxury" hotel, the three-and-a-half-story Stoneman House, built in 1886.[22]

Mackinac National Park underwent a similar assault. The park was created for the "benefit and enjoyment of the people," and was further dedicated as a "national public park, or grounds" for the people's "health, comfort, and pleasure"—the public enjoyment factor receiving even more emphasis than it had in the Yellowstone legislation. Accordingly, this small park underwent heavy resort development. Construction of summer homes, cottages, and hotels in and adjacent to the park (including the impressive thirteen-hundred-bed Grand Hotel, which opened in 1887) made Mackinac a popular destination for vacationers from midwestern and eastern cities.[23]

Yellowstone, however, provided the most striking example of resort-style development in a national park. Its potential was recognized not only in the Madison Junction campfire discussion, but in public statements prior to passage of the Yellowstone Park Act. To Congress it was claimed that the park would become a "place of great national resort" and should be dedicated to "public use, resort and recreation." The *New York Times* editorialized that "in all probability" the mineral springs "with which the place abounds" would soon prove to "possess various curative powers," and claimed that physicians believed the park would "become a valuable resort for certain classes of invalids." Yellowstone could become a spa rivaling those in Europe and attracting people from "all parts of the world to drink the waters, and gaze on picturesque splendors." Such potential fostered the declaration in the Yellowstone Park Act that the area was to be a "public park or pleasuring-ground for the benefit and enjoyment of the people," and the provision allowing the secretary of the interior to lease park lands for "building purposes" and for "accommodation of visitors."[24]

Although eschewing private ownership of Yellowstone, the Northern Pacific anticipated profits from its virtual monopoly on travel into the park. Extensive development did not occur as quickly as the railroad company hoped, however. The national financial crisis of 1873 forced the company to postpone construction of its rail line across Montana. But even as early as 1871, before the park was established, small, primitive hotels (some including thermal-water bathing facilities) were in place near Mammoth Hot Springs and the Lower Geyser Basin. Soon a few crude log structures sprang up near other park attractions. Precisely the kind of development that Jay Cooke disdained, these meager efforts ultimately failed. Not until 1883 did the Northern Pacific rails penetrate to within a few miles of Yellowstone's northwestern boundary. There tourists could transfer to stagecoaches and be driven into the park. Within the year, a consortium backed by the Northern Pacific opened the park's first large hotel at Mammoth Hot Springs.[25]

A parsimonious and often indifferent Congress gave Yellowstone minimal support during its earliest years. Then, in 1883, army engineers began to oversee construction of park roads. Shortly thereafter, to better organize and strengthen park operations, the army was assigned overall management of Yellowstone, its troops arriving in August 1886. (In the 1890s the army also would be placed in charge of Yosemite, Sequoia, and General Grant national parks.) The engineers soon began construction of permanent buildings for Fort Yellowstone, adjacent to the new hotel at Mammoth Hot Springs. The major construction effort was the Grand Loop Road, a 152-mile system routing visitors from one spectacle to another—Mammoth Hot Springs, Norris Geyser Basin, Old Faithful, Yellowstone Lake, the Grand Canyon of the Yellowstone River, and others. By the early part of the twentieth century a system totaling approximately four hundred miles of "mountain roads" (primitive to improved) was nearly complete in Yellowstone.

With the development of the road system and with the backing of the Northern Pacific, large, imposing hotels were built near scenic wonders such as Yellowstone Lake, Old Faithful, and the Grand Canyon of the Yellowstone. Tourists could thus travel safely through the park's vast wilderness landscapes to enjoy civilized pleasures in a variety of grand hotels featuring the kinds of amenities already familiar to the traveling public in the East and Midwest. To promote its investments, the Northern Pacific advertised its route as the "Yellowstone Park Line." By 1910 expenditures for tourist-facility improvements reached a million dollars; and by about 1912 the facilities had produced an equivalent amount of revenue. The federal government also paid its share: by 1906 it too had invested one million dollars in the road system.[26] Hotel and road construction in Yellowstone—far and away the primary management accomplishment during the early decades—essentially paralleled nineteenth-century American resort development.

Other national parks soon experienced the kind of development under way in Yellowstone. Indeed, the enabling legislation for subsequent national parks provided for leasing land to be used for public accommodation, in some instances with wording taken verbatim from the 1872 Yellowstone Act. Roads, trails, public accommodations, and administrative facilities were constructed in the new parks. Usually primitive at first, such developments were followed by well-engineered and architecturally impressive construction. For instance, before the creation of Glacier National Park in 1910, several small tourist accommodations opened in the area. Soon after, the Great Northern Railway Company (principal lobbyist for the park) began construction of large rustic-style hotels and smaller mountain chalets. At a cost of about half a million dollars each, the Great Northern built the Glacier Park Lodge and the Many Glacier Hotel. Its chain of attractive chalets enabled visitors to sleep comfortably overnight while on their way by horseback across the mountainous park.[27]

Clustered village-type developments, as at Yellowstone's Mammoth Hot Springs, emerged as the norm. Typically located near favored scenic attractions, these developed areas featured splendid hotels. By 1915, just over a decade after establishment of Crater Lake National Park, a large rustic hotel, the Crater Lake Lodge, opened to the public. Along with other facilities, this stone, log, and frame hotel was perched on the crater rim, overlooking the deep, sapphire-blue lake. Mount Rainier's picturesque Paradise Inn, with gabled roof and rustic lobby, was completed in 1917 and became a showpiece of the park. Beginning as a very modest accommodation, Sequoia's Giant Forest Lodge, located in the Giant Forest Village, would be enlarged and modernized in the early 1920s.[28] Prior to the establishment of Platt National Park in 1906, that area's thermal springs had already spawned a popular health and recreation resort. Soon after it gained national park status, Platt was further developed, in an architecturally picturesque style, with roads, trails, pavilions, landscaped grounds, and quaint bridges.[29]

Parkwide planning gradually emerged, guiding the placement of roads, trails, tourist accommodations, and administrative facilities. Construction of the Yellowstone road system marked the earliest broad-scale approach. Other parks soon followed, and in 1910 Secretary of the Interior Richard Ballinger called for "complete and comprehensive plans" for national parks. The importance of carefully controlled tourism development was underscored by the 1914 appointment of Mark Daniels as first "general superintendent and landscape engineer" for the national parks.[30] Daniels, a landscape architect and designer of subdivisions in San Francisco, became extensively involved in park planning in Sequoia, Mount Rainier, Crater Lake, Glacier, and especially Yosemite.

In remarks to a 1915 national park conference, Daniels stressed the need for systematic planning. Tellingly, he explained how the implementation of park plans depended in part on the successful promotion of tourism. He commented that the parks "can not get a sufficient appropriation at present from Congress to develop . . . plans and put them on the ground as they should be, therefore we are working for an increase in attendance which will give us a justification for a demand upon Congress to increase the appropriations that are necessary to enable us to complete these things." Daniels' comments suggested a kind of perpetual motion that would become a significant aspect of national park management, where tourism and development would sustain and energize each other through their interdependence.

Already, increasing tourism meant to Daniels "the inevitableness of creating villages in the parks." He stated that the Yosemite Valley was almost in "the category of cities," and that it needed "a sanitary system, a water-supply system, a telephone system, an electric light system, and a system of patrolling." It was clear to him that several national parks would soon "absolutely demand some sort of civic plan" to take care of their visitors.[31]

In the early part of the century, with the rising power of the newly created U.S. Forest Service the need to develop the national parks gained a particular sense of urgency. It was vital to ensure the parks' popularity and prevent their transfer to the Forest Service, which stressed extraction and consumption of natural resources rather than protection of natural conditions or scenic landscapes.[32] Furthermore, as the automobile era rapidly advanced, the national parks would face demands for use and enjoyment from a public more mobile than ever. This situation would foster the continuation of development trends begun by early park management.

The Management of Nature

With park development simulating resort development elsewhere in the country, perhaps the most distinguishing characteristic of the parks was their extensive, protected backcountry. The location of roads, trails, hotels, and other recreational tourism facilities only in selected areas meant that much of the vast park terrain escaped the impact of intensive development and use. Offering the only real possibility for preservation of some semblance of natural conditions, these relatively remote areas would constitute the best hope of later generations seeking to preserve national park ecological systems and biological diversity.

In contrast to tourism development, no precedent existed for intentionally and perpetually maintaining large tracts of land in their "natural condition," as stipulated in the legislation creating Yellowstone and numerous subsequent parks.[33] (The 1916 act creating the National Park Service would require that the parks be left "unimpaired"—essentially synonymous with maintaining "natural conditions.") Moreover, the early mandates for individual parks were not so much the ideas of biologists and other natural scientists, but of politicians and park promoters. There seems to have been no serious attempt to define what it meant to maintain natural conditions. This key mandate for national park management began (and long remained) an ambiguous concept related to protecting natural scenery and the more desirable flora and fauna.

Management of the parks under the mandate to preserve natural conditions took two basic approaches: to ignore, or to manipulate. Many inconspicuous species (for example, small mammals) were either little known or of little concern. Not intentionally manipulated, they carried on their struggle for existence without intentional managerial interference. The second approach, however, involved extensive interference. Managers sought to enhance the parks' appeal by manipulating the more conspicuous resources that contributed to public enjoyment, such as large mammals, entire forests, and fish populations. Although this manipulation sometimes brought about considerable alteration of nature (impacting even those species of little concern), park proponents did not see it that way. Instead, they seem to have taken for granted that manipulative management did not seriously modify natural conditions—in effect, they defined natural conditions to include the changes in nature that they deemed appropriate. Thus, the proponents habitually assumed (and claimed) that the parks were fully preserved.

Most national parks came into existence already altered by intensive human activity, Yellowstone being the least affected. All had experienced some impact from use by Native Americans, whose exclusion from lands they had long utilized was, in effect, reinforced by the establishment of national parks as protected natural areas to be enjoyed by tourists. (Attempts to understand Indian influences on prepark conditions would not begin until the final decades of the twentieth century.) Before their designation as parks in 1890, both Sequoia and Yosemite had been subjected to mining, lumbering, and widespread grazing, with summer herds of sheep and cattle thoroughly cropping some areas. Prospectors had worked on the slopes of Mount Rainier before it became a park, and the initial legislation allowed their activity to continue. In addition to the construction of homes, lodges, and camps, the area to become Glacier National Park had been subjected to mining activity and even oil exploration.[34]

Going well beyond mere protection of flora and fauna, early park managers manipulated natural resources at will. In order to increase sportfishing opportunities, for example, fish populations were extensively manipulated through stocking, which became a common practice in the early national parks. Stocking at Yellowstone began in 1881, less than a decade after the park's establishment, when native cutthroat trout were moved to fishless waters from other areas of the park. Eight years later, nonnative brook trout and rainbow trout were placed in park waters, the army captain in charge of the park at the time stating his hope that stocking would enable the "pleasure-seeker" to "enjoy fine fishing within a few rods of any hotel or camp." These initial efforts soon led to widespread stocking programs, supported by hatchery operations both inside and outside Yellowstone's boundaries.

At Oregon's Crater Lake, William Gladstone Steel, the chief advocate for national park designation, initiated fish stocking in 1888, fourteen years before the park was established. Steel placed rainbow trout in the previously fishless, nearly two-thousand-foot-deep lake. Stocking was uninterrupted by establishment of the park in 1902. Similarly, beginning in the 1890s, native and nonnative fish were stocked throughout Yosemite. Other parks, among them Sequoia and Glacier, developed stocking programs, establishing an early and explicit precedent for extensive manipulation of national park fish populations.[35]

Although the early national parks were set aside principally for the enjoyment of special scenery rather than for wildlife preservation, wildlife quickly became recognized as a significant feature of the parks. Game species, highly prized by hunters, also proved to be the most popular for public viewing. Spokesmen for sporting organizations, particularly the Boone and Crockett Club, and George Bird Grinnell, the editor of the outdoor magazine *Forest and Stream,* encouraged public interest in national park wildlife, and in the 1880s began promoting Yellowstone as a refuge wherein bison and other large mammals should be protected.[36] Such factors helped crystalize early national park wildlife policy, as managers focused on protecting populations of bear and ungulates (the hoofed grazing animals such as elk, moose, bison, deer, and bighorn sheep). Yellowstone, with its impressive variety of large, spectacular mammals (today caricatured as "charismatic megafauna" or "glamour species") would remain the most notable wildlife park in the contiguous states, dominating the formulation of wildlife policy in the national parks.

As they did with fish populations, early national park managers manipulated the populations of large mammals. They sought, for example, to protect favored wildlife species from predators. Native park fauna such as wolves, coyotes, and mountain lions (cougars) were perceived as threats to the popular ungulates and were hunted—the parks were not to be "shared" with such predators. Park rangers and army personnel trapped or shot these animals, or permitted others to do so. Yellowstone's predator control program began very early, accelerated when the army arrived, and continued for decades. Other parks, such as Mount Rainier, Yosemite, and Sequoia, followed suit. Well before the Park Service came into being, predator control had become an established management practice. This effort would ultimately reduce wolves and mountain lions to extinction in most parks.[37]

Park managers also sought to protect favored wildlife species from poachers, who ignored boundaries and hunted the big-game species inside the new preserves—a problem from earliest times in most national parks. The park with the greatest wildlife populations, Yellowstone suffered serious poaching problems, with large numbers of elk, bison, and other mammals taken during the early years. Having virtually no staff, the park could not effectively combat poaching, a situation that changed substantially after the army's arrival in 1886. The military would soon increase attempts to control poaching in Sequoia and Yosemite, while civilian staffs contended with the problem in other parks. In the 1890s poaching threats to bison sparked a campaign led by George Grinnell to strengthen protection of Yellowstone's wildlife. Grinnell helped bring about passage in 1894 of the Act to Protect the Birds and Animals in Yellowstone National Park, establishing penalties and law-enforcement authority to protect animals and other natural resources—measures that had not been provided by the legislation creating the park. This important act set a precedent for similar protection to be extended to other parks.[38]

Although protection of popular large mammals from poachers and predators gradually became more effective, several of the popular species were themselves directly manipulated. Early park managers in Yellowstone employed methods akin to ranching. Fearing the extinction of bison in the United States, the park initiated a program in 1902 that included roundups, winter feeding, and culling of aged animals. To prevent starvation when heavy snows made foraging difficult, winter feeding was extended in 1904 to elk, deer, bighorn sheep, and other ungulates.[39] Bear feeding in Yellowstone began almost spontaneously, along roadsides and at hotel garbage dumps, where the public soon realized that bears could be viewed close up. Feeding at the dumps evolved into a more formalized evening program (soon known as "bear shows") with bleachers for visitors, who were protected by armed rangers. Elsewhere in the parks, bears that threatened the public were often shot or shipped to zoos around the country.[40]

In the early decades of the national parks, forests and grasslands both became special management concerns. In line with accepted policies on other public lands (and on private lands), suppression of forest fires in the parks quickly emerged as a primary objective. As with efforts to prevent poaching, army manpower in Yellowstone, Yosemite, and Sequoia ensured some success with the suppression policy. Disagreement with this policy was occasionally voiced by a few who believed that continuous suppression would allow too much dead, fallen debris to accumulate on the forest floor and eventually fuel unnaturally large, destructive fires. However, because this idea was expressed only intermittently and there was no sustained attempt to put it into practice, it had no real impact. Fire suppression became a deeply entrenched policy in the national parks.[41]

Like fighting poachers and fires, protecting the parks from grazing by domestic livestock was challenging and dangerous. Local ranchers, taking advantage of the remoteness of many park lands, drove their livestock to summer grasslands in the High Sierra Nevada—a practice begun before parks in that area were created and continued after they came into being. John Muir's famous denunciation of sheep as "hoofed locusts" reflected the anger he felt about the threats to native flora and fauna from grazing and trampling. As with its attempts to curtail poaching and fires, the army made a special effort to prevent encroachment of both sheep and cattle in the parks it oversaw. Usually a formidable presence in the parks only during the summer months (which coincided with the grazing season), the troops detained livestock drovers, confiscated their weapons, and sometimes herded their cattle out of the parks at an inconvenient distance from where the drovers were forced to exit.[42] This firm antigrazing policy would at times be compromised by the political influence of western stockmen, who angrily objected to restrictions on grazing public lands and who would form a hard core of resistance, even to the very concept of national parks.

The treatment of nature in the early national parks set precedents that would influence management for decades. Later referred to as "protection" work, activities such as combating poaching and grazing, fighting forest fires, killing predators, and manipulating fish and ungulate populations constituted the backbone of natural resource management. These duties fell to army personnel in parks where the military was present and ultimately, in all parks, to the field employees who were becoming known as "park rangers." As their efforts to curtail poaching and livestock grazing required armed patrol, the rangers rather naturally assumed additional law-enforcement responsibilities. In addition, they assisted the park superintendents by performing myriad other tasks necessary for daily operation of national parks, such as dealing with park visitors and with concessionaires. Deeply involved in such activities, the park rangers were destined to play a central role in the evolution of national park management.[43]

That the national park idea embraced the concept of mostly nonconsumptive land use did not mean that the parks were nonutilitarian. On the contrary, the history of the early national park era suggests that a practical interest in recreational tourism in America's grand scenic areas triggered the park movement and perpetuated it. With Northern Pacific and other corporate influence so pervasive, it is clear that the early parks were not intended to be giant nature preserves with little or no development for tourism. Products of their times, the 1872 Yellowstone Act and subsequent legislation establishing national parks could not be expected to be so radical. Only with the 1964 Wilderness Act would Congress truly authorize such preserves—three-quarters of a century after John Muir had advocated a similar, but not statutory, designation for portions of Yosemite.

Still, it is important to recognize that, although extensive manipulation and intrusion took place in the parks, fundamentally the national park idea embraced the concept of nurturing and protecting nature—a remarkable reversal from the treatment of natural resources typical of the times. Yet with the parks viewed mainly as scenic pleasuring grounds, the treatment of fish, large mammals, forests, and other natural resources reflected the urge to ensure public enjoyment of the national parks by protecting scenery and making nature pleasing and appealing; and it was development that made the parks accessible and usable. Even with legislation calling for preservation of natural conditions, park management was highly manipulative and invasive. "Preservation" amounted mainly to protection work, backed by little, if any, scientific inquiry.

The National Park Service would inherit a system of parks operated under policies already in place and designed to enhance public enjoyment. The commitment to accommodating the public through resort-style development would mean increasing involvement with the tourism industry, a persistently influential force in national park affairs as the twentieth century progressed. Management of the parks in the decades before the advent of the National Park Service had created a momentum that the fledgling bureau would not—and could not—withstand.

CHAPTER 2

Codifying Tradition: The National Park Service Act of 1916

Economics and esthetics really go hand in hand.—MARK DANIELS, 1915

Following a few tentative efforts early in the twentieth century, a campaign to establish a national parks bureau began in earnest in 1910 and continued for six years. In June 1916, as the effort neared success, an article entitled "Making a Business of Scenery" appeared in *The Nation's Business*. Written by Robert Sterling Yard, in charge of the campaign's promotional literature, the article championed the scenery of America's national parks as an "economic asset of incalculable value" if managed in a businesslike way. Yard wrote that, as an example, Switzerland "lives on her scenery," having made it a "great national business" (although diminished by the war ongoing in Europe). The Canadians too had entered "the scenery business" with businessmen in charge of their national parks. It seemed high time that Americans developed such a business. Yard wrote:

> We want our national parks developed. We want roads and trails like Switzerland's. We want hotels of all prices from lowest to highest. We want comfortable public camps in sufficient abundance to meet all demands. We want lodges and chalets at convenient intervals commanding the scenic possibilities of all our parks. We want the best and cheapest accommodations for pedestrians and motorists. We want sufficient and convenient transportation at reasonable rates. We want adequate facilities and supplies for camping out at lowest prices. We want good fishing. We want our wild animal life conserved and developed. We want special facilities for nature study.[1]

The rule rather than an exception, "Making a Business of Scenery" reflected the pervasive utilitarian tenor of the drive to establish the National Park Service. Proponents saw the parks as scenic recreation areas that should be vigorously developed for public use and enjoyment to help the national economy and improve the public's mental and physical well-being, thereby enhancing citizenship and patriotism. The various widely scattered parks and monuments had no centralized, coordinated management. National park supervisors officially reported to the secretary of the interior, but in reality to a "chief clerk," who was involved with diverse bureaus in the Department of the Interior and paid scant attention to the parks. To many, it seemed obvious that a new bureau was needed to manage these areas in an efficient, businesslike way.

Concluding a long period of aggressive politicking, Congress created the National Park Service in August 1916. Analysis of the "legislative history" of the National Park Service Act (referred to as the Service's "Organic Act") illuminates the rationale that has ever since underlain national park management. The act established a fundamental dogma for the Park Service—the chief basis for its philosophy, policies, and decisionmaking.

Repeatedly since passage of the National Park Service Act, critics of various management practices in the parks have cited the act's principal mandate: that the parks be left "unimpaired for the enjoyment of future generations." Often they have asserted that the Park Service violates the spirit and letter of the act by not preserving natural conditions. Particularly since the environmental era of the 1950s, 1960s, and 1970s, they have contended that the Service's primary mandate has always been the preservation of nature, and that the Park Service has misunderstood the congressional mandate to leave the national parks unimpaired.[2]

But in fact, the legislative history of the Organic Act provides no evidence that either Congress or those who lobbied for the act sought a mandate for an exacting preservation of natural conditions. An examination of the motivations and perceptions of the Park Service's founders reveals that their principal concerns were the preservation of scenery, the economic benefits of tourism, and efficient management of the parks. Such concerns were stimulated by the boosterism prevalent in early national park history, and they in turn greatly influenced the future orientation of national park management.

Advocates and Opponents

The drive to establish a national parks bureau was led by four individuals: a horticulturalist, a landscape architect, a borax industry executive, and a young lawyer. The campaign began through the efforts of J. Horace

McFarland, a nationally prominent horticulturalist, urban planner, and leader of the "city beautiful" movement to improve the attractiveness of America's growing cities. McFarland's career was built on his passion for landscape aesthetics and the social benefits to be derived from parks and other professionally landscaped areas. From 1904 until 1925 he served as president of the American Civic Association, an organization that promoted intelligent planning and development to make, as McFarland described it, "American cities, towns, villages and rural communities clean, more beautiful and more attractive places in which to live." McFarland and the association had participated in the move to preserve Niagara Falls and had supported shade-tree planting, city parks, and recreation areas, while opposing the growing billboard blight along the nation's roadsides. Under his guidance the American Civic Association became the leading professional organization supporting the national park legislation. The association would be instrumental in drafting the Organic Act's statement of the parks' principal purpose, and, in the winter of 1911–12, would recommend that the proposed new bureau be designated the National Park Service.[3]

McFarland's contacts extended to cabinet officials and to President William Howard Taft, through whom he initiated the legislative campaign. Alarmed about the proposal to create a reservoir in Yosemite National Park's Hetch Hetchy Valley to supply water to San Francisco, McFarland suggested to Secretary of the Interior Richard Ballinger in May 1910 that the national parks needed a "general, intelligent and logical supervision." McFarland believed that strong, coordinated oversight could best defend the parks against threats such as the damming of Hetch Hetchy, one of Yosemite's outstanding scenic areas.[4]

In December 1910, when Secretary Ballinger formally recommended a national parks bureau to President Taft, he employed a statement prepared by McFarland and reflecting utilitarian goals. Ballinger proposed a bureau of "national parks and resorts," to include a "suitable force of superintendents, supervising engineers, and landscape architects, inspectors, park guards, and other employees." Subsequently, Taft incorporated these views into his message to Congress, advocating that the parks be preserved for the public's "edification and recreation." He called for sufficient funds to "bring all these natural wonders within easy reach of the people"—a means of improving the parks' "accessibility and usefulness."[5]

That same year, at the suggestion of Secretary Ballinger, McFarland recruited the nationally known landscape architect Frederick Law Olmsted, Jr., to the campaign. Son of the principal founder of American landscape architecture, Olmsted, on graduating from Harvard, became an apprentice and then a partner in his father's firm. Adding to his credentials, the younger Olmsted had helped found Harvard's academic program in landscape architecture and served as president of the American Society of Landscape Architects, a professional organization he had helped establish.[6] He had, as well, served on the executive board of the American Civic Association. The year 1910 marked the beginning of Olmsted's long association with the national parks, one that would last until the 1950s, when he became involved in the momentous Echo Park controversy in Dinosaur National Monument.

In line with McFarland's views, Olmsted believed that national park management lacked coordinated leadership and was "mixed up and rather inefficient." Consequently, the parks were in poor condition, without an "orderly or efficient means" of being protected. This "chaotic" situation could, however, be addressed through the "proper businesslike machinery" of sound management. Good national park leadership, Olmsted judged, could be found in a "Western man"—one familiar with the country where all of the national parks were then located, and a man "of really large caliber, of executive ability . . . with the instincts of a gentleman."[7]

Early in 1915 such an individual appeared on the scene when Stephen T. Mather, a Chicago businessman, joined the campaign for a national parks bureau. Mather had political instincts and strategic abilities that complemented those of McFarland and Olmsted. Polished and at ease with the rich, powerful, and famous, he displayed ardent enthusiasm—his biographer referred to him as the "Eternal Freshman," who was "almost pathologically fraternal."[8] In 1917 Mather would be officially appointed as the National Park Service's first director. But beginning in early 1915, after a friend, Secretary of the Interior Franklin K. Lane, asked him to serve as his assistant in the national park legislative drive, Mather devoted his impressive talents and much of his own money (he had amassed personal wealth as head of a borax company with mines in the West) to boosting the national parks. As a chief goal, Mather sought public acceptance and political support for the parks through opening them to greater use. Along with his politicking, he helped finance the purchase of the Tioga Pass Road to make Yosemite's high country accessible to the automobile-touring public. For a while he even paid the salary of the national parks' chief publicist, Robert Sterling Yard.[9]

Serving at Mather's side was his assistant, Horace M. Albright, a young graduate of the University of California and the Georgetown University Law School, who shared Mather's enthusiasm for the parks and gave energetic, intelligent support to the legislative campaign. Albright proved highly effective within the Washington political system, and his skills were crucial to the passage of the act.[10] The youngest of the founding fathers, Albright would resolutely proclaim the founders' concepts of national parks to succeeding generations.

During the legislative drive, from 1910 to 1916, the Department of the Interior sponsored three national park conferences. The first general meetings to be held in the decades since Yellowstone's establishment, these conferences brought together influential people from inside and outside the federal government. Especially because the Organic Act's legislative history includes few official congressional hearings and reports, the conference proceedings provide important evidence of the intentions behind the act. Repeatedly during these conferences, supporters depicted the parks as scenic places for public recreation, enjoyment, and edification—indeed, one participant described the national parks movement as a "campaign for natural scenery." At the first conference (1911, in Yellowstone), Secretary of the Interior Walter L. Fisher's opening remarks drew attention to the crucial need for the parks to attract more visitors; he directed that, in addition to park administration, the meeting should be devoted to concession and transportation matters related to accommodating tourists.

Significantly, the lists of conference participants and agendas reflected what had already become a major factor in national park affairs: the various interest groups that sought to generate business in or near the parks and thus to apply political and economic leverage to shape the character and direction of national park management. The conferences were absorbed with the concerns of these groups. For instance, building on their long involvement with the parks, railroad companies sent numerous spokesmen to the meetings, as did smaller-scale concessionaires who operated facilities in the parks. Representing the industry that would ultimately have the greatest impact on national parks, the fledgling automobile associations were especially prominent at the 1912 conference in Yosemite and the 1915 conference in Berkeley and San Francisco. To one or both of these meetings, officials of the American Automobile Association, the Southern California Automobile Association, and the Automobile Dealers Association of Southern California, among others, came to promote increased public use of the parks.[11]

From within the government came national park superintendents, engineers, landscape architects, and other officials of the Interior Depart-

ment. Even the secretary attended the 1911 and 1912 conferences, and Mather officially represented the secretary at the Berkeley meeting in 1915. Foresters and entomologists represented the scientific professions. At the 1911 meeting, for example, an "expert lumberman" and an "expert in charge of forest insect investigations" advised how to protect forests from fires and insects. Forests, the participants were told, form the "attractive feature in a landscape," and damage to trees "must be considered . . . on the basis of the commercial value" as well as the "aesthetic and educational value."[12]

Most prominent among the railroad delegates at the 1911 conference was Louis W. Hill, president of the Great Northern Railway Company and enthusiastic promoter of the newly established Glacier National Park. Hill's company already had plans for extensive tourist accommodations in and adjacent to Glacier. His remarks to the conference attested to the railroad industry's clear profit motive in its concern for the national parks: the railroads were "greatly interested in the passenger traffic to the parks" and, with lines already built nearby for "regular traffic," each passenger to the national parks represented "practically a net earning." Because his railroad operated in the northern tier of states, Hill was much aware of Canada's aggressive national park promotion, which he claimed diverted many tourists from United States parks. Echoing a prevailing theme in the conferences, he encouraged more advertising of American parks, arguing that such publicity would divert visitors otherwise bound for Canada or Europe.[13]

Throughout the meetings, proponents urged that the parks no longer be abandoned to the haphazard supervision of an Interior Department clerk burdened by other responsibilities. At Yosemite in 1912, Secretary Fisher acknowledged that the Interior Department had "no machinery whatever" to deal with the national parks. He noted that the department lacked the expertise to handle matters such as engineering, park development, landscape management, forestry, sanitation, and construction. Indeed, his office and that of the chief clerk had "never really been equipped to handle these matters, [even] if it had been possible to give them the necessary time and attention."[14]

At the 1912 conference, Richard Watrous, secretary of the American Civic Association, supported maintenance of the parks as "playgrounds," and introduced a resolution supporting creation of a national parks bureau. He believed the bureau could provide the parks with a "definite, systematic, and continuous policy" to improve efficiency of administration. Watrous stated that concern for efficiency was being brought "very promi-

nently" before business leaders and the people, because the White House was giving more attention to the "general subject of economy and efficiency than ever before."[15]

To accommodate visitors, the scenic parks needed to improve accessibility and facilities—practical requirements that put the skills of engineers and landscape architects in demand, as repeatedly emphasized during the conferences. At the 1912 meeting, John Muir recommended utilization of these professions in the parks, a reflection of an increased (but wary) tolerance of tourism late in his life. And Robert B. Marshall, a geographer with the U.S. Geological Survey, who would later serve briefly as chief administrator of the national parks, believed that the proposed bureau should have an engineer as director and that park superintendents should also be engineers, or at least have a substantial knowledge of engineering. Such individuals could ensure "proper maintenance of the great recreation and playgrounds."[16]

Secretary Fisher's successor, Franklin Lane, shared Marshall's views, and in the spring of 1914 created the position "general superintendent and landscape engineer," to provide administrative leadership for the national park system. Initially held by San Francisco landscape architect Mark Daniels, this position replaced the chief clerk as the department's coordinator of parks. Daniels remained in the job until December 1915. He was succeeded by Robert Marshall, whose title became "general superintendent of national parks."[17] These positions were forerunners of the National Park Service directorship.

Addressing the 1915 conference as general superintendent, Daniels declared an urgent need to develop national parks for tourism: "There are roads to be built, and there are bridges to be built, and there are trails to be built, and there are hotels to be built, and sanitation must be taken care of." Earlier he had told the same conference that the only two justifications for the national parks were "economics and esthetics." These factors, he claimed, "really go hand in hand" and were "so intimately related that it is impossible to disassociate them." For Daniels, the function of national parks was like that of city, county, and state parks, because all required the "supplying of playgrounds or recreation grounds to the people."[18]

Daniels spent much of his time as general superintendent seeking to increase public accommodations in the parks with what one observer described as "artistic development" and the "adaptation of the town-planning method." Daniels informed the 1915 conference that he had planned and designed new development for the Yosemite Valley and other national park "villages," where tourist and administrative facilities were to be concen-

trated, and that he had planned new roads and other developments in several of the larger parks.[19]

Robert Marshall, his successor, shared Daniels' eagerness to develop the national parks for tourism. At the 1911 conference Marshall had advocated tennis, golf, and skiing facilities as means of improving the "national playgrounds" and competing with Europe for American dollars. He also recommended that firebreaks be cut throughout the parks, and stated that thousands of cattle could graze the parks each season without doing harm.[20] Marshall elaborated on these ideas at hearings before the House Committee on the Public Lands in the spring of 1916, claiming that the number of visitors to the national parks could be greatly increased—that through businesslike management the parks could pay for themselves: "In a few years we will have an enormous population in the national parks. It is worthwhile. It does not cost much money, and eventually the people will pay for the pleasure we give them."[21]

When the National Park Service Act finally passed in 1916, nearly half a century had elapsed since the Yellowstone Act of 1872. In part, the delay in creating a parks bureau stemmed from concerns about increasing the size and cost of the federal government. Strongly favoring a central national parks office, participants at the conferences scarcely considered the possibility of managing the parks *without* creating a new bureau. Yet Secretary Fisher cautioned the 1912 conference that there was "considerable sentiment" among congressmen to avoid creating a bureau; instead, they would simply designate within the Interior Department an office having as its sole responsibility the management of national parks. As one congressman later put it, an aggrandizing parks bureau might expand and spend ever larger sums of money—it would "start in a small way and soon get up to a big appropriation." Congressman William Kent of California reiterated such concerns in early 1916 when he wrote to Richard Watrous of the American Civic Association that the "most difficult bump to bump is the proposition so blithely entered into of obtaining another bureau," a matter that should be "approached with fear and trembling."[22]

Away from the conferences, the U.S. Forest Service voiced objections calculated to impede passage of the Organic Act. As a bureau of the Department of Agriculture created to manage the already expansive national forest system, it recognized the proposed national parks bureau as a competitor. Forest Service attitudes reflected bureaucratic territorialism and the belief that management of the parks and national forests involved

similar principles. Gifford Pinchot, the first director of the Forest Service and a premier power in natural resource politics, steadfastly opposed the concept of a parks bureau. Earlier he had received support from Secretary of the Interior James R. Garfield, who reported in 1907 that development and maintenance of the parks and the forests were "practically the same," and that roads and trails, fire protection, and game management were all problems that were "being studied in a broader and better way in the Forest Service" than within the park system. Garfield's recommendation that the parks be placed under the Forest Service was rejected by park proponents, who insisted more vehemently than ever that a bureau be established specifically to manage national parks.[23]

With the Park Service legislative campaign under way in earnest, Pinchot wrote to Olmsted asserting that the national forests already provided recreation for about as many people as did the national parks, and that the methods of protecting the parks and forests were similar. To Pinchot, both were "great open spaces," essentially the same except that certain uses were not allowed in the parks. Thus a parks bureau would involve "a needless duplication of effort." Henry S. Graves, director of the Yale School of Forestry before succeeding Pinchot as head of the Forest Service, took a conciliatory stance, agreeing to the establishment of a national parks bureau. However, Graves sought to maintain a clear distinction between national parks and national forests. He wrote to Horace McFarland in March 1916 that he hoped to avoid "hybridizing" through the establishment of "so-called parks" where (just as in the national forests) lumbering, mining, grazing, and water-power developments were allowed. Very likely Graves had in mind parks such as Glacier, where the enabling legislation permitted railroad rights-of-way and water reclamation projects. True national parks, Graves wrote, should be set aside exclusively for the "care and development of scenic features and . . . for the enjoyment, health and recreation" of the people.[24]

Indeed, Graves agreed with Pinchot that duplication between forest and park management would be inevitable, and he wrote that he absolutely opposed any attempt to "dismember the National Forests." He recommended strict qualifications for national parks to resist park proposals on lands that had value for "other purposes," a strategy that would prevent many public lands from becoming parks. Graves would not only keep the national park system smaller, but also place the new bureau within the Agriculture Department, where the Forest Service could exert greater influence. As he described it, this arrangement would promote a close relationship with the Biological Survey, the Bureau of Entomology with its

"experts in insects," and the Bureau of Public Roads with its "corps of trained road engineers."[25]

In contrast, Horace McFarland informed Graves early in the national park campaign that he saw a distinct difference between park and forest management. To McFarland, a national forest was "the nation's woodlot," while a national park was "the nation's playground." He fervently believed the two kinds of management did not mix well—it was unwise for a bureau that managed forests on a sustained-yield commercial basis also to manage national parks. The parks should not be the "secondary object" of the agency overseeing them; this would make park management, as he explained to Pinchot, "incidental, and therefore inefficient." McFarland had no confidence in Pinchot's sense of Forest Service "harmony" with the "economic and sociological purpose" of the national parks. He asserted that there was "very good reason to suppose" that the attitude of the Forest Service was "inimical to the true welfare of the national park idea as serving best the recreational needs of the nation."[26]

McFarland's apprehension about Forest Service opposition remained strong. As congressional hearings on the legislation proceeded in the spring of 1916, he wrote to Olmsted on the difficulty of overcoming the Forest Service's attempt to "emasculate this Park Service proposition." He pointed out that Stephen Mather believed "there is a constant and continual hostility in the Forest Service against the whole idea of National Parks as such."[27]

As the legislative campaign progressed, opposition also arose from western livestock ranchers, concerned about permanent loss of grazing privileges in present and future parks. William Kent, an influential congressman who would soon introduce the national park bill in the House, had a ranch of his own in Nevada and a number of rancher constituents and friends. He backed their cause, arguing that grazing had a beneficial effect on parks by preventing forest fires (a generally accepted belief at the time). Kent would allow grazing, yet ensure that public use areas were preserved "so far as their beauties are concerned."[28]

Although privately opposed to grazing livestock in the parks, Stephen Mather's public stance was influenced by his need for Kent's support in the legislative campaign. Thus, Mather compromised with the ranchers and told Congress in April 1916 that permission to graze was a "very proper" amendment to the bill. In accord with Kent's views, his chief concern was to prevent grazing in areas frequented by park visitors. Mather recalled that the parks' general superintendent, Robert Marshall, had asserted that "a certain amount of grazing in those areas where it will not interfere with the campers' privileges is perfectly proper." Mather testified that he concurred with this assessment, noting also the hazards of allowing grasses and other plants to build up to the point where they could ignite and feed destructive fires. Although initially the Senate would vote against grazing in the parks, inclusion of the provision helped secure House support for the legislation. Mather, Albright, and others found it expedient to agree to the provision despite their private opposition.[29]

The Statement of Purpose

In 1917, looking back on the campaign to establish the National Park Service, Horace McFarland commented that the Organic Act's statement of the national parks' basic purpose was the only item that proponents of the act could not have done without. It was, he said, the "essential thing" in the legislation, and "the reason we feel that [the Organic Act] is worthwhile." Even as the campaign first got under way, Frederick Law Olmsted, Jr., who would author the statement of purpose, had believed it would be of "vital importance" and urged that the purpose of the parks be defined in "broad but unmistakable terms."[30]

In its final form, the statement declared the parks' "fundamental purpose" to be

> to conserve the scenery and the natural and historic objects and the wild life therein and to provide for the enjoyment of the same in such manner and by such means as will leave them unimpaired for the enjoyment of future generations.[31]

Despite its ambiguities, especially in regard to potential conflicts between preserving parks and opening them to public use, this mandate would become the National Park Service's touchstone—its chief point of reference for managing parks. The charge to leave the parks "unimpaired" in effect perpetuated the charge to preserve "natural conditions" as stated in the 1872 Yellowstone legislation and subsequent national park enabling acts. And as "unimpaired" set the 1916 mandate's only actual standard, it became the principal criterion against which preservation and use of national parks have ever since been judged.[32]

The earliest draft of the statement of purpose was prepared by McFarland, Olmsted, and others during a December 1910 meeting of the American Civic Association. Somewhat vague, the draft merely stated that the parks would not be used "in any way detrimental or contrary to the purpose for which dedicated or created by Congress."[33] But the acts by which

Congress had established the different national parks were sufficiently varied and ambiguous that Olmsted was concerned. Later that month he urged an explicit statement that would "safeguard or confirm" the purposes of all of the parks. He recommended a declaration that the parks were

> agencies for promoting public recreation and public health through the use and enjoyment . . . [of the parks] and of the natural scenery and objects of interest therein.[34]

Olmsted expressly recommended that "scenery" be included in this statement—the attribute of parks that was his greatest concern and that contributed most to their public appeal. Writing to Henry Graves of the U.S. Forest Service, McFarland declared this statement to be "for the first time, a declaration of the real purpose of a National Park." Echoing Olmsted, he believed it to be "of extreme importance that such purpose be declared in unmistakable terms." With such remarks, his early 1911 correspondence to both Graves and Gifford Pinchot suggests how important it was to McFarland to distinguish national parks from national forests.[35]

It is significant that for nearly five years—from December 1910 until November 1915—this working version of the statement of purpose defined the national parks primarily as agencies to promote "public recreation and public health" through "use and enjoyment" of the scenery and its special features. This strongly utilitarian concept of the fundamental mission of the parks was not only in accord with the attitudes expressed at the national park conferences and elsewhere, but also may have served to counter the Forest Service's brand of utilitarianism.

Even with its emphasis on recreation and health, Olmsted's statement of purpose raised concern that it might tie the hands of the proposed new bureau. In late 1911 Secretary of the Interior Walter Fisher wrote to McFarland that the statement had the potential to curtail managerial discretion in the parks; it could "embarrass the proposed bureau" and cause questions to be "constantly raised as to the character of each act undertaken." He added that "any one who claimed that any particular action would be detrimental to the value of the parks might undertake to restrain the bureau from the proposed action."[36] (Indeed, Fisher's concern foreshadowed criticism of national park management that would recur many times through the ensuing decades, and that would use as its justification the Organic Act's statement of purpose.)

McFarland responded to Fisher that without the statement even the new bureau itself might not understand the basic purpose of the parks it was being created to manage. To McFarland, the national parks needed a

"Gibraltar"—a statement of their "true and high function"—in order to defend the parks against those who would damage them. He asserted that if Yosemite National Park had had a proper definition of purpose in its legislation, the threat to inundate the Hetch Hetchy Valley might not have arisen. Although he preferred wording that would "avoid the difficulties of too great restriction upon administrative discretion," Secretary Fisher acquiesced and agreed to retain the statement as written.[37]

In the autumn of 1915, after several years of failure to get the national park bill enacted, the American Civic Association redrafted the legislation and arrived at wording close to what would appear in the final bill. At first, the review draft that Richard Watrous of the association forwarded to Olmsted in October 1915 contained essentially the same utilitarian, recreation-and-health definition of parks as before, except that it called for "*conservation* of the scenery and of the *natural and historic* objects" found in the parks (emphasis added)—a more protective statement than before, and more explicit about the kinds of objects (or resources) to be conserved.[38]

But Olmsted concluded that a different version was in order. He reviewed Watrous' draft and responded on November 1, 1915, with a revised statement of purpose that omitted reference to the parks as agencies for public recreation and health. Retaining the commitment to conservation of natural and historic features and to use and enjoyment of the parks, he strengthened the statement by adding that the parks should be left "unimpaired for the enjoyment of future generations." Within two weeks, his new version appeared in the working draft of the bill, the only substantive change being the addition of "wild life" to the short list of resources to be conserved.[39] This revision would be incorporated into the final legislation with only minor alterations.

In recommending wording that required the parks to be left unimpaired, Olmsted did not indicate that he considered the intent of the new version to be a particularly significant departure from that expressed in the bill's earlier public recreation and health statement. In fact, he suggested the new wording almost offhandedly in the last paragraph of a three-page letter, asking "would it not be better to state [his proposed new version]?"[40] Olmsted's original statement of purpose had emphasized public recreation and health needs, and his final version seems to have been intended to further similar goals. Mentioning "enjoyment" twice, the final statement provided for the enjoyment of the parks, but required that parks be left unimpaired so that future generations *also* could enjoy them. These goals could be met by essentially the same means as those of the earlier public health and recreation mandate—by maintenance of the parks' scenic landscapes, which would help ensure continuance of public enjoyment of the areas. Olmsted could have perhaps strengthened the preservation aspects of the statement by plainly requiring the parks to be left "unimpaired for future generations" rather than "unimpaired for the *enjoyment* of future generations" (emphasis added). But given the legislative history's repeated focus on parks as scenic pleasuring grounds, even without the specific references to "enjoyment" the act would not necessarily have called for rigorous preservation of the parks' natural conditions.

Anticipating public use, Olmsted sought to protect the beauty, dignity, and nobility of national park landscapes from commercial blight. Aside from his desire to make management more efficient, prevention of excessive commercialism in the parks was the one concern that Olmsted repeatedly emphasized. Early in the legislative campaign, he expressed fear that some would attempt to "make political capital" out of the parks by developing them for tourism; later he envisioned that a statement of purpose would provide a "legal safeguard" against "exploitation of the parks for commercial and other purposes." In early 1915 he worried about General Superintendent Mark Daniels' eagerness to make the national parks, in Olmsted's mocking words, "accessible to the 'Pee-pul.'" Daniels, he feared, was more concerned with securing an array of "improvements" than with maintaining "the perfect conservation of the quality of the landscapes."[41]

The new mandate was very much a reflection of Olmsted's professional interests. A landscape architect who had developed parks and other public places across the country, he made his living designing outdoor areas for aesthetic appeal, enhancing their scenic beauty for the enjoyment of the people. Indeed, in an unsuccessful effort to include in the legislation authorization of a special board to provide advice and assistance to national park management, the only profession for which Olmsted specifically sought inclusion on the board was landscape architecture.[42] Protecting the majestic national park landscapes through restricted, judicious development was Olmsted's primary concern. His final statement of purpose—against which so much national park management would be both justified and criticized—was thus in accord with the widely held concept of national parks as scenic pleasuring grounds.

A Utilitarian Act

In 1915, with strong support from Secretary of the Interior Franklin Lane, Stephen Mather reenergized the campaign for a new bureau, courting prominent writers, publishers, businessmen, and politicians. Mather and

Horace Albright worked steadily with their key congressional contacts, particularly California congressmen John Raker and William Kent and Utah Senator Reed Smoot. Mather also gained widespread media attention for the national parks, encouraging two highly popular magazines, the *Saturday Evening Post* and the *National Geographic,* to give the parks special coverage. The latter publication devoted its April 1916 issue to the "See America First" theme, praising America's scenic landscapes and tourist destinations and presenting photographs and text on the national parks. With funds from the railroads and from Mather himself, Robert Sterling Yard produced the *National Parks Portfolio,* which illustrated the beauty of the parks, promoting them as tourist destination points. Yard distributed this literature to influential people across the country.[43]

In the spring of 1916, Congress studied the proposal for a national parks bureau. A House report released in May gave its own definition of the purpose of the national parks—that they were set aside for "public enjoyment and entertainment." In hearings before the House Committee on the Public Lands, Richard Watrous of the American Civic Association explained at length his conviction that Canada, having established a national parks office in 1911, was ahead of the United States. He quoted its parliamentary mandate—that the parks were to be administered "for the benefit, advantage, and enjoyment" of the people. This purpose was to be achieved, in the words of an official Canadian government report, "not only by providing for the people of Canada for all time unequaled means of recreation in the out-of-doors under the best possible conditions, but by producing for the country an ever increasing revenue from tourist traffic."[44]

After six years of campaigning, proponents of a new bureau prevailed. Responding to their political strategy and persuasive promotional efforts, Congress passed the bill establishing the National Park Service within the Department of the Interior, and President Woodrow Wilson signed it into law on August 25, 1916.[45]

Among the most important supporters of the legislation were Secretary Lane and congressmen William Kent and John Raker, who less than three years before had been principal players in gaining congressional authorization for a dam in Yosemite's Hetch Hetchy Valley. It was the Raker Act of December 1913 by which Congress authorized the dam that was destined to bring about massive "impairment" to Yosemite through total destruction of natural conditions beneath the dam and reservoir, thus raising the specter of similar havoc in other national parks. Kent took a thoroughly utilitarian view of Hetch Hetchy, believing that a reservoir would be "the highest form of conservation," making the valley more accessible and useful for recreation. Indicating that such gigantic man-made features as dams and reservoirs could be acceptable intrusions that would not "impair" the parks, Kent insisted that the "creation of a lake would not impair the beauty of this wonder spot [Hetch Hetchy], but would, on the other hand, enhance its attractiveness."[46]

The support of Kent, Raker, and Lane for the National Park Service Act represented an accord between the aesthetic and utilitarian branches of the late-nineteenth- and early-twentieth-century conservation movement. Indeed, the national parks themselves constituted both an aesthetic and a utilitarian response to portions of the public domain through the promotion of public use and enjoyment of especially scenic areas. Originally used in reference to the management of reservoirs and of grazing on public lands in the West, the term "conservation" had by the early twentieth century come to identify the nationwide movement for efficient and rationally planned use (often referred to as "wise use") of natural resources. It implied, as one contemporary observer stated, "foresight and restraint in the exploitation of the physical sources of wealth as necessary for the perpetuity of civilization, and the welfare of present and future generations."[47] Creation of the National Park Service had been urged partly on the basis of need for efficient management of the parks; and, efficiently run, the parks (with majestic scenery as the basis of their economic value) could be the essence of "foresight and restraint" in the use of natural resources to benefit future generations.

The Organic Act's statement of purpose called for the Park Service to "conserve" the scenery and other resources, while most early national park enabling acts (including, for example, Yellowstone, Sequoia, Yosemite, Mount Rainier, and Glacier) called for "preservation" of resources—a blending of the related concepts of conservation and preservation.[48] In its broader sense, conservation included preservation as one of many valid approaches to managing resources. The conservation movement comprised a wide array of concerns, of which the wise use of scenic lands in the national parks to foster tourism and public enjoyment was very much a part.

Expressing the hopes and aspirations of McFarland, Olmsted, Mather, Kent, and many others, the Organic Act's "plain language" provided for public use and enjoyment of the parks, and was clearly utilitarian. The act even allowed consumptive use of certain park resources—evidence that the founders intended "unimpaired" to mean something quite different from the strict preservation of nature. Section 3 of the act authorized the leasing of lands in the parks for the development of tourist accommodations, thereby perpetuating the commercial tourism that was ongoing in all national parks, in most instances predating their establishment. The nominal restrictions placed on the leases (twenty years per lease, and no interference with the public's free access to "natural curiosities, wonders, or objects of interest") provided virtually no protection against potentially destructive impacts on the parks by the lessors. The primary statutory restraint on leasing remained the act's statement of purpose.

Section 3 allowed perpetuation of other established park practices by authorizing the National Park Service to destroy animal and plant life if "detrimental to the use" of parks. Already a routine means of protecting the game species most favored by the public, destruction of predatory animals was allowed to continue under this provision. Also, the Park Service was authorized to dispose of timber, particularly if necessary to control insect infestations that might affect the appearance of scenic forests. (Fishing, a consumptive use of park resources, was not mentioned in the act, although it would continue as an exceptionally popular park activity.) In response to pressures from cattle and sheep ranchers, section 3 allowed continuation of livestock grazing in all parks save Yellowstone, when not "detrimental to the primary purpose" of the affected parks. This provision meant, as Mather had testified to Congress, that the parks could serve "different interests without difficulty."[49] In the Organic Act, Congress permitted sheep, cattle, and tourists to use the national parks.

Section 4, the act's final provision, had strong potential to affect natural resources in specified parks and probably, like grazing, was another expedient to gain support of California congressmen. The section affirmed a February 1901 act authorizing the secretary of the interior to permit rights-of-way though Yosemite, Sequoia, and General Grant national parks for, among other things, power lines, pipelines, canals, and ditches, as well as water plants, dams, and reservoirs, to "promote irrigation or mining or quarrying, or the manufacturing or cutting of timber." Before granting permits, the secretary was mandated to determine that such proposed development projects were not "incompatible with the public interest." Although Congress would withdraw this authority in 1920, section 4 demonstrated that, as with livestock grazing, public use of the national parks was in certain instances intended to extend beyond recreation and enjoyment of scenery toward clearly consumptive resource uses.[50]

Together, sections 3 and 4, permitting manipulation of plants and animals and fostering certain consumptive uses, (1) did not modify any natural resource management practices begun in the parks prior to passage of the

Organic Act; (2) slanted the Organic Act toward "multiple use" of the parks' natural resources; (3) moved the definition of national parks further away from any close approximation of pristine natural preserves; and (4) substantially qualified what Congress meant when it required the parks to be left unimpaired.

Even though the Organic Act implied the preservation of nature in its mandate to conserve natural objects and leave the parks unimpaired, the founders gave no substantive consideration to an exacting biological preservation. Olmsted's correspondence, for instance, rarely even alluded to preserving natural conditions. He seems never to have seriously considered the parks as having anything like a mandate for truly pristine preservation.

The absence of explicitly articulated interest in preserving natural resources of all types throughout the parks suggests that the founders assumed that, in effect, *undeveloped* lands were *unimpaired* lands—that where there was little or no development, natural conditions existed and need not be of special concern. The ongoing manipulation of the parks' backcountry resources, such as fish, forests, and wildlife seems not to have been viewed as impairing natural conditions.

Still uninformed on ecological matters, the founders did not advocate scientific investigations to improve understanding of the parks' flora and fauna and ensure their preservation. Rather, as at the national park conferences, they sought advice from foresters and entomologists on how to prevent fire and insects from destroying the beauty of the forests. With threats such as fires, insects, and predators under attack, and with properly limited development, park supporters promoted the parks both as pleasuring grounds and as unimpaired natural preserves.

Although Olmsted had sought a declaration of the national parks' fundamental purpose in "unmistakable terms," the Organic Act decreed what Stephen Mather would call the "double mandate": that the parks be both used and preserved. In truth, the act did not resolve the central ambiguity in national park management—the conflict between use and preservation of the parks. Not defined, the principal concept, to leave the parks "unimpaired," was left open to sweeping interpretation that would allow extensive development and public use (reinforced by section 3 of the act), but would later justify scientific attempts to preserve (and even restore) ecological integrity in parks. Nothing in the act specifically authorized scientifically based park management; but nothing precluded it when it later became a matter of concern.

Following passage of the Organic Act, the National Park Service, as the sole bureau responsible for the act's implementation, would provide its "administrative interpretation" of the new law. During the legislative campaign, the enthusiasm expressed for opening the parks to public use and enjoyment reinforced the urge toward resort-style development. And for its first seventeen years, the Park Service was in fact run by two of its founders, Stephen Mather and Horace Albright—who, because of their personal involvement in passage of the act, never questioned their understanding of the act's intent and the statement of purpose. Under their supervision, park management was set in a direction that would continue with little change for at least the next half-century, thus fundamentally affecting the conditions of the parks and the attitudes and culture of the National Park Service itself.

CHAPTER 3

Perpetuating Tradition: The National Parks under Stephen T. Mather, 1916–1929

> In the administration of the parks the greatest good to the greatest number is always the most important factor determining the policy of the Service.—STEPHEN T. MATHER, 1920

In September 1916 Joseph Grinnell, head of the University of California's Museum of Vertebrate Zoology in Berkeley, coauthored an article in *Science* magazine entitled "Animal Life as an Asset of National Parks." A close observer of the parks (particularly Yosemite), Grinnell, along with his coauthor, Tracy I. Storer, also at the University of California, reflected on the various uses of the parks, from recreation to "retaining the original balance in plant and animal life." Regarding their concern for nature, they warned that "without a scientific investigation" of national park wildlife, "no thorough understanding of the conditions or of the practical problems they involve is possible." They also predicted that, with settlement of the country causing alterations to nature, the national parks would "probably be the only areas remaining unspoiled for scientific study."[1] This article, published less than a month after passage of the National Park Service Act, sounded an early cautionary note that national park management should have firm scientific footing.

Under Stephen Mather's direction from its founding until early 1929, the Park Service ignored Grinnell and Storer's counsel. In November 1928, shortly before the ailing Mather resigned as first director of the Service, his soon-to-be successor, Horace Albright, wrote him about possible new positions for forest, fish, and wildlife management. After more than a decade of enthusiastic development of the national parks for tourism, Albright stated that it was "highly essential" to begin hiring staff in "other than . . . landscape architecture and engineering, both of which have been pretty well provided for." Influenced by an emerging interest in science within Park Service ranks, he urged that the bureau not set itself up for charges of having provided "thousands for engineering of one kind or another and hardly one cent for experts to look after our fish resources, wild life and forests."[2]

Indeed, the Park Service had continued the management practices of its military and civilian predecessors. Rather than altering the direction of park management, the Organic Act's immediate outcome had been administrative and political gains for the national park system. The act consolidated park management, enabling it to focus on the needs of the entire system and giving it a voice with which to promote the national park idea to Congress and the public. National park leadership was elevated to a fully visible and aggressive new bureau within the Department of the Interior, and was backed by leading proponents of outdoor recreation, tourism, and landscape preservation. The fact that by the time Mather resigned he had become an institutional hero within the Service and commanded respect in broader conservation circles suggests that his persistent expansionist and developmental policies met with widespread approval.

Building Park Service Leadership

Because the various national parks had previously been independent of one another, with no effort at a cooperative approach to management policy and practice, very little organization building had taken place within the system. Thus, Mather did not face a powerful, cohesive managerial clique. Even though the U.S. Army had held responsibility for three of the most complex parks in the system, it had not sought to build a national park empire. Prior to withdrawal, its leaders urged that park duties were costly and inappropriate for the army and should be terminated.[3] The military's departure from Yosemite, Sequoia, and General Grant national parks in 1914, and Yellowstone in 1918, left a significant void in park management. Moreover, Mather judged many of the civilian superintendents of the other parks to be ineffectual, and would soon replace them with his own men.[4] Enjoying considerable discretion as director, he could determine what kinds of expertise were most needed to run the parks under the new mandate. Furthermore, within funding limitations, he could select the Service's directorate, the park superintendents, and professional support with little if any interference.

Although the Organic Act was passed in August 1916, it was not until the following spring that Congress appropriated funds for the Park Service.

Mather oversaw interim operations; and with a staff of six, the Park Service's headquarters in Washington, D.C., officially opened on April 17, only eleven days after the United States entered World War I.[5] Taking the place of general parks superintendent Robert Marshall, the Service's directorate assumed its leadership role. Next to Mather, Horace Albright was the most powerful individual in the directorate, serving first in Washington, and then, from 1919 to 1929, as Yellowstone superintendent, with continuing directorate responsibilities. Moreover, during Mather's long periods of absence due to severe stress and nervous conditions, he made Albright, his protégé and closest advisor, acting director of the Park Service.

As Mather staffed the new bureau, two groups assumed positions of special power and influence: one group consisted of landscape architects and engineers—professionals who oversaw park development; the other consisted of park managers—the superintendents and their rangers who were in charge of day-to-day operation of the parks. Under Mather's direction, each group coalesced, attaining a bureaucratic status that would flourish under succeeding directors.

As the Service matured into a sizable and highly successful bureau, it would develop a strong sense of identity and purpose and, concurrently, a sense of working together as a kind of close-knit family—the "Park Service family," as it would become fondly known by many employees. Together with the Service's ever-powerful directorate, the landscape architects, engineers, superintendents, and park rangers formed the core of an emerging "leadership culture"—in effect, the dominant family members. Under their guidance the Mather era locked in place the utilitarian tendencies of the pre–Park Service years and crystallized the business-capitalist predisposition for continual development, growth, and expansion. With continuous reference to the Organic Act's mandate as fundamental dogma, the Service's leadership groups defined the values and principles of the new bureau and established its managerial traditions—the leadership culture itself became locked in place. Policies developed and honed during the Mather era would exert an enduring, pervasive influence on national park history.

Applicable to National Park Service evolution, sociologist Edgar H. Schein, in his study of organizational culture and leadership, discusses how organizational cultures "begin with leaders who impose their own values and assumptions on a group." Such cultures come to be defined by the "shared, taken-for-granted basic assumptions held by members of the group or organization." Around these, the culture will develop a "basic design of tasks, division of labor, organization structure, [and] reward and incentive systems." Schein states further that if an organization is successful and its assumptions "come to be taken for granted," then its culture will "define for later generations of members what kinds of leadership are acceptable." Thus, "the culture now defines leadership"—it will "determine the criteria for leadership and thus determine who will or will not be a leader."[6]

In such regards, it can be argued that, in line with the values and objectives set by the Park Service founders (especially Mather and Albright), the perceived needs of the national parks and the intended purpose of the Service have always been reflected in the bureau's organizational arrangements. Such arrangements reveal a hierarchy of goals and functions and disclose the professions that controlled policy formulation and decisionmaking and formed the Service's leadership culture.

The first true professions to appear in the National Park Service—engineering and "landscape engineering" (later designated landscape architecture)—made up two of the four divisions in the Service's organizational chart dated July 1, 1919.[7] As developmental professions capable of overseeing the planning, design, and construction of park facilities, they fit very naturally into Mather's plans for implementing the Organic Act. The extensive involvement of these professions initially sprang from the public understanding of national parks as pleasuring grounds and soon worked to perpetuate this perception.

The emerging bureaucratic strength of landscape architecture no doubt benefited from the profession's having been so well represented among Park Service founders. Especially prominent were leaders of the American Society of Landscape Architects and the American Civic Association, including Fredrick Law Olmsted, Jr., and Horace McFarland (a horticulturalist deeply involved with aspects of landscape architecture), whose influence and support continued well after the Service was established. Mark Daniels, the national parks' first general superintendent, was a landscape architect. Mather himself was a longtime member of the American Civic Association; following his resignation, the landscape architects awarded him an honorary life membership in their national society.[8]

Mather believed that landscape architects filled a "serious gap" in his organization; and in 1922, seeking to ensure that new construction "fit into the park environment in a harmonious manner," he required their approval on "all important plans" for the parks. This authority was also extended to park development undertaken by concessionaires.[9] In developing the parks in cooperation with architects and engineers, landscape architects sought not only to avoid intruding on scenery, but also to display scenery to its best advantage with the proper placement of roads, trails, and buildings. They designed plantings to screen unattractive development from view, and planned intensively developed areas, with parking lots, sidewalks, buildings, lawns, and gardens. The resolve to blend new construction with natural surroundings—to develop the parks without destroying their beauty—formed the basis of landscape architecture's central role in national park development.

The authority of the landscape architects did not mean that their decisions went unchallenged; rather, they frequently skirmished with superintendents, concessionaires, and others over the details of plans and designs. In September 1922 a dispute over the design of two bridges in Yosemite caused Arno B. Cammerer, then an assistant director of the Service, to defend the landscape architects' approval authority. Cammerer wrote confidentially to Olmsted that, regarding such disagreements, some superintendents were "bucky in the matter" and needed to be better educated in park design and development concerns. He pressed the issue later that year at the superintendents conference, and again in the 1923 conference, when he reiterated that the superintendents must cooperate with the landscape architects.[10]

The pervasiveness of landscape architecture in the national parks encouraged some in the profession to argue for it to have even greater authority within the Service. Landscape architect Paul Kiessig wrote to Horace Albright in 1922 that national parks are "primarily a landscape thing," that "scenery is the attribute that sets a park aside to be conserved and protected for all generations," and that a park's "original charm" must be protected. Claiming that the superintendents had a "perennial resistance" and a "basic aversion" to the ideas of the landscape architects, Kiessig advocated that not only national park superintendents, but also an assistant director of the Service, should be men trained in landscape architecture.[11]

Later, in May 1929, while seeking to gain dominance in the "Field Headquarters" (a recently established office located in San Francisco to improve coordination among the mostly western national parks), landscape architect Thomas C. Vint asserted that his profession deserved the central role in park development. Writing to Albright to express concern about engineers having too much influence in the San Francisco office, Vint asked rhetorically if the parks were to be developed on a "landscape or engineering basis." Predictably, his choice was a landscape basis, which would put the parks under a kind of umbrella profession, combining architecture, engineering, and horticulture, with a strong focus on "the element of beauty." Vint believed that all employees should "think landscape," and that no matter what the organizational structure of the San Francisco office was, it would still become a "landscape organization."[12]

Albright later recalled that the lack of "integrated planning" in the parks led Mather to begin hiring landscape architects. As much as any other factor, the emergence of a formal, parkwide planning process gave the profession its powerful, enduring role in national park affairs. In February 1916, during the campaign to establish the National Park Service, James S. Pray of the American Society of Landscape Architects had called for "comprehensive general plans" for each park. This idea was endorsed two years later by Interior secretary Franklin Lane, who required that all park improvements be "in accordance with a preconceived plan developed with special reference to the preservation of the landscape," a plan that would require knowledge of "landscape architecture or . . . proper appreciation of the esthetic value of park lands." Lane stated that these comprehensive plans were to be prepared as soon as funds were available. Mather did not get systemwide planning under way until 1925, when he authorized preparation of five-year plans for the parks. By late 1929 the Service employed nine landscape architects, a number that increased to twenty by 1932.[13]

In the early 1930s the Service would expand its long-range planning and prepare comprehensive, parkwide plans (which became known as "master plans"), supplemented by more detailed plans for areas to be intensively developed. By this time, planning and landscape architecture had come under the command of Thomas Vint.[14] And in February 1931, landscape architect Conrad L. Wirth joined the Park Service, rising quickly to assistant director. Under Vint and Wirth—probably the two most influential landscape architects in National Park Service history—landscape architecture became firmly established as one of the Service's most powerful professions, a status it has not relinquished to this day.[15]

Although never acquiring the bureaucratic strength that landscape architects wielded, Park Service engineers nevertheless gained considerable influence. Mather hired his first engineer in 1917, the year before he employed the first landscape architect; engineering remained a vital part of the organization throughout his directorship. Chief among the engineers' responsibilities was the construction of park roads. Designed for horse traffic, the early roads needed widening, realigning, and paving to accommodate automobiles, which had begun to be allowed in the parks just before establishment of the Park Service. Mather aggressively lobbied

Congress for funds for road rehabilitation and new construction, and in 1924 Congress funded the Service's first large road program. Two years later the Service concluded a formal agreement whereby the Bureau of Public Roads would oversee the building of major highways and bridges in the national parks. Park Service engineers coordinated the work with the bureau and also oversaw other development, such as park buildings, water and sewage systems, electrical systems, trails, and campgrounds.[16] With their projects often creating massive intrusions on park landscapes, engineers had to coordinate regularly with landscape architects on matters of aesthetics and scenery. As the key link between construction and the preservation of majestic park scenery, landscape architects had the bureaucratic advantage over engineers.

In 1927 Frank A. Kittredge, who had impressed Mather during the initial planning and construction of Glacier National Park's spectacular Going to the Sun Highway, transferred from the Bureau of Public Roads to become the Park Service's chief engineer. With congressional increases in construction funds in the late 1920s and into the New Deal era, the engineering office grew in size and influence. In a time of such expansive development of the national parks, the engineers mixed easily with park management and attained membership in the Service's leadership circles. Indeed, many of Mather's new superintendents were former engineers. Indicative of the engineers' ability to cross over into park management, Kittredge himself would later become head of the newly created regional office in San Francisco, overseeing parks in much of the area from the Rocky Mountains west. In time, he would serve in superintendencies at Grand Canyon and Yosemite before returning to engineering.[17]

Under Mather, field management began to develop a genuine professionalism, with identifiable duties and standards of operation. As one of his principal objectives, Mather wanted the new bureau to have organizational strength and durability—what Horace Albright later called a "strong internal structure."[18] The heart of this structure was to be the park rangers and superintendents. By the time Mather resigned in early 1929, the rangers and superintendents had coalesced as a distinctive group with a strong sense of identity and a common understanding of how national parks should be managed. Proudly wearing the dark-green field uniform, they became the chief bearers of Park Service family tradition and the forerunners of today's "green blood" employees.

The national park ranger corps had slowly evolved during the late nineteenth and early twentieth centuries. In 1914, while attempting to establish a "ranger service"—a distinct corps of rangers—general superintendent Mark Daniels drew up regulations to coordinate and standardize ranger work. Without a strong national office to oversee this effort, Daniels' ranger service did not succeed.[19] His regulations, however, issued to all parks by Secretary Lane in January 1915, reflected the essentially frontier skills expected of a ranger. In addition to age requirements, appointments and promotions, salary scales, and uniform and equipment standards, the regulations called for rangers to have "experience in the outdoor life" and to be able to endure hardships, ride and take care of horses and mules, shoot a rifle and a pistol, cook simple meals, build trails, and construct cabins. These types of skills would enable them to patrol park backcountry for poachers and unauthorized livestock, kill predators, fight fires, and undertake other park protection activities.[20] In time, those rangers most deeply involved in such natural resource management activities would become known as "wildlife rangers."

Secretary Lane's regulations also directed rangers to be "tactful in handling people," a requirement that foretold an increasingly significant responsibility during the Mather era. With rapidly increasing automobile travel after World War I, the rangers had greater contact with park visitors who were not poaching or trespassing, but instead were enjoying the parks. The need to assist visitors brought about establishment of "ranger naturalist" positions, which, under the supervision of a "park naturalist," had duties including staffing park museums, leading hikes, and giving nature talks.[21] Like the wildlife rangers, the ranger naturalists needed a serviceable understanding of their park's natural history.

Mather believed the success or failure of the national parks depended on the rangers. Albright saw them as the "core of park management" (as he later put it) and recognized that the public's impression of the National Park Service came primarily from contact with these uniformed personnel.[22] In his effort to build ranger *esprit de corps,* Mather always wore his official uniform and mixed with the rangers during his many park visits. Symbolic of his concern for the rangers' welfare and morale, in 1920 Mather himself paid for construction of the Yosemite "Rangers' Club," which became famous throughout the Park Service as a gathering spot for rangers, superintendents, and the Service directorate. Mather also authorized the first conference of chief rangers in 1926. Held in Sequoia, and chaired by veteran Yellowstone chief ranger Sam Woodring, the conference was designed to expose rangers to the variety of issues faced by the Service, in order to broaden their understanding of park management.[23]

Perhaps most important for morale building was Mather's effort to improve the rangers' status as government employees. When the Park Service was established, employment was tied to individual parks, rather than to the park system. Thus rangers had no official "transfer rights" and had to resign from one park and pay their own moving expenses to the next location.[24] For low-salaried rangers, such fragmented employment opportunities severely restricted chances for career advancement. Furthermore, they fostered a provincial view, causing rangers to focus only on the parks they served, rather than the park system as a whole. Mather encouraged the rangers to consider their national park work as a career rather than a mere job; and his lobbying won salary increases and transfer rights (including moving costs) and ultimately brought rangers under the Civil Service's competitive examination system.[25]

The park rangers developed a natural alliance with the superintendents, based on mutual goals and perceptions as well as common career paths. Organizationally, the link between superintendents and rangers was through the chief ranger—usually the second most powerful position in the park, the incumbent of which acted for the superintendent during his absence.[26] The bonds that developed between rangers and superintendents during the Mather era became a fundamental aspect of park management and the internal politics of the Service.

In 1924 Horace Albright recalled believing that many of the superintendents on board when Mather took charge had been "incompetent men appointed as politicians." Seeking loyal, qualified employees, Mather hired new superintendents whom he trusted, and who could help build a close-knit, mutually supportive organization. He tended to choose men who had out-of-doors experience and who were engineers (particularly topographical engineers) or had served with the army or the U. S. Geological Survey. (Only in the 1970s would women begin to attain leadership roles in the Park Service.)[27] Mather's early superintendency appointments included Roger W. Toll, an engineer and former army officer, to Mount Rainier and later to Rocky Mountain and to Yellowstone; Washington B. ("Dusty") Lewis, a Geological Survey engineer, to Yosemite; "Colonel" Thomas Boles, an engineer, to Carlsbad Caverns; John R. White, a British-born, Oxford-educated soldier of fortune and former U.S. Army officer, to Sequoia and General Grant; and J. Ross Eakin, a Geological Survey engineer, to Glacier and later to Grand Canyon and to Great Smoky Mountains.[28]

The park rangers constituted another source from which to select superintendents, a factor that helped bond the two groups. For example, following the army's departure from Sequoia and General Grant, Walter

Fry, a longtime ranger in those parks, was chosen to be superintendent. When Fry resigned in 1919, Mather replaced him with John White, who had worked briefly as a ranger at Grand Canyon before his elevation to the Sequoia position.[29] Most of Mather's early superintendency appointments did not come from the ranger ranks, however, perhaps because he did not have much confidence in the small rangers corps that was in place when the Park Service began operations. But as the Service recruited and trained rangers, they increasingly became obvious choices to fill superintendency positions.

Mather's most significant appointment came in 1919, when he named Horace Albright to the Yellowstone superintendency. Albright was to manage Yellowstone and was also to serve as Mather's field assistant (in effect, his deputy), in direct charge of all parks and all offices not located in Washington.[30] By placing his most trusted Park Service friend and confidant in the premier national park superintendency and in charge of field areas, Mather reinforced the bonds between the superintendents and the Park Service directorate.

To strengthen his organization and develop common solutions to management problems, Mather held superintendents conferences about every two years. He considered these meetings to be a continuation of the national park conferences begun in Yellowstone in 1911, which he had first attended at Berkeley in 1915. Albright recalled that they served as "forums for spreading the best ideas and tackling the biggest problems throughout the system." Mather also used the conferences to develop camaraderie among the superintendents, often staging large, festive dinners, with singing, horseplay, practical jokes, and other group activities. And at times he insisted that the superintendents travel to the conferences in automobile caravans (such as to Mesa Verde in 1925), in order to visit parks along the way and discuss various management issues. The conferences provided the superintendents with opportunities not only to form lifelong friendships, but also to become more aware that they were part of a national organization.[31] Through Mather's conferences, they began to comprehend the parks as a system and to influence policy on a systemwide basis.

A Formal Policy and a Bureaucratic Rivalry

In the winter of 1917–18, as the Park Service neared completion of its initial year of operation, Horace Albright drafted a comprehensive statement of national park management policies. After a thorough review (by prominent conservationists, among others), Secretary Franklin Lane issued the policies in the form of a letter to Director Mather. Albright later recalled that the Lane Letter, as it became known, was "a landmark" and the Service's "basic creed."[32] As the new bureau's first formal statement of its responsibilities under the Organic Act, the letter reflected the founders' emphasis on the parks as scenic pleasuring grounds. It was also an affirmation of management practices long under way in the parks.

The letter opened with a reference to the Organic Act's statement of purpose, declaring that the parks were to be maintained in "*absolutely* unimpaired" condition. This statement, and one in the next paragraph that all activities were subordinate to the duty of preserving the parks "in essentially their natural state," constituted a formidable commitment to preservation. However, the letter then explained that national parks were set aside for the "use, observation, health, and pleasure of the people" (sounding much like Olmsted's early but discarded statement of purpose). It declared the parks to be a "national playground system," to be made accessible "by any means practicable," including through construction of roads, trails, and buildings that harmonized with park scenery. It also encouraged educational use of the parks and appropriate outdoor sports—including winter sports. And the letter urged the Service to "diligently extend and use" the cooperation offered by tourist bureaus, chambers of commerce, and automobile associations to increase public awareness of the parks.

The Lane Letter authorized cattle grazing in "isolated regions not frequented by visitors" and where "natural features" would not be harmed. It forbade sheep in the parks, however. It also forbade hunting, limited timber cutting to that which was most necessary (including thinning to "improve the scenic features"), and called for elimination of private holdings in parks. In the single specific reference to science, the letter recommended that the Service *not* develop its own scientific expertise, but that it seek assistance from the government's "scientific bureaus."[33]

As a "landmark" and "basic creed" for the National Park Service, the Lane Letter delineated the values and assumptions of the bureau's emerging corporate culture, thereby setting the tenor and direction of park management during the Mather era and far beyond. In 1925 the Service prepared a second major policy statement, signed by Secretary of the Interior Hubert Work and subsequently known as the "Work Letter." It conveyed essentially the same concerns—some verbatim—as had the Lane Letter.[34]

The utilitarian values expressed in both policy letters were probably stimulated in part by rivalry with the U.S. Forest Service. It had taken six years of campaigning to convince Congress to create a national parks bureau; as director, Mather realized that the parks still might not be politically secure. Aside from possible public indifference, probably the greatest threat to the Park Service's future came from the Forest Service, whose proponents resented not having gained control of the national parks. About the time the Park Service was established, some Forest Service public recreation programs began to appear; and in 1924 the bureau's New Mexico office designated the first "wilderness" area, a virtually roadless portion of the Gila National Forest.[35] Initially these were regional initiatives, rather than national. Yet such moves increased the Forest Service's range of land management practices and encroached on responsibilities the Park Service claimed for its own—thus helping to perpetuate the bureaucratic rivalry.

Although the National Park Service may not at first have been aware of the Forest Service's administrative designation of a wilderness area, the beginnings of national forest recreation programs did cause consternation among Park Service leaders. In a 1925 paper on the issue, Mather argued that no overlaps existed between the Park Service functions and those of other bureaus, particularly the Forest Service. He quarreled with attempts to confuse the duties of the two bureaus and with continuing claims by Forest Service advocates that it could operate the national parks at little extra cost beyond that of managing the forests. Placing the parks under the Forest Service, which was engaged, as Mather put it, in "commercial exploitation of natural resources," would, he believed, destroy the parks. As if seeking to prove that his own utilitarian biases were as strong as those of Forest Service leadership, he stated that outdoor recreation responsibilities belonged to the Park Service—that the parks were "more truly national playgrounds than are the forests." In order to meet the dictates of the Organic Act, Mather believed the Service was obligated to develop the parks—to "grant franchises for the erection of hotels and permanent camps, operation of transportation lines, stores and other services, etc."[36] Rivalry with the utilitarian Forest Service stimulated Mather's bent for recreational tourism management.

Appropriate and Inappropriate Park Development

Espousing strong democratic ideals and believing in the high social value of the national parks, Mather once wrote that the "greatest good for the greatest number" was "always the most important factor" in determining Park Service policy. As the individual with primary responsibility for implementing the Organic Act, he urged that the Service develop the parks for tourism by providing "such imperative necessities as new roads, improved roads, trails, bridges, public camping facilities, and water supply and sewerage systems."[37] In addition, Mather personally sponsored the creation in 1919 of a support organization—the National Parks Association—headed by his friend Robert Sterling Yard, who had helped greatly in the campaign to establish the Park Service. Reflecting the goals of its parent organization and its sponsor, the association's principal objectives were to protect the parks, enlarge the national park system (through such significant additions as would make the system "an American trademark in the competition for the world's travel"), and promote public enjoyment without impairing the parks.[38]

Soon after Mather first became associated with the national parks, he and Horace Albright had made a tour of the parks, seeking to assess the situation in the field. Mather noted that park facilities were inadequate. Only Yellowstone and Yosemite had extended road systems, but the roads had been designed for horse traffic, now being replaced by the automobile; many park hotels and campgrounds were primitive. The following year Mather claimed in his annual report that the parks had been "greatly neglected." Repeatedly urging park development, Mather got results. By the time his health problems forced him to resign early in 1929, the parks had undergone extensive development involving virtually every type of facility needed to support recreational tourism and park administration.[39] Shortly after Mather's resignation, Albright, as the new director, summed up the park development that had occurred before and during Mather's tenure by reporting that the Park Service was responsible for "1,298 miles of roads, 3,903 miles of trails, 1,623 miles of telephone and telegraph lines, extensive camp grounds, sewer and water system[s], power plants, buildings," and more.[40]

During Mather's directorship, the railroad companies continued to promote their hotels in or near Yellowstone, Glacier, Mount Rainier, and other parks. More important for future park development, the emerging automobile age meshed perfectly with Mather's desire to make the parks popular. A member of both national and local automobile associations, he worked closely with them to encourage tourism. In 1916 he advocated preparing the parks for the "great influx of automobiles by constructing new roads and improving existing highways wherever improvement is necessary."[41] The previous year he had helped found the National Park-to-Park Highway Association. This organization promoted highway improvement and new construction designed to connect the major western national parks and enable tourists to make a giant circle through the West, visiting the parks. Mapped and signposted, but only partially paved, this route of approximately six thousand miles was officially dedicated in 1920.[42]

In 1919 Mather recommended to Secretary Lane that the Service establish a "travel division" or "division of touring" in its Washington office to assist in advertising the parks. Rather than create a new division, however, he kept this responsibility largely with his publications and public relations office. He regularly and enthusiastically reported to the secretary on tourism to the parks, noting, for instance, in 1925 that "it is again my pleasure" to report a large increase in numbers of visitors, who would bring with them, he claimed, a "great flow of tourist dollars."[43]

In overseeing the burst of park development that took place under Mather, Park Service leadership viewed specific development proposals in light of whether they were appropriate or inappropriate in a national park. The appropriate development generally was that which supported the traditional needs of recreational tourism, such as roads, trails, hotels, and park administrative facilities. In many cases designed to harmonize with park landscapes, this type of development generally was not considered to "impair" the parks—although disagreements arose over numerous specific proposals.

In contrast to most tourism-related construction, developments such as dams or mines were considered inappropriate. Outside the realm of park tourism needs and not intended to harmonize with the landscape, they were viewed as serious threats to the scenic qualities of the parks—impairments that could indeed undermine the Park Service's ability to meet its basic mandate from Congress.

Roads, when not built in excess, were accepted as appropriate development. Mather was convinced that each of the large parks should have one major road penetrating into the heart of the scenic backcountry, and he wrote in 1920 that the "road problem" (the need for more and better roads) was "one of the most important issues before the Service."[44] After using his own money to help purchase the Tioga Pass Road, which cut across Yosemite's high country and passed through the scenic Tuolumne Meadows and along the shore of Tenaya Lake, he convinced automobile associations to improve the road. Mather predicted that tourists would soon use "every nook and corner" of Yosemite; and, recognizing that the Tuolumne Meadows had already become popular, he anticipated adding automobile camps to avoid "insanitation and other evils." In 1919, realizing that the Yosemite Valley had become crowded, he advocated a new road to take visitors through the upper end of the valley, passing near Nevada and Vernal falls, and connecting with the Tioga Pass Road—a proposal that was never implemented.[45]

Despite such efforts, Mather declared that he did not want the parks "gridironed" with roads. He would limit road development to leave large areas of each park in a "natural wilderness state," accessible only by trail. This concern was echoed in Secretary Work's 1925 policy letter (drafted by the Park Service), stating that excessive construction of park roads should be "cautiously guarded against."[46] Two years later, for example, Mather adamantly objected to the "Sierra Highway," proposed to cross Sequoia National Park; he wrote to Superintendent John White that the idea of the road "does not appeal to me in any way." Mather wanted it "distinctly understood" that the Service was "not attempting to over-build from the road standpoint."[47]

Overall, though, the Mather era brought extensive road building. Aware that tourists were bypassing Glacier National Park because it had no east-west road, Mather pushed for construction of the Going to the Sun Highway, to cross the continental divide, traversing miles of previously undisturbed areas. Even in Alaska's remote Mt. McKinley (now Denali) National Park, the Service undertook a road program, contracting with the Alaska Road Commission in 1922 to build a highway into the park's interior. With the railroad from Seward to Fairbanks nearing completion, Mather sought to accommodate the anticipated increase in tourism; by the time of his resignation, highway construction extended about forty miles into the park.[48] In 1924 Mather's aggressive lobbying brought congressional increases in national park road appropriations, followed in 1927 by a ten-year, $51-million program to improve existing roads and build new ones.[49] He looked forward to improving roads in parks such as Lassen Volcanic and Hawaii, which previously had had no significant road development. Also, Mather's engineers and landscape architects laid plans for building Trail Ridge Road, to cut across Rocky Mountain National Park's high country, in places at elevations of more than twelve thousand feet above sea level.[50]

In addition to roads, the Service designed and built hundreds of miles of foot and horse trails and encouraged visitors to get out of their automobiles to see the parks. Writing in the *Sierra Club Bulletin* in 1920, Mather advocated that the parks' remote areas be "rendered accessible by trails and public camps" (a clear reflection of the Sierra Club's own creed, to "render accessible" the California mountains).[51] Two years later the park superintendents recommended a policy of building ten miles of trail for every mile of automobile road, to avoid the "cheapening effect of easy accessibility" with the automobile. Every year Mather reported progress in trail construction in the parks, and in many cases these trails did indeed make some remote areas accessible. For example, with enthusiastic Sierra Club support, the John Muir Trail and the High Sierra Trail (both part of an extensive Sierra trail system) were built through Sequoia National Park's backcountry. The planning, design, and construction of these major trail projects covered the entire span of Mather's career with the Service, and beyond.[52]

Mather also backed construction of administrative and tourist facilities throughout the national park system—another type of acceptable development. As before 1916, many of these structures were built in a "rustic" architectural style, designed to harmonize with the grand scenery surrounding them. Increasingly, the appearance of national parks reflected the influence of the landscape architects, and the carefully designed landscapes sometimes competed with the parks' scenic features for the public's attention. During Mather's directorship, the log and stone structures and winding scenic roads designed by men like Vint and Kittredge helped establish the classic appearance of national park development, which would not be seriously modified until the midcentury "Mission 66" construction program brought about more modern designs.

The 1920s saw completion of headquarters buildings in, for example, Sequoia, Grand Canyon, Glacier, and Mount Rainier. "Ranger stations" were built in many parks, some in a "trapper cabin" style suggestive of the fur trade era and particularly favored by Mather. To enhance its interpretation of natural history in the parks, the Service erected museums, two of the most prominent early examples being in Yosemite and Grand Canyon, built with donations from the Laura Spelman Rockefeller Memorial. In Mather's last year as director, the Service accepted more funds from the Rockefeller memorial to build rustic-style museums in Yellowstone, at Norris Geyser Basin, Madison Junction, Fishing Bridge, and Old Faithful. In addition, Mather approved construction of hotels built by park concessionaires, including Sequoia's Giant Forest Lodge, the Phantom Ranch in the depths of the Grand Canyon, and Yosemite's impressive Ahwahnee Hotel.[53]

Although much of the new construction was designed to be in harmony with the natural settings, the development of national parks also involved less aesthetically pleasing facilities, including parking areas, campgrounds, water storage and supply systems, electrical power plants, sewage systems, and garbage dumps. And tourism development often went beyond such basic accommodations to include, for instance, a golf course and zoo in Yosemite, as well as a racetrack for the "Indian Field Days" celebration held during the summers. Mather personally encouraged construction of golf courses in Yosemite and Yellowstone, believing that tourists would stay longer in the parks if they had more to entertain them.[54]

In line with the Lane Letter's endorsement, the director believed that where feasible the parks should be developed for winter sports—especially Yosemite, which he hoped could become "a winter as well as a summer resort."[55] The winter use concept reached a peak in early 1929, at the very end of Mather's career, when Horace Albright led an aggressive campaign to host the 1932 Winter Olympics in Yosemite, a proposal that would have required extensive development. Albright's enthusiastic support for this project was evidenced in his February 1929 letter to Yosemite assistant superintendent James V. Lloyd, applauding him for doing a "magnificent job" of "stirring up the San Joaquin Valley towns to support the plan to secure the winter sports of the 1932 Olympiad for Yosemite." Although the Service lost out to Lake Placid, Yosemite would soon initiate a winter sports carnival to attract off-season tourists, using facilities developed during Mather's time, such as a toboggan run and an ice rink. Albright would also promote winter resort facilities in Rocky Mountain, arguing that it "has been done in other parks, and we will have to find a place for the toboggan slide, ski jump, etc., where it will not mar the natural beauties of the park."[56]

Earlier, Albright had found himself in disagreement with Mather over the proposal to build a cable-car tram across the Grand Canyon. Albright supported the tram as a means of enhancing the public's enjoyment of the canyon, but Mather opposed it as an inappropriate intrusion. Following a lengthy debate and analysis, the proposal was defeated.[57]

In fact, the Mather era witnessed a continual debate over the degree and types of tourism development to be allowed in the parks. Without definitive guidelines, and with the most substantial precedents being the early development of national parks and resorts elsewhere in the country, Mather and his staff groped to determine what was indeed appropriate. Following their November 1922 conference in Yosemite, and in response to negative comments in the press regarding development, the superintendents drafted a statement analyzing the role of park development. They noted that without facilities to accommodate the public, a national park would be "merely a wilderness, not serving the purpose for which it was set aside, not benefitting the general public." Yet they recognized that there was "no sharp line between necessary, proper development and harmful over-development." Seeking a cautious golden mean, they stated that the parks needed "more adequate development" but that "over-development of any national park, or any portion of a national park, is undesirable and should be avoided." To this, Sequoia superintendent John White added his view that the Service's "biggest problem" was to develop the parks "without devitalizing them; to make them accessible and popular, but not vulgar; to bring in the crowds and yet to maintain an appearance of not being crowded."[58]

The superintendents, in their 1922 recommendations against overdevelopment, urged that the parks be kept "free from commercial exploitation" and argued against industrial uses such as dams, power plants, and mining.[59] Indeed, while rapidly developing the parks for tourism as it deemed appropriate, the Service gained early acclaim through its opposition to commercially motivated development proposals considered inappropriate in a national park setting. To a large degree, it was the actions of others (not the Park Service itself) that the Service viewed as threatening the parks.

Perhaps its most difficult confrontation during the Mather years came with the fight against water development proposals in Yellowstone. In 1919 Idaho's congressional delegation, intending to irrigate lands in the southeastern part of the state, sought legislation permitting several dams to be built in Yellowstone, including those proposed at the outlet of Yellowstone Lake and on the Falls and Bechler rivers. Water from these sources in the park would be tapped to supply Idaho farms. Montana also lobbied for a dam at the outlet of Yellowstone Lake as a means of flood control and irrigation.[60]

Among the founders of the National Park Service, perhaps the most notable disagreement on how the parks should be protected involved dam proposals. The Yellowstone proposals had the backing of Interior secretary Lane, who had joined other founders in support of the Hetch Hetchy dam and continued to promote reclamation projects in the West. A firm believer in utilitarian management of the nation's natural resources, Lane claimed that the dams in Yellowstone would "improve the park instead of injuring it."[61] Mather vehemently opposed the secretary on this issue, threatening to resign rather than support the dams.

In his 1919 annual report, Mather argued briefly against the "menace of irrigation projects" and the following year reported extensively on a number of recommended dams and the threats they posed to Yellowstone and other parks. Because of the precedents that the dams could set, Mather saw the proposals as putting the entire national park system in a state of "grave crisis." They would not only destroy the beauty of the lakes and streams, but also flood meadows, forests, and other feeding grounds for wildlife.[62] Mather and his staff lobbied vigorously against the Yellowstone dam proposals. Their efforts were boosted when Secretary Lane resigned suddenly in 1920 and his successors, John Barton Payne (a strong conservationist) and Albert B. Fall (who was no conservationist, but liked Mather and the national parks) both agreed to oppose the dams. With secretarial support withdrawn, the bills ultimately failed, and the potentially massive intrusions in the parks were averted.[63]

The Service was not always successful in opposing dams—for example in Glacier, where the park's enabling legislation specifically allowed use of park streams for irrigation and power. Construction of a dam at the lower end of Sherburne Lake outraged Mather, but he wrote in 1919 that the dam provided one consolation: it was a "glaring example of what is to be avoided in national parks having lakes still untouched."[64] The dam may have provided park supporters with a glaring example, but it did not faze pro-development groups. Soon irrigation associations in both Montana and Canada sought to tap the waters of Lake St. Mary, also in Glacier. Robert Yard of the National Parks Association urged that the lake's scenery be used in its defense, writing to the Park Service to "play up St. Mary Lake as one of the scenic marvels of the world." The Service successfully opposed this project; still, Glacier's legislation would encourage reclamation groups to continue seeking water projects in the park that the Service considered unacceptable.[65]

Among the most pervasive threats of inappropriate development were potentially unsightly uses of privately owned lands (today known as "inholdings") situated within national parks. The result of patents being issued on public lands before establishment of a park, inholdings were anathema to the Service. Excluded from Park Service control, use and development of inholdings could cause serious intrusions, potentially scarring the landscape and crippling the Service's efforts to leave the parks unimpaired. Chief among the threats resulting from the inholdings were mining, timbering, and uncontrolled commercial development. Experience had shown what could happen on private lands. For instance in Sequoia, before the Park Service was established, the Mt. Whitney Power Company dammed two rivers and built roads, flumes, and a power plant on its lands.[66] In Glacier, more than ten thousand acres were in private hands when the park was created in 1910, some of this acreage along the shore of Lake McDonald, where summer cottages and resorts had been built.[67]

Indeed, from the first, the Service made acquisition of private lands a high priority. Consolidation of all lands within park boundaries would allow control over development in the parks. To reduce the threat of inappropriate development, the Park Service continually sought to acquire inholdings, accepting them as direct donations, purchasing them, or swapping them for federal lands elsewhere.[68]

The 1918 Lane Letter declared that privately owned lands "seriously hamper the administration of these reservations" and advocated their elimination. Those in "important scenic areas" had the highest priority for acquisition. But for nearly a decade Congress failed to appropriate funds for buying inholdings, thereby forcing the Service to rely on private donations for such purchases. Mather himself contributed substantially to land acquisition in Sequoia and other parks, such as Yosemite and Glacier. Under the Service's prodding, Congress in 1927 and 1928 began to make regular appropriations for inholding purchases, but with the requirement that these funds be matched by private donations. In 1929, shortly after Mather's resignation, Director Albright predicted that reliance on private funds would not be satisfactory because potential donors felt that acquisition of park lands was the government's responsibility. Although Mather had secured some congressional funding, the inholdings remained, in Albright's words, one of the Service's "greatest problems"—a threat to the parks' integrity, and a "distinct menace to good administration and future development."[69] Albright's remarks foreshadowed a long, still-ongoing struggle to control inholdings.

Deletions and Additions of Park Lands

In many instances, development considered inappropriate was in fact allowed to take place when Congress simply revised national park boundaries to *exclude* the proposed development from parks. An ambitious early effort to adjust a park's boundary had come in the 1880s with the attempt by mining interests to remove from park status a huge tract in the northeastern part of Yellowstone, through which they wished to build a railroad to mines near Cooke City, Montana. This attempt proved unsuccessful; however, a similar effort succeeded with the removal in 1905 of several hundred square miles from Yosemite National Park. The Yosemite deletions involved lands with many inholdings and with potential for timber and mineral production.[70] Such tradeoffs occurred on a smaller scale during the Mather years. For example, in Rocky Mountain National Park lands in several areas near the park's boundary and including private holdings were desired for irrigation reservoirs and thus legislatively removed from the park. In one instance the Service divested itself of an entire park—Sully's Hill, located in eastern North Dakota and deemed unworthy to be in the national park system. The area was turned over to the Biological Survey as a game preserve in 1931.[71] Although such deletions meant the loss of lands and natural resources, they also meant that the parks proper would not be impaired by the proposed development, which would occur outside the adjusted park boundary—thereby effectively sidestepping the protection issue.

In contrast, at the time of Mather's resignation in early 1929 the Service was seeking boundary *extensions* for Sequoia (the Kings Canyon area), Yellowstone, Bryce Canyon, Zion, Rocky Mountain, and Glacier national parks.[72] In proposing changes to park boundaries, the National Park Service was in effect demarcating the resources it wished to manage—the lands it hoped to leave "unimpaired for the enjoyment of future generations." These efforts constituted a fundamental form of natural resource management in that they helped determine whether certain lands would be managed under national park policies or under the more consumptive policies of other land management bureaus or private interests.

The Service's efforts to determine which lands it would manage involved the larger question of supporting or opposing new park proposals—and the factor of scenery influenced decisions in all instances. The Lane Letter encouraged extensions of existing parks to complete their "scenic purposes." It mentioned in particular the High Sierra peaks bordering Sequoia National Park, and stated that expansion to include the Teton Mountains represented Yellowstone's "greatest need."[73] In this vein, Albright later recalled that when he and Mather first viewed the Tetons in 1916 they remarked that they had "never seen such scenery" and that they were "flabbergasted." During this initial visit they agreed that the "whole magnificent area" should become part of Yellowstone National Park. (In 1929, the Tetons—but very little of the Jackson Hole valley just east of the range—were established as a park separate from Yellowstone.)[74]

For new parks, the Lane Letter advised the Service to seek only those areas of "supreme and distinctive quality or some natural feature so extraordinary or unique as to be of national interest and importance"—only "world architecture" should be included in the national park system. As with the Tetons, Albright's enthusiasm for Utah's spectacular Zion Canyon was instantaneous. Before visiting Zion, he had heard it described as "Yosemite painted in oils"; and during his first visit to Zion he quickly determined it should become a national park. Shortly after seeing the area, he telegraphed Mather to tell him of its incredible beauty and to urge that they "do something about it."[75] Similarly, "world architecture" had been an obvious factor in the February 1917 establishment of Mt. McKinley National Park, the first national park created after the advent of the Service. Not only did the park include the highest mountain in North America, but it also was seen as a "vast reservoir of game," including caribou, Dall sheep, bear, and moose. More than with other early national parks, the spectacular wildlife was the chief motivating factor behind the park's creation, which Mather and Albright aggressively sought.[76]

During Mather's directorship, the Service gained some of the most spectacular national parks in the system—not only Mt. McKinley, Zion, and Grand Teton, but also Grand Canyon and Bryce Canyon. Moreover, the establishment of Acadia, Great Smoky Mountains, Shenandoah, and Mammoth Cave national parks meant that the park system (and thus the Park Service) gained greater representation in the more populous and more politically powerful eastern states—a factor of considerable importance to Mather.[77] In addition, presidential proclamations created numerous national monuments, including two vast natural areas in Alaska: the Katmai volcanic area at the top of the Alaska Peninsula, and Glacier Bay, a region of immense glaciers in southeast Alaska. All of these units added enormously to the Park Service's reputation as protector of places of majestic natural beauty throughout the country.[78]

Although the Service was highly successful in expanding the system during Mather's tenure, it also fought against proposals that it believed did not meet its standards. And the question of "national park standards" (that is, which areas had clear qualifications to become national parks) became a significant issue. Most outspoken was the National Parks Association, which, even though it generally did not criticize the Service's management practices once a park was established, nonetheless sought to keep "inferior" areas out of the system. This concern intensified in the early 1920s when Secretary of the Interior Albert Fall proposed the "All-Year National Park," a group of small areas in southern New Mexico (close to some of Fall's own ranch land) that the association believed would make decidedly inferior national parks. With strong opposition from the association and the Park Service, and with Fall's political disgrace in the Teapot Dome Scandal, the proposal did not succeed.[79]

At about the time Mather left the directorship, Horace Albright reported that Congress was considering more than twenty bills for new parks, but that most of these "lacked merit." He noted further that there were "few worthy candidates for parkhood remaining"—a statement that did not anticipate the later growth of the park system.[80] The Service's chief objection invariably seemed to be the lack of sufficient scenic qualities. For instance, Mather opposed creating a park out of the colorful, eroded lands of southwestern North Dakota (much later to become Theodore Roosevelt National Park) because, echoing the Lane Letter, he believed they lacked the "quality of supreme beauty required by National Park standards."

Mather recommended that the area in North Dakota instead become a state park—a recommendation the Service made frequently for areas it did not want in the system.[81] Indeed, largely as a means of relieving pressure on the national parks, Mather convened a conference on state parks in 1921 to encourage growth of these local systems. Advocating a "State Park Every Hundred Miles," the National Park Service agreed to assist the states in their park programs.[82] The 1921 conference signaled a significant step toward involvement with affairs external to the national parks, which efforts would grow dramatically during the New Deal era.

Nature Management

The values and perceptions of National Park Service leaders were reflected in the treatment of nature during Mather's directorship. From the first, the Service did not see itself as a scientific bureau. Its leadership assumed that its unique mandate to leave parks unimpaired did not require special scientific skills and perceptions different from those used in more explicitly utilitarian land management. Biologists were not part of the bureau's emerging leadership circles, and had very little voice in its rank and file. Instead, as Mather claimed in 1917, scientific assistance from other bureaus could be "had for the asking"—and the Service borrowed scientists as well as their resource management strategies.[83] Secretary Lane's 1918 policy letter stated that the "scientific bureaus . . . offer facilities of the highest worth and authority" for addressing national park problems, and the Service should "utilize [their] hearty cooperation to the utmost." The 1925 Work Letter reaffirmed this policy.[84]

In the same year that the Work Letter appeared, Mather, attempting to prove that the Park Service was avoiding needless duplication of government functions, listed the federal bureaus on whose expertise the Service relied. At least six bureaus, representing the departments of Interior, Agriculture, and Commerce, were named as substantial contributors to natural resource management in the parks. The bureaus that the Service, as Mather put it, "calls upon . . . for help" included the Geological Survey (conducting topographical surveys and gauging streams), the Forest Service (preserving trees along roads approaching parks and protecting "park areas for which no funds are available"), the Bureau of Biological Survey (managing wild animals, including reducing predatory species), the Bureau of Animal Industry (vaccinating Yellowstone's buffalo herds and controlling hoof-and-mouth disease), the Bureau of Entomology (fighting insect infestations in parks), and the Bureau of Fisheries (maintaining hatcheries and stocking park lakes and streams).[85]

It is important to note that while the Park Service was steadily building up its landscape architecture and engineering capability, it was content to borrow scientists from other bureaus to manage national park flora and fauna—a telling reflection of how much greater the Service's interest was in recreational tourism than in fostering innovative strategies in nature preservation. With the Park Service borrowing scientific expertise, its natural resource management programs under Mather were to a large extent imitative rather than innovative.

Moreover, natural resource management was an adjunct to tourism management. The Park Service sought to present to the public an idealized setting of tranquil pastoral scenes with wild animals grazing in beautiful forests and meadows bounded by towering mountain peaks and deep canyons. Mather described the parks as having a kind of primeval glory, "prolific with game" grazing in "undisturbed majesty and serenity." Albright shared this idyllic view of the parks, once commenting on the great public appeal of seeing "large mammals in their natural habitats of mountain forests or meadows."[86] Such suggestions of peace and tranquility did not allow for violent disruptions like raging, destructive forest fires blackening the landscapes, or flesh-eating predators attacking popular wildlife. To Park Service leadership, the vision of a serene, verdant landscape seemed to equate to an unimpaired park. Maintaining such a setting amounted to facade management—preserving the scenic facade of nature, the principal basis for public enjoyment.

As before the Park Service was established, natural resource management focused on husbanding certain flora and fauna: forests, fish, and large grazing mammals, primarily the species that contributed most to public enjoyment of the parks. Indeed, the Park Service conducted a kind of ranching and farming operation to maintain the productivity and presence of favored species. Those species that threatened the favored plants and animals had to be sacrificed—eradicated, or reduced to a point where they would not affect populations of the more desired flora and fauna.

The management of nature in the national parks took into account the benefits to be reaped outside park boundaries as well. For example, both Mather and Albright approved predator control programs partly to protect livestock on lands adjacent to the parks. In the manner of wildlife refuges being established by the Biological Survey, Mather noted in his annual report of 1924 that wildlife moving out of the parks to adjacent lands where they could be hunted provided "one of the important factors wherein the national parks contribute economically to the surrounding territory."[87] Thus, tourists would be drawn into the parks, while hunters would use nearby lands. In this regard, Horace Albright told the 1924 conference of the American Game Protective Association that if state governments were cooperative in game conservation, "there will always be good hunting around several of the big parks." Later, in an article on the parks as wildlife sanctuaries, Albright tellingly observed that although animals in the national parks were referred to as "wildlife," once they left the park they were called "game."[88]

As wildlife moved in and out of the parks, so did poachers, who imperiled some large-mammal populations. By the time the Service began operating, illegal hunting of the parks' wildlife had diminished owing to aggressive protection efforts, especially in parks controlled by the army. Nevertheless, poaching remained a serious concern, and rangers patrolled park boundaries and interior areas, scouting for evidence of illegal hunting. The most difficult poaching problems occurred in the recently created Mt. McKinley National Park. Not until 1921, four years after the park's establishment, did the Service obtain funds to hire a superintendent and an assistant to protect the huge numbers of wildlife. Given the vastness of McKinley and its proximity to a railroad and to mining villages, and with road construction ongoing in the park, the park's wildlife continued to be subjected to poaching, especially in remote areas.[89]

The Predator Problem

Of all of the natural resource management efforts in the parks, the most controversial was the killing of predators in order to protect more popular species. Predator control efforts in the parks were in accord with the ongoing, nationwide campaign to control carnivorous enemies of domestic livestock, as demanded by farmers and ranchers and promoted by the Biological Survey. Inherited from army and civilian park management, the programs attained legal justification through the Organic Act's authorization to destroy animals considered "detrimental to the use" of parks.[90] Determined to keep the national parks unimpaired, the Service acted as though the predators themselves were impairments—threats to be dealt with before they destroyed the peaceful scenes it wished to maintain. Mather believed predator control helped increase the populations of the "important species of wild animals," and he once stated that the national parks offered sanctuaries to all wildlife "except predatory animals."[91] Shortly after he succeeded Mather, Horace Albright defined predators as those species that preyed on "animals that add so much to the pleasure of park visitors"—clearly tying predator reduction to public enjoyment. Albright saw predator control as a means of protecting those "species of animals desirable for public observation and enjoyment," and declared that the "enemies of those species must be controlled."[92]

Bloodthirsty predators seemed to have no place in the beautiful pastoral parks, at least not in large numbers. From its beginning the Service practiced predator control "with thoroughness" (as an internal report later put it) and developed an expanded list of undesirable predatory animals—at times including the cougar, wolf, coyote, lynx, bobcat, fox, badger, mink, weasel, fisher, otter, and marten. For a time during the 1920s, rangers destroyed pelican eggs in an effort to reduce the numbers of pelicans in Yellowstone and protect trout populations to enhance sport fishing.[93]

As before 1916, implementation of predator control programs varied from park to park and remained largely at the discretion of the superintendents—depending on, in the words of a Park Service report, "local conditions and the Superintendents' ideas." Generally, the rangers were responsible for predator control; as a means of augmenting salaries and encouraging predator hunting, they were often allowed to sell for personal profit a percentage of the hides and pelts of the predators they killed. In addition, the parks sometimes hired predator hunters.[94] Perhaps the most noted was Jay Bruce, "official cougar killer" for the State of California (whom Mather once had entertain visitors to Yosemite with tales of killing mountain lions). Yosemite superintendent Washington ("Dusty") Lewis reported in 1919 that in the previous three or four years, Bruce had killed more than fifty cougars in or near the park. This had prevented, Lewis stated, the mountain lion's "slaughter" of Yosemite's deer.[95]

Reflecting the policy of borrowing expertise from other agencies, Mather commented in 1926 that most predator control in the parks was conducted by rangers or by the Biological Survey.[96] In parks such as Zion, Rocky Mountain, Glacier, and Grand Canyon, the Biological Survey supplied its own hunters or supervised contract hunters. Its classification of which animals were harmful predators was generally accepted as a guide by the Park Service.[97] The survey further influenced the Service in the means by which predators were exterminated: not only shooting, but poisoning, trapping, and tracking with dogs. Furthermore, the Park Service obtained support from state game and fish offices, especially in California, which supplied hunters (like cougar killer Jay Bruce) and information on predator and prey species.

The predator programs came under increasing criticism beginning in about the mid-1920s. Critics focused on the methods of control (especially the use of poisons and steel traps); the lack of scientific information to justify the programs; and, most fundamentally, the very idea of killing predators in the national parks. Moreover, by the early 1920s some parks had begun to report that the largest predators (wolves and cougars) were disappearing. Glacier, Yellowstone, and Rocky Mountain indicated in 1923 and 1924 that wolves and cougars were reduced to the point of extinction, making it unprofitable to hire special hunters. It is possible that both species were eradicated from these parks by the mid-1920s.[98]

Even though extinction of large predators was taking place in some parks, official and unofficial pronouncements of the Service began to maintain that it was only *reducing* predator populations, not eliminating them. In their 1922 conference, the superintendents stated that some nuisance animals such as the porcupine and the pelican should be reduced in number, but not eliminated. In all cases they agreed that predators should be killed only when they threatened "the natural balance of wild life." Each superintendent was to study conditions in his park and determine if and when any one species was becoming "too powerful for the safety of another." Already some superintendents (at Mount Rainier and Sequoia, for example) had largely ended control in their parks. Yet others continued their programs, as at Rocky Mountain National Park, where in 1922 Superintendent Roger Toll initiated a cooperative effort with the Biological Survey to poison predators or track them with dogs. Toll later stated that he wanted the predators reduced to the "lowest practicable numbers" and that the park had too many predators "for the good of the game."[99]

Throughout the remainder of the 1920s, the Service's basic predator policy included reducing rather than eliminating predators; killing mainly wolves, coyotes, and cougars, with declining emphasis on other predators; and allowing superintendents broad discretion in defining and implementing their predator programs. These policies were affirmed at superintendents conferences of 1923 and 1925. Still, even the smaller predators continued to be hunted, and not infrequently. In September 1926 a particularly striking example of elimination of smaller predatory animals occurred when otters at one of Yellowstone's lakes were killed because they were eating trout—some of which were probably nonnative species introduced for sportfishing.[100]

Criticism of predator control in the national parks intensified in the late 1920s, when organizations such as the Boone and Crockett Club, the New York Zoological Society, and the American Society of Mammalogists protested the Service's policies.[101] These groups and their allies expressed concern that predator control did not have an adequate scientific basis, and stressed their views that predators had a natural place in the parks. University of California biologist Joseph Grinnell spoke to the 1928 superintendents conference (the last one held during Mather's tenure), emphasizing the need for scientific research to guide national park policy and the necessity to leave predators alone.[102]

In March 1929, two months after Mather left office, Director Albright reported to the secretary of the interior that predators were being "controlled but not eliminated." Although a gradual reduction in predator killing had taken place, the Service's overall policy had not changed substantively. The superintendents' continuing discretionary authority was apparent in that, even though they had adopted a strongly worded anti-steel-trap resolution at their 1928 conference (they voted to "forbid absolutely" the use of traps in national parks), such use continued in Grand Canyon until 1930 and in Yellowstone until 1931. Although parks such as Mount Rainier and Sequoia had largely discontinued predator control, Yellowstone aggressively killed coyotes throughout the Mather era and beyond.[103]

Even Albright's official declaration of a new predator policy, published in the May 1931 issue of the *Journal of Mammalogy* (and almost certainly influenced by newly hired Park Service wildlife biologists), kept open the option of killing predators "when they are actually found making serious inroads upon herds of game or other mammals needing special protection." Yet the new policy statement did help move the Service away from the traditional views that had dominated the bureau during Mather's time. Predators, in Albright's words, were to be "considered integral parts of the wild life protected within the national parks."

Albright summed up the new policy by underlining what he saw as the difference between the parks and other lands—that predators not tolerated elsewhere were to be given "definite attention" in the national parks. He pledged the Park Service to maintain "examples of the various interesting North American mammals under natural conditions for the pleasure and education of the visitors and for the purpose of scientific study."[104] But Albright's modification of predator control came only after populations of major predators had been eliminated or seriously reduced in some parks. In the years ahead, the new policy would be observed in varying degrees by different park superintendents. Indeed, Albright himself would staunchly advocate the continued killing of certain predators.

Popular Wildlife Species

The Service's treatment of large-mammal populations did not follow a policy of letting nature take its course; rather, it involved frequent and sometimes intensive manipulation, such as killing predators or nurturing favored species. In his annual reports to the secretary of the interior, Mather regularly and enthusiastically commented on wildlife management activities in the national parks. Yet the 1918 Lane Letter mentioned wildlife only in passing, merely stating that for the "care of wild animals" the Service would use experts from other bureaus. (The Biological Survey, while routinely providing assistance in national park wildlife management activities, actually had full responsibility for "game preserves" in two national parks—Wind Cave in South Dakota and Sully's Hill in North Dakota.)[105]

Ranching techniques applied to national park wildlife were particularly evident with Yellowstone's bison. The Park Service inherited the bison program that the U.S. Army had established in the early twentieth century in an effort to save the species from extinction in the United States. With activities centered at the Buffalo Ranch in the Lamar Valley of northeastern Yellowstone, the Service continued the practice of treating bison much like domestic livestock. The park's "chief buffalo keeper" allowed the animals to roam the valley under the watchful eye of a herdsman during the months when forage was readily available, then rounded up and corralled the herd at the Buffalo Ranch for a time during the winter. There the bison were fed hay raised on approximately six hundred acres of plowed, seeded, and irrigated park land—the most intensive of several haying operations in the park (and one that lasted until 1956). In the corrals, the keepers separated the calves, then castrated many of the young bulls to reduce the number of intact males, who were usually difficult to manage. Despite problems with disease, by early 1929 the herd had reached a population of about 950, up from fewer than 50 when the program began in 1902.[106] (Bison populations in other areas of the park existed at population levels usually not of concern to management, particularly given their more remote locations and lack of contact with the public. In comparison with the Lamar Valley herd, they were easier to ignore.)

By the late 1920s, the Service came to believe that the areas most practical for growing hay for winter feeding were used to capacity, and that the Buffalo Ranch and rangelands in the Lamar Valley could support only about a thousand bison. Already, as a means of thinning the herd, the park sold bison to be slaughtered for market. Superintendent Albright's suggestion in 1924 that the park establish a pemmican plant as possibly "the only feasible way of dispensing of buffalo meat in large quantities" was never realized. The first slaughtering occurred in 1925, when Mather gave permission for seventeen animals to be killed in conjunction with "Buffalo Plains Week," a local summer celebration. This was followed by several years of slaughtering for market. For instance, Chief Ranger Sam Woodring noted in 1929 that the park had "disposed of 100 [bison] steers this fall, the slaughter contract being awarded to the highest bidder." The park also began donating meat to local Indian tribes.[107]

In another practice that helped control herd size throughout most of Mather's tenure and well beyond, the Service obtained authority from Congress to ship "surplus" bison to state and federal preserves and to local parks and zoos across the country. Those bison remaining in Yellowstone constituted a major tourist attraction, with the greatest spectacles being roundups and stampedes, the latter held for a few years in the 1920s, especially to entertain distinguished visitors.[108]

Following precedents set by army and civilian park managers and by the Biological Survey, the Park Service gave constant attention to other ungulates, such as deer, antelope (pronghorn), bighorn sheep, and especially elk. Attempting to maintain "ideal" wildlife populations, the Service usually sought to increase herds—or to reduce them when they were believed to be too large. The desire to control populations led managers to concentrate on the animals' food supply—mainly the condition and sufficiency of their range, particularly in winter when snow cover limited the areas available for grazing and browsing.[109] It seems the Service valued park grasslands mainly as pasturage for ungulates, rather than as areas biologically important for plants and other life forms.

To augment the food supply when the range appeared to be insufficient, the army had begun winter feeding of antelope, deer, and bighorn sheep in Yellowstone in 1904. Later the Park Service undertook supplemental feeding of hay in several other western parks during harsh winter weather. In Yellowstone, according to Superintendent Albright, the winter feeding program used fifteen hundred tons of hay by 1919. Ten years later, the park reported that in addition to obtaining hay from neighboring sources, it had three hay "ranches" for winter feeding of elk.[110] This was in addition to hay raised for bison and for large numbers of horses used by park rangers and by concessionaire-operated trail rides. In Jackson Hole, south of Yellowstone, the Biological Survey had managed a similar winter feed program since 1912 at its National Elk Refuge, feeding hay to thousands of elk that migrated to the south end of Jackson Hole from Yellowstone and other nearby high country. By this means, the Biological Survey helped care for Yellowstone's "southern elk herd."

The winter feeding program was also intended to entice the animals to stay in the parks. Human settlement and domestic livestock grazing in adjacent areas (whether public or private lands) greatly complicated wildlife park management by impeding the ability of wildlife to use winter grazing areas in lower valleys outside the parks. In addition, migration out of the parks usually meant that the animals were subjected to hunting. Thus, as a further means of protecting elk, Yellowstone's rangers sometimes sought to block winter migration (except to Jackson Hole) by herding the animals back into the park, but apparently this approach met with little success.[111]

Even though Mather and Albright promoted national parks as a source of game for hunting on adjacent lands, they frequently objected to the slaughter that took place outside parks during fall migration. Basically, they wanted well-controlled hunting, not wanton killing of the animals. Despite the Service's protests to the appropriate state governments, excessive hunting often occurred on adjoining lands.[112] Yellowstone chief ranger Sam Woodring reported in January 1929 that he had seen hunters north of the park so thick that they looked like a "skirmish line of troops." Woodring recalled that several years earlier he had watched while Montana hunters just outside Yellowstone surrounded a herd of elk at 7:00 A.M. and held them for an hour, at which time state laws permitted hunting. Then half the herd was "shot down in less than thirty minutes."[113]

With care and feeding, populations of the large grazing and browsing animals appeared to increase in the national parks during the 1920s. Yet at times under such protected conditions, the animals' population fluctuations seemed much more pronounced. Lacking systematically obtained data, it is probable that population and mortality counts by the parks were not very accurate. Still, park management concluded that Yellowstone experienced an unusually high die-off during the winter of 1919–20, when it estimated that perhaps as many as fourteen thousand elk had starved to death. It also believed that another elk population crash occurred in the winter of 1927–28.[114]

The greatest controversy involving animal populations took place in the Kaibab National Forest, on the plateau north of Grand Canyon and adjacent to the new national park created from Forest Service lands in 1919. The protected deer population on the Kaibab Plateau increased rapidly until by 1920 the Forest Service believed the herd was far too large for the forage available. Concerned about the importance of the deer to the visitors at Grand Canyon National Park (and not convinced that there were too many deer), Mather objected to the Forest Service proposal to shoot about half the herd. His objections helped delay the kill. With no artificial reduction, the deer population in the Kaibab crashed during the winter of 1924–25, when thousands died from starvation and cold. Although the very high mortality rate continued through the 1920s, Mather—without well-researched data—never accepted that there had been too many deer.[115] (In the 1930s the Kaibab situation would directly influence Park Service ungulate reduction policy.)

Having fostered an apparent increase in big-game populations in the parks, the Service obtained specific authority from Congress to ship "surplus" animals to state and federal preserves and to city parks and zoos in different parts of the country. In addition to the bison shipments, by the early 1920s Yellowstone had become, in Mather's words, a "source of supply" and a "distributing center" for elk, with the park having shipped the animals to twenty-five states, plus various locations in Canada.[116] Some of the elk went to other national parks; for instance, prior to the establishment of Rocky Mountain National Park, the Forest Service began introducing Yellowstone elk to replenish that area's dwindling elk population. Besides elk and bison, Yellowstone shipped out bear. Mather noted in 1921 that at times it had even been possible "to give away families of beaver, and these interesting animals have been captured and shipped with encouraging success."[117]

Building on precedent, the Service set up zoos in the parks to ensure that tourists would have a chance to see the more popular animals. Mather approved a zoo in the Wawona area of Yosemite, and for a while the park maintained caged cougars that had been captured as cubs. In the early 1920s, "for the pleasure and education of the visitors," as Mather stated it, the Park Service imported from the San Joaquin Valley to Yosemite a small herd of Tule elk, which were not native to the park and which were kept behind fences.[118] In Yellowstone, visitors to the Buffalo Ranch often had the opportunity to see bison feeding in fenced areas. And in 1925 the park set up a zoo at Mammoth, with bears, coyotes, and a badger exhibited in captivity with bison. Foretelling an even greater effort to exhibit bison, Horace Albright wrote in 1929 that the bison should be made "more readily accessible to the visiting public." He argued that the Park Service needed to solve the problem of "how to handle this herd under nearly natural conditions and at the same time get it near the main highways where it can be easily and safely observed."[119]

Shipping the large mammals out of the parks, placing them in park zoos, or slaughtering them for market did not adequately solve the perceived problem of excessive animal populations. As early as 1921, Mather noted that the demand for bison to be shipped out had not "kept pace with the supply" available in Yellowstone.[120] By the late 1920s, the pressure that large numbers of animals were believed to be placing on their rangelands had become a matter of widespread concern. The increased deer population in Yosemite, for example, caused Albright to comment at the 1928 superintendents conference that "it looks as if when we protect the deer we are going to lose the flowers." Rocky Mountain National Park, which earlier had built up its elk herd through importation and winter feeding and protected them by killing predators, also had begun to worry about excess populations.[121] The Mather era ended with populations of most ungulates believed to have increased, and with gathering alarm about overgrazed, deteriorating rangelands burdened by an apparent surplus of animals.

The management of bears in the parks presented fewer problems than that of ungulates, and, perhaps more than with any other large mammal, was most directly aimed at public enjoyment. Virtually all bear management during Mather's time had to do with controlling their interaction with park visitors, rather than manipulating their population to a desired level. On the one hand, the Service sought to bring bears and people together, particularly with the bear shows at garbage dumps and by allowing roadside feeding. On the other hand, it removed "problem" bears that threatened visitors or their possessions.

From early in national park history, the public could feed black bears along park roadsides. Even though it officially frowned on such activity, the Park Service allowed roadside bear feeding to continue in Yellowstone and other parks as automobile travel stimulated public interest.[122] Also popular were the bear shows at garbage dumps, where visitors could enjoy the bears at close range. A major attraction well before 1916, the shows continued without interruption under Park Service managers aware that bears were especially popular animals. Horace Albright noted in 1924 that bears "seriously compete with geysers and waterfalls and magnificent canyons" for the public's attention, and attract greater interest than any other animal in the parks.[123] In a number of national parks, the shows became a regular feature, as in Sequoia, where on summer evenings hundreds of tourists gathered at Bear Hill to watch black bears feed. In Yellowstone the larger and more dangerous grizzly bears came to dominate the feeding shows.

Although bears are predatory, they seem not to have been subjected to predator control programs, perhaps largely because of the public's fascination with these animals. However, coupled with the Park Service's efforts to let the public view bears was the need to remove the animals when they became troublesome—and this often meant killing them. As tourism increased, so did the conflicts between people and bears, in campgrounds, near hotels, and along roadsides. The Service frequently shot the most recalcitrant animals, as in Sequoia, where by the 1920s the practice had become common. The parks also shipped problem bears to zoos.[124] With roadside feeding and bear shows being feature attractions, bear management continued essentially unchanged through Mather's administration and beyond.

More extensively than any other wildlife, the Park Service manipulated fish populations. It aggressively promoted sportfishing, building on pre-1916 precedent to make fishing a premier national park attraction and a major aspect of tourism management. Although the Service sought to halt the poaching of mammals in the parks, it enthusiastically sanctioned not only the regulated taking of fish but also the introduction of numerous nonnative species.

Mather reported annually on fish management, noting in 1922 that "fish planting on a grand scale" in Yellowstone had resulted in a public catch of about sixty thousand trout in one year. That same year the park planted 1.2 million fingerlings of trout and almost 7.4 million fry and eyed eggs, and installed a hatchery at Fish Lake. Mather stated that the "magnificent results" of these efforts proved the necessity of establishing hatcheries in other parks; thus Yellowstone, Glacier, Mount Rainier, and Yosemite, for example, got new hatcheries. In 1925 a study of Glacier's "fisheries needs" determined that the park had a "great deal more fish food . . . than was formerly believed." Based on this finding, Mather believed that more intensive stocking of Glacier's streams and lakes was justified.[125] The following year the park planted close to 3.3 million trout fry, all from its own hatchery. Also in 1926, Yellowstone planted 5 million eyed native cutthroat trout eggs, and the Oregon state game commission shipped "several carloads" of trout to Crater Lake. Although many of the fish planted in the parks were native species, others were not. For example, fish introduced in Yellowstone Park included nonnative rainbow, brook, brown, and lake trout.[126]

For staffing and operating the hatcheries and for stocking lakes and streams, the Service continually relied on the Bureau of Fisheries, a commodity-oriented and production-oriented bureau devoted to assisting the nation's commercial fishing industry and sportfishing enthusiasts. Similarly, game and fish commissions in states such as California, Colorado, Washington, and Oregon worked closely with the Park Service in managing fish. The state commission in Oregon, for example, assisted the Service in the stocking of Crater Lake.[127] Reflecting a continuing dependence on outside expertise, Horace Albright reported in 1929 on the detailing of a fisheries expert from the Bureau of Fisheries to supervise "fish operations" in the parks. Albright viewed this as the "first step" in coordinating National Park Service fish management. He anticipated that the Service could be sure that the employee so detailed would "take our point of view in developing the sport of fishing in the National Parks, and see to it that the parks are stocked with fish."[128]

Earlier, the Service's fish programs aroused concern on the part of the Ecological Society of America, which passed a resolution in late 1921 strongly opposing the introduction of nonnative species. The resolution referred not just to fish but to all plants and animals not native to the parks. The society urged that such introductions be "*strictly forbidden* by the park authorities." Albright responded in regard to Yellowstone's fish programs, disingenuously stating that his park was "averse" to planting nonnative species. He added that the park was continuing to stock nonnative fish, but was limiting them to streams where these species had previously been introduced. Moreover, he asserted that the Service was adhering to a policy that "foreign plant and animal life are not to be brought in."[129]

In actuality, the Park Service had developed no stringent prohibition of such practices. Planting of fish, including nonnative species, continued on a "grand scale," along with introduction of nonnative trees, shrubs, and grasses for landscaping developed areas. Also, nonnative grasses (for instance timothy, now recognized as an aggressively invading species) were planted in Yellowstone's irrigated fields to supply hay for winter; and Tule elk were displayed behind fences in Yosemite.[130] The Park Service's attitude toward nonnative species was equivocating. Indeed, neither of the major policy statements of the Service's first decade (the Lane and Work letters) prohibited introduction of nonnative flora and fauna. Both specifically approved the propagation and stocking of fish, without limiting these activities to native species.[131]

In March 1929, shortly after Mather's resignation, Director Albright repeated his claim (this time to the secretary of the interior) that "exotic plants, animals, and birds, are excluded" from the parks. In truth, there were still exceptions to this policy, such as the continued introduction of nonnative fish.[132] On the other hand, the Service had started trying to *eliminate* certain nonnative species, fearful that they were harming native flora and fauna. In 1924 it began eradicating feral burros from Grand Canyon, killing more than twelve hundred in the first five years. In the late 1920s, rangers in Hawaii National Park began a campaign to eradicate feral goats from the park.[133] Many years later, both of these programs would bring the Service into heated public controversy.

Forest Management

Throughout Mather's directorship, the Park Service maintained a steadfast policy of protecting the forests from two major threats, fire and insects, of which fire—the "Forest Fiend," as Mather called it—seemed the greater. For assistance, Mather turned to the Forest Service, which had developed expertise in fire fighting. Moreover, many national parks were adjacent to national forests; thus the two bureaus shared miles of common boundary and recognized fire as a common enemy threatening their different interests and purposes.[134]

Even though the Forest Service had a fundamentally different mandate for land management—providing for the harvest of a variety of resources—the Park Service readily accepted the Forest Service's total-suppression fire policy (this was, after all, a continuation of the army's policy for the parks). When Sequoia superintendent John White proposed the alternative of "light burning" of forest debris and understory as a means of avoiding larger conflagrations, Park Service leaders stayed with tradition and supported full suppression, as practiced in the national forests.[135] In accord with the thinking of the time, and seeking to keep its forests green and beautiful, the Park Service viewed suppression of all park forest fires (it did not differentiate between natural and human-caused fires) as fully compatible with its mandate to preserve the national parks unimpaired. Under Mather's prodding, Congress in 1920 began making annual appropriations for fire control in the parks.

A prime example of fire's threat to adjacent lands administered by both services occurred in the summer of 1926, when major fires broke out in Glacier National Park and in neighboring national forests. For most of the summer the Park Service and the Forest Service fought these fires, at huge expense.[136] The Glacier fire was especially important in that it inspired the Park Service to create an office of forestry—the first formal organizational designation specifically for natural resource management. Forestry management had begun as simply another duty placed under forester Ansel F. Hall, who was chief of the recently created Division of Education as well as the Park Service's "chief naturalist." But in the summer of 1928, Hall hired John Coffman from the Forest Service, and the division's designation was changed to "education and forestry."

Going beyond the customary use of generalist rangers and other available personnel to fight park fires, Hall's and Coffman's positions marked the first use of in-house, professionally trained foresters. In addition, the Park Service in 1927 had joined the Forest Protection Board, an interagency organization that fostered cooperative fire suppression, and on which Coffman served as national park representative.[137] The forestry office was located in Hilgard Hall, on the University of California campus in Berkeley. A den of forestry expertise, Hilgard Hall also housed the Forest Service's California Forest and Range Experiment Station and the university's forestry study programs.[138] Such close proximity to other foresters surely encouraged an even stronger commitment to strict fire suppression in national parks.

Coffman went to work energetically, overseeing fire protection for the Park Service "as thoroughly as a Fire Chief in the U.S. Forest Service," as he later put it. His duties were reflected in the "Forestry Policy," a comprehensive, systemwide statement on national park fire management prepared by Hall and Coffman in the fall of 1928. The new policy called for the Service to prepare fire plans for all parks for the "prevention, detection and suppression" of fires; train firefighting personnel; cooperate with other federal and state agencies with lands near national parks; implement "hazard reduction" (such as the removal of combustible dead trees) in areas with high fire potential; and establish a fire reporting and review process for each fire season and for individual fires. This statement became official Park Service policy in 1931.[139]

The Forest Protection Board also planned to fight forest diseases and insect infestations. Participating on the board, the Department of Agriculture's Bureau of Plant Industry had the primary responsibility for assisting the Park Service and other participating bureaus in disease control. Of particular concern was white pine blister rust, a nonnative fungus. Albright reported in 1929 that the bureau had checked for blister rust in Acadia, Mount Rainier, and Glacier. This disease eradication program would lead to extensive control efforts in the 1930s.[140]

Although the Park Service's new forest policy statement contained no reference to forest diseases, it did have a section on insect control, which received increased attention during Mather's time. By the mid-1920s Congress was appropriating funds for the control of insects in the parks. Principal targets at the time included the lodgepole sawfly and spruce budworm in Yellowstone, the needle miner and bark beetle in Yosemite, and the pine beetle in Crater Lake (where Mather feared the infestation was "utterly beyond control"). To combat such attacks, the Service used chemical sprays and also felled infested trees, peeled off their bark, and burned them.[141] The Park Service relied on the Bureau of Entomology, which trained personnel and frequently supervised the control work. The bureau was, Mather noted, very supportive in combating insect "depredations" that threatened to cause "injury of the scenic beauty."[142]

Significantly, the Service concentrated its insect control effort in scenic areas important to the visiting public, such as along road and trail corridors and in zones of special appeal. Mather reported in 1925 that in Yellowstone spraying had been increased "along the roads and at places of the most scenic importance." Insect control at Crater Lake the following year focused on protecting "a beautiful stand of yellow pine, one of the three finest forests in the park."[143] Indeed, the policy of concentrating on areas important to the public would be reiterated in the Service's 1931 forest policy statement: unless there were extenuating circumstances, "remote areas of no special scenic value and not of high fire hazard, little used or seen by the public and not planned for intensive use within a reasonable period of years, may be omitted from insect control plans."[144]

Like fire, hordes of insects threatened to damage forests over vast tracts of public land, no matter what mandates governed their management. Thus, a February 1928 meeting of the Forest Protection Board, attended by several national park superintendents, stressed multibureau cooperation in insect control. The insect spraying program was, in the words of a Forest Service representative at the meeting, usually carried out "regardless of boundaries" between the parks, forests, and other public lands. Horace Albright agreed to the cooperative effort, stating that the Service's forestry office stood ready to participate.[145]

In protecting forests and other areas of the national parks, the Service faced the problem of livestock grazing, which impacted native flora and fauna in many parks. During the campaign to create the Park Service, Mather had supported grazing in the parks as a means of securing congressional support. Authorizing livestock grazing in all parks but Yellowstone, the Organic Act was followed by Secretary Lane's policy letter of

1918. Diverging from the act's general authorization of livestock grazing, Lane declared that sheep would not be allowed in national parks. The Service was seriously committed to fighting this use, more damaging than cattle grazing; and its fight against sheep would prove more successful than against cattle.

Regarding cattle, the Lane Letter declared they could graze in "isolated regions not frequented by visitors, and where no injury to the natural features of the parks may result from such use."[146] In Sequoia the army had terminated grazing, but the Service revived it in 1917, under pressure from ranchers who—emboldened by the Organic Act's authorization of grazing—claimed they needed park lands to ensure sufficient beef supplies because America's entry into World War I seemed probable. Even after the armistice negated this purported patriotic rationale, the ranchers fought to continue grazing (a pattern that occurred in other parks during the World War II era). Although Mather opposed increased grazing in the parks during World War I, in his public pronouncements he sometimes showed a willingness to compromise, noting after the war that grazing exemplified "the principle of *use*" of the parks and would be allowed where it did not interfere with tourist use.[147]

In stark contrast, scientific judgments on grazing recognized the extensive impact on the national parks. Charles C. Adams, a well-known biologist with Syracuse University, surveyed several parks and commented on overgrazed conditions in a 1925 article in *Scientific Monthly*. In Sequoia, for example, Adams reported that areas in the northeast part of the park had possibly "suffered more . . . than any other overgrazed area" he had seen in a national park or even a national forest. And on Grand Canyon's south rim, the "extreme overgrazing" was so bad in both the park and the adjacent national forest that Adams believed it was impossible to tell by the range condition whether the land was "in the park or in the forest." He added that, partly as a result of earlier livestock grazing, Grand Canyon had already been "greatly modified from a natural wild [area]" before becoming a national park in 1919.[148] Although Mather strongly opposed repeated attempts by ranchers to increase grazing privileges, cattle grazing continued in many national parks and would prove an enduring vexation for the Park Service.

Ecological Concerns and Mather's Leadership

During the Mather years, objections to the Service's management of national park flora and fauna were infrequent (except for complaints from ranchers and others who basically opposed parks and wanted access to their resources). The Ecological Society of America's 1921 resolution against introduction of nonnative species was similar to resolutions by the American Association for the Advancement of Science adopted in 1921 and 1926.[149] Yet these statements hardly represented sustained criticism. The protest that built up in the middle and late 1920s against the killing of predators was perhaps the most severe criticism of the Service's natural resource management encountered during Mather's time. As was true of professional organizations, few individuals criticized the Service's treatment of nature in the parks. Joseph Grinnell, who had a special interest in Yosemite, may have been the most consistently vocal advocate for managing the parks on a more scientific basis.[150]

Probably the most penetrating critique came from biologist Charles Adams, who in his 1925 analysis of the parks urged that park management align itself with the emerging science of ecology. Adams' examination of such programs as fire, wildlife, and fish management led him to conclude that the Service must develop an ecological understanding of its natural resources. As he put it, if the Service is to preserve the parks "in any adequate manner . . . there must be applied to them a knowledge of ecology." He referred to the "theoretical" policy of maintaining the parks as wilderness—a policy to which the Park Service had "not adhered." The Service was not meeting what he believed to be its true mandate, the preservation of natural conditions in the parks.[151] Adams' critique was important as an early effort to promote ecologically based management of the national parks (and thus to interpret the Organic Act in that regard). Yet even more important was that it had little if any effect on national park policies: the Park Service under Mather was firmly set on a different course.

Adams noted also that naturalists in the parks were not "devoted to technical research, but in the main to elementary educational work with the park visitors."[152] Indeed, in addition to its manipulation of flora and fauna, the Service's natural history concerns focused on ensuring public enjoyment, not preserving biological integrity. Establishment of Ansel Hall's Education Division in 1925 confirmed the naturalists' duties as an important part of park operations. Hall, the chief naturalist, advised parks on museum planning and operation, and on hiring ranger naturalists and giving nature walks and evening campfire talks, among other programs. (The naturalist and education functions were forerunners of today's "interpretation" activities.)[153]

In the late summer of 1928, Hall's division rather suddenly moved toward generating a scientific base for natural resource management when

Yosemite's assistant park naturalist, George M. Wright, an independently wealthy biologist, offered to fund a survey of national park wildlife. From his observations in Yosemite and other parks, Wright understood that the Service had no scientific understanding of its wildlife populations.[154] Presented with his proposal, the Park Service, after some deliberation, agreed to its first systemwide research designed to enhance management of natural resources. The project was to be conducted by an expanded educational division. Moreover, Wright's proposal prompted Albright's November 1928 suggestion to Mather that the Service develop its own scientific expertise, as it had already done with landscape architecture and engineering. As Albright saw it, the Service needed "a few specialists with scientific training who have strictly the National Park point of view."[155]

Under Mather the Park Service had established itself as a national leader in recreational tourism, but had done nothing in the way of research-based preservation of natural resources. Only at the very end of Mather's directorship—and with the promise of private funding—did the Park Service move to develop in-house scientific expertise to address natural resource management issues. Yet this shift toward ecologically informed management would have to contend with the emphasis on recreational tourism that Mather had firmly established, building on the policies of earlier park managers.

Little concerned about science and ecology, Stephen Mather was a promoter, builder, and developer of the national park system. Conservationists of later generations would question his devotion to tourism and the treatment of nature in the parks during his tenure, but his efforts greatly advanced the formation of a system of national parks that are today highly valued both for their scenery and for their biological richness. Assuming leadership at a propitious time in national park history and backed by highly placed conservation-minded friends, Mather made the parks an enduring feature of the American landscape and a source of national pride. He resigned as director effective January 12, 1929. Through fourteen years of extremely demanding work, interrupted by nervous collapses and culminating in a heart attack and a stroke, he sacrificed his health for what he saw as a truly grand cause. When Mather died in January 1930, a year after his resignation, tributes poured in from Congress, conservation groups, businessmen, officials, and friends across the country.

In organizing the Park Service and giving it direction, Mather imparted his vision of what national parks should be—ideas which the new bureau's emerging leadership readily accepted. In effect, Mather envisioned the national parks as "nature's paradise," a kind of rugged, mountainous version of peace and plenty. The Service sought to present the parks as a paradise of beauty and richness, free of fires and predators. Underlying Mather's vision were strong social concerns. He had thoroughgoing democratic and patriotic tendencies: to him, the national parks were places where American people, through "clean living in God's great out-of-doors," could renew their spirits and become better citizens. Furthermore, the parks were "vast schoolrooms of Americanism," where people could learn to "love more deeply the land in which they live."[156]

It is important to note that the management traditions firmly set in place by Mather and his emerging leadership cadres flowed quite logically from the founders' vision of the parks as scenic pleasuring grounds. Moreover, throughout Mather's career, Congress did not challenge his management and development of the parks. Rather, through creation of numerous new parks and through increased funding for development especially during his last years as director, Congress clearly indicated its approval of Mather's policies.

Utilitarian Aesthetics and National Park Management

Mather and other supporters of the National Park Service have sometimes been identified as "aesthetic conservationists," concerned about preserving lands for their great scenic beauty—as opposed to the "utilitarian conservationists" exemplified by Gifford Pinchot and the Forest Service, who sought sustained consumptive use of natural resources.[157] Certainly, through its determined efforts to preserve the scenic facade of nature, the Park Service under Mather focused on aesthetic conservation. But as practiced during the early decades of the Park Service, the nurturing of forests and certain animal species that contributed most to public enjoyment had a strongly utilitarian cast. It was, to a degree, even "commodity" oriented, as with fish management and the ranching and farming types of operations intended to ensure an abundance of the favored large mammals.

Just as it was virtually impossible to separate the basic idea of national parks from tourism development and economics (a connection dating back to the Northern Pacific Railroad's support of the 1872 Yellowstone legislation), so too was it difficult to separate the treatment of specific park resources (bears, fish, and forests, for example) from the promotion of public enjoyment of the parks, which fostered tourism and economic benefits. In viewing recreational tourism effectively as the highest and best use of the national parks' scenic landscapes, and developing the parks for that purpose, the Service took a "wise use" approach to the parks—an approach reflected in the bureau's capitalist-oriented growth and development rhetoric. Through the promotion of tourism in the national parks, scenery itself became a kind of commodity.

The basic concept of setting lands aside as national parks, the development of the parks for tourism, and the detailed management of nature in the parks—none of these ran contrary to the American economic system. The establishment of national parks prevented a genuine free-enterprise system from developing in these areas and required a sustained government role in their management. But this was done in part as a means of protecting recognized scenic values, which through tourism also had obvious economic value. With regard to national parks, aesthetic and utilitarian conservation coalesced to a considerable degree; frequently the differences between the two were not distinct. The national parks, in fact, represented another cooperative effort between government and private business—notably railroad, automobile, and other tourism interests—to use the resources of publicly owned lands, particularly in the West. Through the Park Service, the federal government collaborated with business to preserve places of great natural beauty and scientific interest, while also developing them to accommodate public enjoyment and thereby creating and perpetuating an economic base through tourism.

With no precedents and no scientific understanding of how to keep natural areas unimpaired, the newly created National Park Service believed that it was truly preserving the parks. During Mather's time the Service seemed to define an unimpaired national park as a *carefully and properly developed* park. With use and enjoyment of the parks being unmistakably intended by the Organic Act, harmonious development of public accommodations became a means of keeping parks "unimpaired" within the essential context of public use.

In comparison with other public and private land management practices of the time that championed consumptive use of resources, the national parks stood almost alone in their orientation toward the preservation of nature. Generally perceiving biological health in terms of attractive outward appearances, the Service seemed to believe that it could fulfill what Mather called the "double mandate" for *both* preservation and public use. It could preserve what it considered to be the important aspects of nature while promoting public enjoyment of the parks. For instance, the 1918 Lane Letter, the principal national park policy statement of the Mather era, embraced these two goals without any suggestion of contradiction. It asserted that the parks were to remain "absolutely unimpaired," but also stated that they were the "national playground system."[158]

The Park Service's faith in the importance of development and its compatibility with maintaining natural conditions in the parks found expression on no less than the bronze plaque honoring Stephen Mather, cast shortly after his death and with replicas in many national parks and monuments. The plaque's inscription noted that in laying the "foundation of the National Park Service" Mather had established the policies by which the parks were to be "*developed* and conserved unimpaired" for the benefit of future generations (emphasis added).[159] This assertion—in effect a restatement of the Organic Act's principal mandate—affirmed the belief that developed parks could remain unimpaired. It would characterize Park Service rationale and rhetoric from Mather's time until at least the end of the first half-century of the Service's history. By that time (the mid-1960s) increased postwar tourism and an improved understanding of ecology would reveal much more clearly the inherent tension between park development and preservation.

Biologist Charles Adams recognized in 1925 that the U.S. Forest Service had been launched with the advantage of a forestry profession already developed in Europe in the late nineteenth century. In contrast, he believed that national parks were a "distinctly American idea," with European precedents limited to "formal park design rather than large wild parks" such as those in the United States. Adams noted also that there had been "no adequate recognition" that "these wild parks call for a new profession, far removed indeed from that of the training needed for the formal city park or that of the conventional training of the forester."[160] In effect, America's national parks required more than "facade management"—more than the customary landscape architecture and forestry practices of the Park Service.

Indeed, during the Mather era the Service built on precedents it found in landscape design and in tourism and recreation management to make the parks enormously inviting. Although operating under a unique and farsighted mandate to keep the parks unimpaired, the newly established bureau relied on precedents of traditional forest, game, and fish management. The Service practiced a selective kind of preservation, promoting some elements of nature and opposing others—altering natural conditions largely in an attempt to serve the other part of its mandate, the public's enjoyment of the parks.

CHAPTER 4

The Rise and Decline of Ecological Attitudes, 1929–1940

We know that it is impossible to keep any area in the United States in an absolutely primeval condition, but there are reasonable aspects to it and reasonable objectives that we can strive for.—GEORGE M. WRIGHT, 1934

The survey of park wildlife initiated in the summer of 1929 and funded through the personal fortune of biologist George Wright marked the National Park Service's first extended, in-depth scientific research in support of natural resource management. The success of this effort inspired the Park Service to establish a "wildlife division," inaugurating a decade of substantial scientific activity within the Service. During this period, the wildlife biologists under Wright developed new perspectives on natural resources, opening new options for park management. They promoted an ecological awareness in the Service and questioned the utilitarian and recreational focus that dominated the bureau.

Yet in January 1940, little more than a decade after the survey began, the Park Service biologists were transferred to the Interior Department's Bureau of Biological Survey.[1] Although the biologists remained responsible for national park wildlife programs, their administrative separation symbolized the diminishing influence of science in the Service by the late 1930s. The decade of the 1930s thus witnessed a rise—and then a decline—of ecological thinking in the National Park Service. It also saw a vast diversification of Park Service programs, which expanded responsibilities beyond management of mostly large natural areas and drew attention to matters other than nature preservation.

Park Service Leadership and the Wildlife Biologists

In addition to efforts to make tourism development harmonious with scenic park landscapes, the Service during the Mather era tended to measure its success in leaving the parks unimpaired by the degree to which it restricted physical development. The undeveloped areas (the vast backcountry of the parks) were considered to be pristine, evidence that park wilderness had been preserved. For example, in the fall of 1928 Yellowstone superintendent Horace Albright (soon to succeed Mather as Park Service director) published a *Saturday Evening Post* article entitled "The Everlasting Wilderness," in which the absence of physical development was equated with pristine conditions. Responding to fears that the Service might "checkerboard" the parks with roads, Albright noted the relatively small percentage of lands impacted by road and trail construction in the parks. He argued that Yellowstone's roads affected just ten percent of the park, leaving the remaining ninety percent accessible only by trail—a huge backcountry of "everlasting wilderness" with flourishing wildlife and excellent fishing streams. Comparable statistics were given for Yosemite, Grand Canyon, Mount Rainier, and other parks. All national parks, he wrote, were to be "preserved forever in their natural state," and the vast majority of Yellowstone's lands remained as "primeval" as before the area became a park.[2]

Albright notwithstanding, virtually the entire scientific effort within the National Park Service during the 1930s contradicted such thinking. A clear and concise statement of the scientists' perceptions came in a 1934 memorandum from Ben H. Thompson, one of the Wildlife Division's biologists, when he wrote to Arno B. Cammerer (who succeeded Albright as director in 1933) about setting aside supposedly pristine park areas solely for scientific study. Thompson bluntly declared that no "first or second class nature sanctuaries are to be found in any of our national parks under their present condition." He cited factors such as the parks' limited size; even a park as large as Yellowstone could not provide "protection and habitat unmodified by civilization" for carnivores and large ungulates.

Thompson then detailed some of the changes that had occurred. He declared that cougar, white-tailed deer, wolf, lynx, and perhaps wolverine and fisher, were most likely "gone from the Yellowstone fauna." Rocky Mountain National Park's "carnivore situation" was much the same, except that it had also lost its grizzly population. At Grand Canyon feral burros had "decimated every available bit of range" in the canyon, and domestic livestock had taken a "heavy toll from the narrow strip of South Rim range." Moreover, Grand Canyon's cougars were "almost extirpated," and bighorn sheep "greatly reduced," while the "entire ground cover and food supply for ground dwelling birds and small mammals" had been altered by cattle grazing. Yosemite National Park had lost its bighorn and grizzly populations, and its cougars were "almost gone." In Glacier the grizzly were "very scarce," the trumpeter swan and bison were missing, and game species in general were "seriously depleted because of inadequate boundaries." Finally, Thompson commented that there was "no need to repeat the story for the smaller parks."[3]

Thompson's views of park conditions were in striking contrast to Albright's depiction of the parks as "preserved forever in their natural state." Albright's ideas arose from essentially romantic perceptions of the majestic landscapes, equating the parks' undeveloped and unoccupied lands with unimpaired conditions—a perception almost certainly shared by Park Service officials and by the public.

But the new cadre of wildlife biologists judged the same landscapes in ecological terms. Although roads and other development had not penetrated many areas of the national parks, other activities had, such as predator control, cattle grazing, and suppression of forest fires. As Thompson indicated, these interferences had greatly altered natural conditions, affecting backcountry well away from developed areas.

The wildlife biologists thus became a minority "opposition party" within the Service, challenging traditional assumptions and practices—in effect reinterpreting in scientific terms the Organic Act's mandate to leave the parks unimpaired. Throughout the 1930s they urged that the Service concern itself not just with scenery and public enjoyment, but also with careful, research-based management of natural resources so as to leave the parks in a condition as near to pristine as possible. Events of the 1930s would reveal the Park Service's response to this new perception of its mandate.

The continuity between the administrations of Stephen Mather and Horace Albright has been seen as remarkably strong.[4] Indeed, Mather's constant reliance on Albright's support and advice resulted in a virtually seamless transition between the two directorships. Albright too was a promoter, builder, and developer of the national park system. As Mather's chief assistant and then as director, he greatly expanded the park system and managed the parks to ensure public enjoyment.

Albright's directorship was brief—January 1929 to August 1933. Reversing the direction taken by Mather, who left mining to work for the national parks, Albright resigned from the Service to become an executive of the United States Potash Company. Throughout the rest of his long life, however, he kept exceptionally close watch on Park Service activities, continually passing judgment on the Service's operations and speaking out with firmly held opinions. As director, he supported the survey and the Wildlife Division—yet he no doubt failed to anticipate the management implications of the wildlife biologists' new policies. A dedicated proponent of recreational tourism in the parks, Albright would remain steadfastly loyal to most management practices of the Mather era, which often would place him at odds with the wildlife biologists. At times, he proved one of their most vocal adversaries and critics.

Albright could criticize with authority. He had been one of the principal founders of the Park Service, Stephen Mather's closest confidant, superintendent of Yellowstone, and the Service's second director. After leaving the Park Service and joining U.S. Potash, Albright relocated from Washington to another hub of power, with an office in midtown Manhattan, high in the new complex known as Rockefeller Center. There he maintained close contact with national park benefactor John D. Rockefeller, Jr.—a relationship of enormous importance to National Park Service interests.

Arno Cammerer, Albright's successor, had been in the Service's directorate since 1919. Although much less dynamic than Mather and Albright (and less prominent in the annals of Park Service history), Cammerer effectively led the bureau during a period of rapid change and expansion. His tenure as director lasted until 1940, when for reasons of poor health (probably exacerbated by his protracted difficulties with Secretary of the Interior Harold L. Ickes) he stepped down to become regional director in the Richmond, Virginia, office. As Park Service director during the New Deal era, Cammerer took advantage of many opportunities, using New Deal money and programs to develop the parks and move the Service much further along in the direction set by Mather and Albright.[5]

Establishment of the Service's scientific programs under Albright and Cammerer marked an important break in continuity from the Mather era. Yet the programs emerged only in a fortuitous, opportunistic way. In more than a decade of ever-expanding operations and expenditures, the Service had not felt it necessary to commit funds for scientific studies to improve its knowledge of natural resources and provide guidance for park management. Had George Wright not offered to fund a survey, the Service might well have waited many more years before initiating its own science programs. Moreover, wildlife biology is the only major management program in the history of the National Park Service to have started as a privately funded endeavor within the Service.

The Service's initial response to Wright's offer reflected the bureau's traditional approach to natural resource matters. For instance, Assistant

Director Arthur E. Demaray (acting for Mather in September 1928) suggested that the survey be done not by the National Park Service but under the auspices of the Biological Survey, in keeping with the Service's established practice of using other government bureaus to do "special work of this kind," as Demaray phrased it. Demaray initiated informal talks with the head of the Biological Survey to implement this proposal. The Park Service directorate was persuaded otherwise, however, most likely by Wright, who was donating the funds and strongly believed that the Service itself should assert responsibility.

In favor of the wildlife survey, yet adhering to traditional Service attitudes, Albright emphasized the benefits that Wright's proposal would bring to national park educational programs aimed at enhancing public enjoyment and appreciation of the parks—the Service's chief concern.[6] In March 1929, two months after becoming director, he reported to the secretary of the interior on the Service's need for scientists—that they should be "attached to the educational division," which could "gather data for museums, for all other educational activities, and for the other divisions as needed." Albright also reported that there were no regularly appropriated funds for scientific research, yet he did not ask the secretary to provide such funds. Still, he approved of the scientific survey that Wright was funding, as did Ansel Hall, head of the Education Division, who saw the survey as urgently needed for both education and wildlife management.[7]

The wildlife survey was, in fact, assigned to Ansel Hall's Education Division, located on the University of California campus in Berkeley, where Wright had studied zoology and forestry. With the encouragement of Mather and Albright (themselves University of California alumni), the university was becoming a center of Park Service activity that included education, forestry, and landscape architecture, in addition to wildlife management. Wright's mentor, Joseph Grinnell, head of the university's Museum of Vertebrate Zoology and a longtime proponent of scientifically based management of the national parks, was close by. Also, Ben Thompson and Joseph S. Dixon, the biologists who had joined Wright on the wildlife survey team, were graduates of the university and former students of Grinnell.[8]

Particularly interested in Yosemite and other Sierra parks, Grinnell was an important figure in the promotion of scientific research in the national parks. In 1924 he and Tracy Storer had elaborated on their earlier thoughts on national parks in an article entitled "The Interrelations of Living Things," stating that the more they studied the parks the more they were aware that "a finely adjusted interrelation exists, amounting to a mutual interdependence" among species. They perceived that each species "occupies a niche of its own, where normally it carries on its existence in perfect harmony on the whole with the larger scheme of living nature." For wildlife management they urged the Service to take into account such habitat-related matters as food supply, shelter from predators, and secure breeding places. Throughout his career, which ended with his sudden death in 1939, Grinnell championed an ecological approach to national park management, and he regularly communicated with the wildlife biologists and Park Service directorate.[9]

Grinnell's ecological views reflected the evolving concepts of nature and natural systems that marked a significant scientific advancement during the period when Wright and his fellow Park Service biologists were launching their careers. Biologists were gaining an increased comprehension of the role of habitat in the survival of species; and an understanding of the importance of the overall environment in which different species existed melded animal and plant ecology and led to studies of food chains, predator-prey relationships, and other interrelationships of animal and plant life.[10] New ecological ideas underlay the growing academic interest in game management, and largely through Grinnell and his students new theories began to be applied to natural systems in the national parks.

Following preparatory work, Wright, Dixon, and Thompson began their field survey in May 1930. By the following spring they had completed a report of more than one hundred fifty pages, including brief analyses of most of the large mammals in the principal natural parks. Formal publication came in 1933, under the title *Fauna of the National Parks of the United States: A Preliminary Survey of Faunal Relations in National Parks* (referred to as Fauna No. 1, since it was planned as the first in a series of wildlife studies). A landmark document, Fauna No. 1 was the Service's first comprehensive statement of natural resource management policies, and it proposed a truly radical departure from earlier practices. The biologists proposed to perpetuate existing natural conditions and, where necessary and feasible, to *restore* park fauna to a "pristine state." Achieving this goal would require not only thorough scientific research but also, the report noted, "biological engineering, a science which itself is in its infancy."[11]

The wildlife biologists recognized a fundamental conflict in national park management: that efforts to perpetuate natural conditions would have to be "forever reconciled" with the presence of large numbers of people in the parks, a situation in land management that, they observed, had "never existed before." This conflict had contributed to a "very wide range of maladjustments" among park fauna. Identified as additional contributing factors were human manipulation of lands prior to park establishment and the "failure" to create parks as "independent biological units" with vital year-round habitats for the larger mammals.[12] To correct the maladjustments, the biologists proposed a number of actions. For example, those species extirpated from certain parks should be restored when feasible. And the species whose populations had been reduced to the "danger point" should receive management's special attention. Similarly, where park habitats had been seriously altered, they should be restored.

In confronting the impacts of public use of the parks, the team remained loyal to traditional attitudes, stating that public use "transcends all other considerations." Still, foretelling the concerns of Park Service scientists in decades to come, they urged that the "most farsighted administrative policy" was to "minimize the disturbance of the biota as much as possible." Alternative development solutions should be sought "even if a larger expenditure of money is thereby involved."[13]

Of all their proposed solutions, the survey team most frequently emphasized the need to expand boundaries to include year-round habitats for protection of wildlife that migrated out of the high-mountain parks during winter. It was, the biologists noted, "utterly impossible" to protect animals in an area they occupied only part of the year. With annual migration patterns having been of no concern in the initial establishment of park boundaries, the parks were like houses "with two sides left open," or like a "reservoir with the downhill side wide open."[14]

Fauna No. 1 recognized that nature had always been in a state of flux; thus, there "is no one wild-life picture which can be called the original one." Yet the biologists identified the "period between the arrival of the first whites and the entrenchment of civilization" in areas later to become parks as the point of reference for purposes of wildlife management. They believed that little could be determined regarding changes that had resulted from earlier, American Indian uses, adding that "the rate of alteration in the faunal structure has been so rapid since, and relatively so slow before, the introduction of European culture that the situation which obtained on the arrival of the settlers may well be considered as representing the original or primitive condition that it is desired to maintain."[15]

The report concluded with a series of recommendations entitled "Suggested National-Park Policy for the Vertebrates," which would, in fact, soon be declared official policy. Two recommendations were fundamental: the Service should base its natural resource management on scientific research, including conducting "complete faunal investigations . . . in each park at the earliest possible date"; and each species should be left to "carry on its struggle for existence unaided" unless threatened with extinction in a park. The remaining recommendations in effect qualified or elaborated on these two basic tenets, with specific statements on concerns such as protection of predators, artificial feeding of threatened ungulates, preservation of ungulate range, removal of exotic species, and restoration of extirpated native species.[16]

As an official policy recommendation from within a government bureau, Fauna No. 1's proposal for perpetuating and even restoring natural conditions was unprecedented in the history of national parks, and in all likelihood in the history of American public land management. George Wright acknowledged the limitations on such a proposal when he told the 1934 superintendents conference that the wildlife biologists realized the impossibility of keeping "any area in the United States in an absolutely primeval condition," but added that "there are reasonable aspects to it and reasonable objectives that [the Service] can strive for."[17]

Fauna No. 1 stands as the threshold to a new era in Park Service history. Its conception of "unimpaired" in essentially ecological terms marked a revolutionary change in the understanding of national parks by Service professionals. Recommendations for scientific research, ecological restoration, protection of predators and endangered species, reduction or eradication of nonnative species, and acquisition of more ecologically complete wildlife habitats were among the many farsighted aspects of this report.

Although he would later take serious issue with some of their proposals, Director Albright lent support to the early work of the wildlife biologists and indicated a broadening concern for their programs, beyond educational purposes alone. Although his policy limiting predator control in the parks (enunciated in the *Journal of Mammalogy* in May 1931) reflected pressure from outside the Service, it almost certainly was also influenced by the wildlife biologists, who would in Fauna No. 1 strongly recommend ending predator control. Very likely the biologists themselves drafted detailed commentaries such as Albright's 1932 "Game Conditions in Western National Parks," an account of various wildlife problems confronting the Service. In a June 1933 article in *Scientific Monthly,* entitled "Research in the National Parks" (again probably drafted by the biologists), Albright stated that it had been "inevitable" that scientific research would become part of national park management. Research, he observed, served not only education in the parks, but was "fundamental" to the protection of their natural features, as required by national park legislation.[18] Albright thus endorsed science as an important element in the Service's management of nature—a position he had not previously taken.

In addition, Albright began to provide fiscal support for the scientists. In July 1931, two years after the wildlife survey had gotten under way, the Service undertook to assume half the survey costs, with the other half still funded by George Wright.[19] And two years later, on July 1, 1933, the director formally established the Wildlife Division, with Wright as division chief and Dixon and Thompson as staff biologists. At this time the Service began to pay all costs. Headquartered at the University of California, the division was made part of the newly created Branch of Research and Education (successor to the Education Division) and placed under Harold C. Bryant, another student of Joseph Grinnell's.

With the Wildlife Division, the Service began to develop its own cadre of scientists who were "park-oriented," as Park Service biologist Lowell Sumner later recalled. Reflecting on the emergence of biological research and management in the 1930s, Sumner also observed that Fauna No. 1 soon became the "working 'bible' for all park biologists." In March 1934, Director Arno Cammerer endorsed Fauna No. 1's recommendations as official National Park Service policy. In a memorandum to the superintendents, Cammerer, who had recently succeeded Albright, pledged the Service to make "game conservation work a major activity." He admonished the superintendents that Fauna No. 1's policy recommendations (quoted verbatim in his directive) were "hereby adopted and you are directed to place [them] in effect."[20]

Cammerer's directive reiterated a recommendation Albright had made two years before, that the superintendents appoint rangers to coordinate wildlife management in each park—"preferably," as Albright had put it, men with "some biological training and native interest in the subject." (He was, in fact, endorsing a procedure already being used to select wildlife rangers.) Cammerer instructed the rangers to conduct a "continual fish and game study program" in each park, and to assist the wildlife biologists when they were in the field.[21] The biologists also received some support from the park naturalists, who, although busy with the growing educational programs, collected plant and animal specimens and provided other field assistance.

In addition to working with biologists, however, the wildlife rangers' natural resource management efforts included established programs such as controlling predator, rodent, and mosquito populations; assisting the foresters in fighting insects and fires; and working with fishery experts to stock park waters.[22] Consistently contradicting Fauna No. 1, these ranger activities represented traditional management practices that did not, as the biologists saw it, preserve natural conditions. Allied with the foresters, the wildlife rangers would quickly find many of their established practices strongly opposed by the biologists.

The biologists' efforts gained momentum with the advent of President Franklin D. Roosevelt's New Deal emergency relief programs, which made money and manpower available to the Park Service. The Service obtained increased support for park development from several relief programs, including the Works Progress Administration, Public Works Administration, and Civilian Conservation Corps (CCC). Of these, the Civilian Conservation Corps most affected the Wildlife Division and the national parks themselves. Authorized by the Emergency Conservation Act of March 1933, the CCC put unemployed young men to work on public land conservation and reclamation projects. Soon becoming one of the New Deal's most acclaimed programs, it remained active until World War II.[23]

Quick to realize the potential of the New Deal programs, Director Albright aggressively sought CCC money and manpower for developing the national parks. However, CCC projects such as road and trail construction, administrative and visitor facility construction, and water and sewage development resulted in the extensive alteration of natural resources. Much of the CCC work conflicted with Fauna No. 1's call for "farsighted" policies to "minimize the disturbance of the biota." Living in camps of two hundred or more men, the CCC crews sometimes vandalized areas and harassed park wildlife.[24] In addition to extensive park development work, the CCC crews undertook many highly manipulative natural resource projects, such as assisting the wildlife rangers in mosquito control, firefighting, and removal of fire hazards.

In June 1933 Albright cautioned his superintendents that the CCC crews must "safeguard rather than destroy" the resources of the national parks. He suggested that the "evident dangers to wild life" resulting from CCC work might be kept at a minimum through consultation with the Wildlife Division.[25] Given such concerns, and at George Wright's urging, the Park Service used CCC funds to hire additional wildlife biologists to monitor CCC and other work in the parks. By 1936 the number of professionally trained wildlife biologists had grown ninefold, from the original three-man survey team to twenty-seven biologists. Most were stationed in the parks or in field offices.[26] Thus, Fauna No. 1 provided policies and the CCC provided funds for the Park Service to develop its own more scientifically informed natural resource management.

Still, overall commitment to the wildlife biology programs was limited.

Just as the Park Service had begun its own scientific efforts only when Wright provided money from his personal fortune, it also took special New Deal funding (rather than the Service's regular annual appropriations) to finance most of the wildlife biology programs in the 1930s. Of the twenty-seven biology positions, the Park Service's annual appropriations (which gradually increased during the Depression) paid for only four; the rest were funded with CCC money.[27] Ironically, then, with most of the Wildlife Division's money and positions coming from the CCC, the bulk of the Service's increased scientific effort was tied to park development programs—which resulted in considerable alteration to the very natural resources that the wildlife biologists sought to preserve.

In 1935, given the growing complexity of the division's work and its need to coordinate activities with other Park Service operations, Director Cammerer transferred the Wildlife Division to Washington, D.C. In its new headquarters and with an expanded staff of biologists located in key parks, the Wildlife Division reached its apex. Then, in February 1936, the Service's wildlife management programs suffered a severe setback with George Wright's sudden death in a head-on automobile accident east of Deming, New Mexico. Although not fully apparent at the time, the loss of Wright's impressive leadership skills marked the beginning of the decline of National Park Service science programs. Through the remainder of the decade the number of wildlife biologists would decrease, thereby diminishing their influence even before they were transferred to the Biological Survey in January 1940.

Conflict over Park Development

The "conservation" aspects of the Civilian Conservation Corps were indeed utilitarian, oriented toward what was in effect "wise use" of the national parks' scenic resources through accommodating public use and enjoyment. Virtually all of the CCC's park development and much of its direct manipulation of natural resources were in one way or another intended to address such utilitarian concerns. The CCC and other New Deal park programs thus represented a continuation of the Service's traditional goals and values. With funds available in unprecedented amounts, it was possible to implement much of the park development envisioned in master plans prepared during Mather's and Albright's directorships. By one estimate, during the New Deal the Service was able to advance park development as much as two decades beyond where it would have been without Roosevelt's emergency relief programs.[28]

For the first time, wildlife biologists became involved in decisions on development, which previously had been the responsibility of landscape architects, engineers, superintendents, and the Washington directorate. Yet the biologists' role was mainly advisory. They reviewed and commented on details such as alignment of roads and trails and placement of facilities, calculating the impacts of development on fauna and flora, and recommending means of keeping impacts at a minimum.

Moreover, the biologists had limited involvement in updating park master plans. Writing to Cammerer in February 1934 on the need to include wildlife management in the plans, George Wright had argued that involvement would help "more than any thing else" to focus attention on wildlife issues.[29] As the biologists' influence reached its peak in late 1935, Wright had reiterated to the director the need for the master plans to include natural resource information—rather than "contemplated and completed physical development only." For example, Mount Rainier's plan should include a "fish sheet," describing the "kinds and distribution of native fishes" before their being affected by modern human activity. It should also comment on the advisability of stocking native or exotic fish species and whether or not the park truly needed a fish hatchery. This kind of information would, Wright asserted, provide help "which the master plans could, but do not, give," and thus would protect against the "honest but sometimes misguided zeal" of superintendents who had to manage the parks without such information.[30] Despite his pleas, there is no indication that the biologists achieved substantial involvement in the Service's master planning process.

The biologists reviewed a variety of park development projects. For instance, reporting from Death Valley National Monument in September 1935, biologist Lowell Sumner recommended approval of a number of proposals, including road and trail construction, campground expansion, and water well and water pipeline development. He consented to a proposed road construction project by noting that it did not appear to endanger bighorn sheep, and urged his fellow biologists to conserve their energy for "curbing less desirable projects." In the same report, Sumner recommended that biologists not only review project proposals, but also closely monitor actual construction whenever natural resources were particularly vulnerable.

Significantly, the wildlife biologists' criticism of development that they considered inappropriate tended to stress ecological factors—a different focus from concerns about visual intrusion into park landscapes. Among the less desirable projects in Death Valley, for example, was the proposed road improvement in Titus Canyon, to which Sumner strenuously objected because it would threaten wildlife habitat—rare plants grew in the canyon and an important watering hole for bighorn sheep lay at the end of the existing primitive road. Such arguments would have been heard rarely, if at all, before the wildlife biologists became involved with project review. Sumner also claimed that it was unsafe for humans to frequent the canyon and pleaded that it remain "unvisited and undisturbed." Declaring that Death Valley was being developed at a rate that "has never been paralleled by any national park or monument," he warned that the park could lose its remaining pristine areas. Instead of road improvement, he urged that the Titus Canyon area be designated a "research reserve," to be set aside for research purposes only, a recommendation that apparently was ignored.[31]

In a report from Glacier National Park in 1935, biologist Victor Cahalane opposed the park's sawmill operation, used to dispose of dead trees considered fire hazards. With an ecological orientation similar to Sumner's, Cahalane recommended against the sawmill and argued for adhering to the Service's stated policies rather than to "a purely utilitarian viewpoint." He concluded with a rhetorical question and a blunt injunction: "Is it not more in keeping with our ideals to leave the dead trees standing than to instigate a logging operation in a national park? The project is not approved." The Wildlife Division regularly received strongly worded field reports like Sumner's and Cahalane's. Following review by Wright and his Washington-based staff, the reports were forwarded to the directorate with comments, some of which did not concur with the field scientists' recommendations.[32]

Sharp conflicts inevitably arose over the reviews, probably exacerbated by the fact that the wildlife biologists were newcomers to the project review process, which was the traditional territory of superintendents, landscape architects, and engineers. Responding to Cahalane's objections to construction of a shelter for campers at Grand Canyon's Clear Creek, Superintendent Minor Tillotson wrote to Director Cammerer in October 1935 that Cahalane's views were "not only far-fetched but picayunish." Tillotson argued that because a trail had been built, provision should be made for use of the primitive area to which the trail led: "objections to the development as proposed . . . should have been voiced before all the money was spent on the trail." Stating that he was "always glad" for the wildlife biologists' advice, the superintendent chided that in this case they had "gone considerably out of their way" to find something to criticize.[33]

Of all national park development, roads—both their initial construction and their improvement to allow increased use—definitely constituted the most severe intrusion. Probably most of the park roads newly constructed during the 1930s were primitive, intended to provide access for firefighting only. But they penetrated the backcountry, inviting further development as tourist roads (for instance, in Titus Canyon) and diminishing wild qualities and biological integrity, as Sumner feared. Thus roads became a major focus of debates over development in the parks. Conflicting attitudes toward national park roads began to crystallize during the 1930s—attitudes that would typify the dichotomy in Park Service thinking for decades.

Improvement of the Tioga Road through Yosemite's high country sparked conflicts in the 1930s (as it would again in the 1950s). During realignment of the road in the mid-1930s, Lowell Sumner objected to plans to use a small unnamed lake along the road as a borrow pit, brusquely depicting the plans as an example of the tendency of road builders to "slash their way through park scenery." Engineers, he wrote, wanted to straighten roads and reduce grades "to spare the motorist . . . the necessity of shifting out of high gear." Such construction practices resulted in more cuts and fills and therefore more borrow pits.[34] In this instance, Sumner objected as much to the "disfiguration" of park scenery as to the impact on natural resources per se.

R. L. McKown, Yosemite's resident landscape architect, reacted angrily to Sumner's barbed comments, writing to the top Park Service landscape architect, Thomas Vint, that such remarks were "derogatory of our Landscape Division" and that Sumner was "misinformed" about the division's principles. McKown claimed the division went out of its way to prevent slashing through scenery. The pressure to straighten park roads came, he asserted, not from the landscape architects but from the Bureau of Public Roads, which was responding to the public's desire for "high speed motor ways in our national parks" similar to what they found elsewhere. McKown also noted that if the lake were not used for borrow, the materials would have to be found at least four thousand feet farther along the route, and to him the added cost seemed unwarranted.[35]

Sumner apologized to McKown, granting that the Landscape Division was actually seeking to reduce the road's intrusion. The division was, in Sumner's words, "the prime guardian of the natural in our parks"—a remark that seemed to contradict the role the Wildlife Division was assuming for itself. Sumner then commented that "even the most skillful camouflaging in the interest of landscaping cannot altogether prevent it from being an intrusion on the wilderness," an indication that he may have believed that the landscape architects' work indeed mostly amounted to camouflaging.[36]

Sumner recognized that limiting visual intrusions into wilderness areas did not necessarily mean that the areas' natural resources would remain free from serious harm. Reflecting on the construction of the Tioga Road, he wrote in October 1936 that it illustrated the "complex, irrevocable, and perhaps partly unforeseen chain of disturbances" resulting from roads. The Sierra Club would later describe road development in national parks as being "like a worm in an apple." Sumner himself characterized park roads as an "infection," bringing on further, gradual development of an area, with gasoline stations, lodges, trails, campgrounds, fire roads, and sewage systems—until the "elusive wilderness flavor vanishes, often quite suddenly." This he feared was happening along the Tioga Road and in other park areas where the superintendents were under unrelenting pressure to develop.[37]

In fact, the potential for greater use of an area subsequent to road improvements was clearly indicated in a final construction report on a portion of the Tioga Road. The report anticipated that the Tuolumne Meadows, through which the road passed, would soon become one of the park's more heavily used recreational areas, particularly attractive for hiking, nature study, fishing, and horseback riding. With each summer season, the report stated, more people were using the area, and a "large increase of cars pulling trailer houses has been especially noticed." Furthermore, the road improvements were likely to attract a substantial amount of transcontinental traffic simply crossing the mountains.[38]

Quite representative of the wildlife biologists' attitudes, Sumner's remarks on the Tioga Road revealed a cautious, pessimistic view of development. He feared widespread park development stemming from New Deal relief and conservation programs, believing that such "improvements" could ultimately lead to the national parks' ecological ruin. In early February 1938, Sumner wrote to his mentor, Joseph Grinnell, expressing concern that true wilderness in the parks would soon vanish if the Service did not halt development. He lamented that although the Park Service should be the leader in wilderness preservation, it "has been more at fault than many other agencies" in destroying such natural values.[39]

In another statement prepared in 1938 and entitled "Losing the Wilderness Which We Set Out to Preserve," Sumner warned against exceeding the "recreational saturation point" in parks with roads, trails, and development for winter sports and other activities. Concerned about modifications to natural resources, he argued that ground impaction affected even minute soil organisms active in maintaining porosity and soil nitrogen—the thinking of Park Service scientists had moved well beyond management's traditional preoccupation with scenic landscapes and large mammals.[40]

With the wildlife biologists questioning traditional practices, Park Service leaders made an earnest effort to rationalize national park development, at times using park preservation as the principal justification. For instance, Director Cammerer declared in a 1936 article for the *American Planning and Civic Annual* that park roads could be used as an "implement of wilderness conservation." Noting that the Service opposed grazing, mining, hunting, and lumbering in parks, the director wrote that the "core" national park idea is "conservation for human use." So, he asked, what forms of park use *should* the Service permit? His answer was to build sufficient roads so that the public could use and enjoy the parks as called for in the Organic Act. Espousing a utilitarian rationale for preservation in the national parks, Cammerer stated that the Park Service must provide an "economically justifiable and humanly satisfying form of land use, capable of standing on its own merit in competition with other forms of land use." He strongly opposed allowing roads to penetrate all areas of a park; but by building roads in a "portion" of a park area so that the public could enjoy it, the Service could save large undisturbed areas for the "relatively few who enjoy wilderness." He commented perceptively that unless "bolstered by definite, tangible returns" such as public use and enjoyment made possible through roads, the preservation of national park wilderness would fall before the onslaught of pragmatic economic needs. Cammerer added that roads were a "small price" to pay, and that they could potentially "make many friends" for the remaining park wilderness because people do not "know what a wilderness is until they have a chance to go through it."[41]

Thomas Vint put forth arguments similar to Cammerer's. In 1938, with the national wilderness preservation movement under way, Vint published an article (also in the *American Planning and Civic Annual*) that clearly tied park development to backcountry preservation. In "Wilderness Areas: Development of National Parks for Conservation," he wrote that the time comes when "it is worthwhile, as a means of preservation of the terrain, to build a path." And with increased traffic, a path must be "built stronger to resist the pressure." There followed a progression of development and improvement: Vint depicted this progression, beginning with paths for foot traffic, then paths for horses and wagons, and ultimately roads for automobiles, which in turn go through "various stages of improvement."[42]

Vint then asked a question fundamental to national park management: at what point does park development "trespass on the wilderness or intrude on the perfect natural landscapes?" Closely restricted development, he believed, was the key to preventing trespass of park wilderness—development that would accommodate people and at the same time control where they went. The lands remaining untouched (in Vint's words, "*all* of the area within the boundaries of the park that is not a developed area") would be saved as wilderness.[43] Reminiscent of Albright's earlier assertions about roads and wilderness in Yellowstone, Vint's comments evidenced a tendency to equate undeveloped areas with adequately preserved wilderness—a perspective that Ben Thompson had challenged a few years before, and that differed substantially from Lowell Sumner's view of roads as "infections," ultimately contaminating large corridors of the parks.[44]

From 1916 on, Park Service leaders had overseen the initial construction or improvement of hundreds of miles of park roads, often through the heart of primitive backcountry. Yet they also opposed road construction in instances when they believed, as Vint put it, that the "trespass on the wilderness or [intrusion] on the perfect natural landscapes" was excessive. A clear example of this came in the 1930s with Superintendent John White's protracted opposition to the "Sierra Highway," proposed to cut through Sequoia National Park's remote backcountry.[45] Giving strong support to White, Acting Director Demaray in 1935 wrote Secretary of the Interior Ickes (himself not enthusiastic about national park roads) that the proposed road was "an unjustifiable and destructive invasion of a great national resource, the primitive and unspoiled grandeur of the Sierra." The highway, he continued, would "destroy the seclusion and a large part of the recreational value of every watershed, canyon, valley, and mountain crest which it traversed"; the proposal was "psychologically wrong and physically wasteful."[46] These words sounded much like Lowell Sumner's; and, indeed, the planned Sierra road was defeated. All the same, such a position stood in contrast to the Service's aggressive promotion of other road projects, such as Glacier National Park's Going to the Sun Highway, Rocky Mountain's Trail Ridge Road, Mt. McKinley's road system, and the Tioga Pass Road in Yosemite.[47]

The wildlife biologists' cautious approach to park development was in accord with ecological concerns, but threatened to inhibit the spending of large amounts of New Deal funds to develop parks. With abundant park development funds available at a time when the wilderness preservation movement was emerging, the rationale that development fostered preservation appears to have been particularly useful to Service leadership.

It is important to note that the idea that national parks must be made accessible for public use in order to secure public support clearly had legitimacy. As Mather and his successors thoroughly understood, the public was hardly likely to have supported undeveloped, inaccessible national parks. National parks were originally intended to be public pleasuring grounds; and proponents of the Organic Act had evidenced an unmistakable interest in the accessibility and enjoyment of park landscapes—as reflected in the act's wording and amplified in, for example, Secretary Lane's policy letter of 1918. In a clear indication of support for the Service's emphasis on recreational tourism in the parks, Congress provided millions for park roads and other development, with funding reaching unprecedented levels during the New Deal era.

The concept of development as a means of ensuring preservation provided the Service with a rationale for believing that it could meet the congressional mandate to provide for public use while leaving large portions of the parks unimpaired. Nevertheless, while park development continued apace, the number of wildlife biologists available to provide an ecological perspective diminished—and dissenting opinions of the remaining wildlife biologists continued to encounter formidable, entrenched Park Service tradition.

Biological Research

At their 1929 conference, the national park naturalists had noted that scientific data on the parks' natural history were "almost infinitesimal." This disheartening situation had begun to change that very year, as field research for Fauna No. 1 got under way, then continued under the Wildlife Division. Lowell Sumner later estimated that during the 1930s about half of the biologists' work involved research and wildlife management, while the other half was devoted to review of and comment on proposed development projects. He calculated that prior to World War II the biologists had produced perhaps a thousand reports. Having joined the Service in 1935, Sumner estimated that he himself had prepared about 175 reports before the war began.[48]

Research focused on subjects such as bison, elk, and bird life in Wind Cave; white-tailed deer and winter birds in Shenandoah; grazing mammals in Rocky Mountain; and deer and bighorn in Glacier National Park. Park naturalists contributed further to the gathering of information, as in Great Smoky Mountains, where specimens of about two thousand plant species were collected by the mid-1930s.[49] Given the large number of documents prepared and the limited number of biologists in the Park Service, only a few of the reports were truly comprehensive.[50]

An important element of the biologists' programs during the 1930s was the establishment of "research reserves," areas within national parks designated to be used for scientific study only. Probably at the urging of the Ecological Society of America and leading biologists such as John C. Merriam of the Carnegie Institution, who feared the disappearance of all unmodified natural areas in the United States, the Park Service in the mid-1920s had gradually begun to develop a research reserve program. In 1927 Yosemite National Park designated approximately seven square miles of high mountain country north of Tuolumne Meadows as a "wilderness reserve," later termed a research reserve—the first of its kind in the national park system.[51]

The park naturalists discussed the reserves at their 1929 conference, advocating that the areas be permanently set aside primarily for scientific study. They were to be, as the naturalists phrased it, "as little influenced by human use and occupation as conditions permit." Park Service director Horace Albright followed up in the spring of 1931 by issuing a research reserve policy to "preserve permanently" selected natural areas "in as nearly as possible unmodified condition free from external influences." In effect, the areas would help meet Fauna No. 1's recommendation for each species to "carry on its struggle for existence unaided." The reserves were to be entered only in case of emergency or by special permit; as a further means of protection, their location was not to be publicized.[52]

The research reserves emerged in the 1930s as the most preservation-oriented land-use category the Park Service had yet devised—an important philosophical and policy descendant of the congressional mandate to leave the national parks unimpaired, and much more restrictive than the traditional policy of allowing park backcountry to be developed with horse and foot trails.[53] In George Wright's view, the greatest value of the reserves lay in providing scientists with the opportunity to learn what certain portions of the parks were like in their original, unmodified condition. This "primitive picture" would provide a basis of knowledge to benefit all future research. He also believed that the reserves would not become "an actuality" until their flora and fauna had been surveyed. To Wright, setting aside the reserves was a "most immediate urgency," which should be accomplished before further biological modifications took place.[54]

The research reserves became an integral part of park management in March 1932, when Director Albright asked that they be formally designated through the cooperation of the park superintendents and naturalists and the Washington office. He requested that the superintendents indicate the location of the reserves in the five-year park development plans (master plans), and he assigned the wildlife biologists responsibility for gathering information and tracking the progress of the program. By 1933, research reserves had been designated in Yellowstone, Sequoia, Grand Canyon, and Lassen Volcanic national parks. Others followed, in Great Smoky Mountains, Glacier, Mount Rainier, Rocky Mountain, and Zion, as well as Yosemite, for a total of twenty-eight designations in ten parks.[55]

The research reserve idea worked better in theory than in practice, however. The wildlife biologists apparently did not participate in the actual selection of many of the reserves, probably because a number of the areas were designated while the biologists were busy completing Fauna No. 1 and because they had not been given a meaningful role in the master planning process. As late as February 1934, the Wildlife Division seemed poorly informed on the exact location and character of many of the reserves; and regarding those they knew something about, Wright noted that some of the areas were not worthwhile research areas—indications that the biologists had had limited input in selection of the reserves. A reserve in Lassen Volcanic National Park was no more than a strip of land three-quarters of a mile wide and about five miles long, whereas two of Grand Canyon's reserves were so close to the park boundary that activities outside the park were certain to affect their biotic makeup. Observing the potentially serious external influences on the reserves, Wright advocated the establishment of "buffer areas" around the parks (including additional winter range for wildlife), rather than "withdrawing further and further within the park" to create reserves.[56] Like the parks themselves, the reserves were not satisfactory biological units.

Expressing deep concern about the reserve program, Victor Cahalane, Wright's assistant division chief, wrote in September 1935 about the problem of selecting research reserves in parks so "artificialized and mechanized." Cahalane believed that the difficulty of finding even relatively small unaltered research areas indicated the extent to which the Service had failed to meet its basic mandate to protect the parks' wilderness character. Reflecting biologist Ben Thompson's earlier comments about alterations to natural conditions in the parks, Cahalane wrote that Glacier National Park had no pristine area worthy of becoming a research reserve. This had occurred "not by reason of a network of roads" in Glacier, but because "all streams now contain exotic species of fish, because the wolverine and fisher have been exterminated from the entire park and the bison and antelope from the east side, and because exotic plants . . . have been carried to practically every corner of the park." Recognizing the existing problems with "pristine" areas in the parks, Cahalane called for a "show-down on this matter of preservation of the greatest resource of the National Park Service—the wilderness."[57]

Beyond the difficulty of identifying minimally altered natural areas to be designated as research reserves, these areas were the product of decisions made wholly within the Park Service and were therefore subject to administrative discretion and vulnerable to the sudden impulses of management. The reserves had no specific mandate from Congress. They could be protected, ignored, or, as happened with Andrews Bald research reserve in Great Smoky Mountains National Park, created and then summarily abolished. Indeed, the "show-down" that occurred over Andrews Bald went directly against the biologists' recommendations and reflected the Park Service's ingrained disregard for scientific research. The outcome was an ominous portent for the science programs overall.

Designated a research reserve in the mid-1930s, Andrews Bald was one of several such areas in Great Smoky Mountains intended to be strictly preserved so that "ecological and other scientific studies" could be conducted on a long-range basis, especially to determine natural plant succession. (The "grassy balds"—open, mountaintop areas of grasses and low-growing shrubs, and without tall trees—were one of the primary scenic features in the Smokies. They were then, and remain today, of special scientific interest.)

In early April 1936, a terrific windstorm killed hundreds of trees in and around Andrews Bald, precipitating a sharp debate in the Service over how to manage the area. Dead and dying trees, some still standing, littered the landscape and, in the minds of the superintendent and most of his staff, constituted a fire hazard that needed to be cleaned up.[58] Superintendent J. R. Eakin wanted a cleanup, as did the park's rangers and foresters; and in a letter to Park Service director Arno Cammerer, Eakin stressed the potential fire problems. Reflecting an ongoing disagreement over what to do with naturally killed trees, the superintendent noted that "again" the Wildlife Division and the naturalists were "not concerned with fire protection" and the danger that might arise if dead trees were left in place. Particularly concerned about scenery, Frank E. Mattson, the park's resident landscape architect, argued for cleanup of the windfall, stating that because the bald attracted so many sightseers, it should be treated "much as a trailside or roadside" area.[59]

By contrast, the wildlife biologists (supported by park naturalist Arthur Stupka) advocated special consideration for the reserves, so that scientific studies could "be started and continued thru the years to come." They urged that the trees be left untouched. Although acknowledging the fire prevention concerns, the biologists argued that the windstorm was a natural phenomenon and that cleanup of the area would "thwart the objectives" of Andrews Bald research reserve. Still, Superintendent Eakin believed the area constituted a serious fire hazard; in an exchange of correspondence with the Washington office, he insisted that the trees should be cleared.[60]

In a stinging reply to Eakin, Acting Director Arthur Demaray finally granted permission for clearance, but added that the Andrews Bald Biotic Research Area was thereby abolished. He further stated that "I wish to call your attention to several factors which you seem to have overlooked": the reserve had been approved by Eakin himself, it was included in the park's master plan, and preservation of such areas was "an established policy of the Service." In the acting director's view, the superintendent's insistence was forcing a change in the official use of the area from research and strict preservation to recreation: "The reason the research area is now abolished is that you have convinced us you made an error in approving its establishment. Its apparent proper use is primarily recreational."[61]

Andrews Bald illustrated the vulnerability of the reserves to administrative discretion and, too, the vulnerability of research in the national parks. An area committed to serve research purposes over a long period of time was subject to sudden modification as a result of internal decision-making. Indeed, the urge to clear the trees was not truly based on whim, but reflected the deep-seated, traditional allegiance of the superintendents, foresters, and landscape architects to preserving national park scenery and accommodating public use, while generally showing little interest in science.

Even though the research reserves were supported by the director's policy pronouncement of 1931 and represented the bureau's strongest commitment to preservation of natural conditions, the Park Service eventually disregarded the entire program. Certainly most reserves did not vanish in as confrontational a way as did Andrews Bald, yet Lowell Sumner later recalled that the research reserve program came to be largely ignored, beginning about the time of World War II.[62]

Although it may seem that ignoring the research reserve program meant that these areas would be left alone with no human interference, this was very likely not the case. With the program untended and the reserves in effect forgotten, these areas of special research value were subjected to alteration through such practices as fire protection (for example, the removal of trees from Andrews Bald), firefighting, forest insect and disease control, grazing, and fish stocking and harvesting. The neglected research reserves were subject to the kinds of modifications that concerned George Wright in the early 1930s when he stressed the "most immediate urgency" of establishing the reserves.[63]

Rangelands and the Grazing Species

In contrast to the research reserve program, which was intended to leave selected natural areas undisturbed, the biologists believed that in other instances it was necessary to interfere with nature and, as stated in Fauna No. 1, assist certain species to combat the "harmful effects of human influence" in order to restore the "primitive state" of the parks. Fauna No. 1 also specifically called for preservation of ungulate range and advocated that a park's "deteriorated range" should be "brought back to [its] original productiveness."[64] Of all the Park Service's attempts to interfere with nature during the 1930s, the manipulation of Yellowstone's northern elk herd received the greatest attention and ultimately became the most controversial.

To many familiar with Yellowstone, the park's northern elk herd seemed to have become so large that it was overgrazing its range (mostly consisting of the Lamar and Yellowstone river basins). The resulting deterioration appeared to limit use of the range by competing ungulates such as deer and pronghorn. The wildlife biologists determined that the northern herd needed to be reduced, in line with Fauna No. 1's recommendations, a proposal that would entail shooting large numbers of elk. For humane reasons, shooting the animals seemed far preferable to allowing them to die of winter kill when heavy snows restricted their range. Furthermore, reduction could bring the population to a specified level.

The biologists concluded that "human influence" had caused the winter range problems in Yellowstone. This understanding in the 1930s (which decades later would become strongly disputed) was based on some fundamental assumptions: prior to Anglo-American settlement of the lower valleys to the north of the park, the herd had wintered in those valleys; and after the park was established, its protected elk population had expanded enormously. The scientists also believed that the elk population had crashed in the period 1917–20, and that this dramatic decline had been caused by range deterioration through overgrazing. With drought conditions affecting the range in the late 1920s and early 1930s, and with elk populations believed to have increased due to protection in the park, a second population crash was seen as imminent—one that the Wildlife Division expected to bring "hideous starvation and wastage."[65]

In 1931 Joseph Dixon and Ben Thompson (who were working with George Wright on Fauna No. 1) had participated in a reconnaissance of the deer population explosion in the Kaibab National Forest, north of Grand Canyon. Reporting that an overpopulation of deer threatened the national forest, they recommended reducing the deer herds. Probably influenced by what seemed to have happened in the Kaibab, the biologists made their recommendation that Yellowstone's elk population also be reduced. In a February 1934 report documented with numerous photographs (and reprinted in Fauna No. 2 the following year) the Wildlife Division announced that, as a result of an overpopulation of elk, Yellowstone's northern range had been overused to the point that it was in "deplorable" condition. The biologists believed that the situation had worsened since they first saw the area in 1929, and that it now threatened the survival of other animals dependent on the range. Arguing that the overpopulated herd was on the "brink of disaster," the report warned that the next hard winter would cause starvation and death for thousands of elk.[66]

The elk reduction program had strong, apparently unanimous support among the Park Service's wildlife biologists. Their statements and reports did not equivocate on the wisdom of artificially lowering Yellowstone's elk population. Commenting in the late winter of 1935 that without reduction the problems of overgrazing and winter starvation would continue—the "old winter range ghost will be walking again"—Wright himself saw the program as critical to the success of the park's wildlife and range management.[67] Olaus Murie, who had overseen the Bureau of Biological Survey's elk management in Jackson Hole, south of Yellowstone, also urged reducing the northern herd, as did his brother, Adolph, a highly respected National Park Service wildlife biologist. In late December 1934, just before the first big reduction began, Olaus Murie wrote to Ben Thompson approving elk reduction, noting that "if carefully handled it will be successful," and adding that he looked forward "with great interest to the outcome of the experiment."[68]

Beyond their own observations, the biologists based their elk policy on research conducted in the region in the 1920s and early 1930s by U.S. Forest Service biologist W. M. Rush, whose work was privately funded with money obtained by Park Service director Horace Albright. Rush's conclusions supported the biologists' views.[69] Also, because they believed that longer hunting seasons and increased bag limits in Montana and on adjacent Forest Service lands would provide only limited help, the biologists recommended that the park itself undertake reduction to ensure that the proper number of elk would be killed each winter. Until the desired population level was reached, Yellowstone must be prepared "to slaughter elk as it does buffalo."[70]

Much more cautious was the opinion of Joseph Grinnell, mentor to numerous Park Service biologists. Asked by Director Cammerer to comment on the proposed reduction, Grinnell observed that the elk situation in

Yellowstone was "truly disturbing from any point of view." He remarked on the "damage" that he believed elk grazing had done to the winter range, and agreed that human influences had been an important factor in bringing on the situation. Although he carefully avoided criticizing the decisions of his former students and close friends, Grinnell withheld support for the reduction program. Rather, he expressed hope that the killing of any park animals, predators as well as elk, would become a thing of the past. In his summation Grinnell advocated "adjustments through natural processes" to restore the "primeval biotic set-up." More than the Park Service biologists of the 1930s, Grinnell expressed faith in allowing "natural processes" to control elk populations, with aggressive measures taken to reduce adverse human influences on the animals.[71]

Reduction began in January 1935, with Yellowstone's rangers shooting the elk and preparing their carcasses for shipment to tribes on nearby reservations. With the intention of reducing elk populations to the range's "carrying capacity," the Park Service's goal of killing 3,000 elk the first winter included animals to be taken outside the park under Forest Service and Montana State Fish and Game Department regulations liberalized to increase the number killed by hunters. During the first reduction effort, hunters on lands adjacent to Yellowstone took 2,598 elk and park rangers killed 667, for a total of nearly 3,300.[72]

Responding to an inquiry from the American Museum of Natural History in March 1935, George Wright expressed relief that the Park Service itself had not had to kill large numbers of elk during the initial reduction; yet he wrote that "we are glad to have established a satisfactory precedent" regarding the "propriety of direct control" in the national parks. Even after further reduction in 1936, biologist Adolph Murie studied Yellowstone's range and found it "undoubtedly worse" than it had been in six or seven years. He recommended that the kill be increased to 4,000 the following winter. A lengthy 1938 report by Yellowstone ranger Rudolph L. Grimm again confirmed the belief that the range was overgrazed, and advocated continued reduction.[73]

With a "satisfactory precedent" established in the mid-1930s, Yellowstone's elk reduction program began its long history, with the policy eventually being applied in other areas, particularly Rocky Mountain National Park. At the end of the decade, the wildlife biologists reported that the "basic and most important problem" at Yellowstone was still the condition of the park's range. "As in the past," they asserted, the abundance of elk "depletes the forage of other ungulates using the same range."[74] Although he did not speak out aggressively against the reduction program, Grinnell continued to oppose it, writing to Director Cammerer in January 1939 that he did not approve of regulating "the numbers of certain animals in certain Parks." Grinnell urged that the Service submit the problem to a group of specially trained ecologists.[75]

Throughout the 1930s, management of Yellowstone's Lamar Valley bison—the herd of most concern to the Park Service—remained more intensive and varied than management of the park's elk. Using domestic livestock ranching methods first developed by the U.S. Army and then expanded during Mather's time, bison management changed little during the decade. Still headquartered at Buffalo Ranch, it continued to involve roundups, winter feeding in the corrals, and removal of surplus animals (including those not wanted for breeding), which were slaughtered or shipped live to other areas.

In Fauna No. 1, the biologists had had little to recommend concerning bison management, stating only that winter feeding of the animals was "absolutely necessary." Regarding park fauna in general, the report's recommendations called for allowing threatened species to exist on a "self-sustaining basis" when such measures as feeding were no longer necessary. Similar counsel was given in Fauna No. 2, which urged returning bison to their "wild state" to the degree that the "inherent limitations" of each park would permit. But such measures as winter feeding and slaughtering would have to continue until "artificial management" was no longer necessary.[76]

Based on recommendations made during the late 1920s and early 1930s, the park sought to keep Yellowstone's Lamar Valley herd limited in size, at first seeking a population level of 1,000 animals, then 800 beginning about 1934—levels believed to be within the "carrying capacity" of the bison range and what the Buffalo Ranch facilities could accommodate. Even by the following year, some concern was being expressed that the population was much too high. Harlow B. Mills, a biologist at Montana State College who had worked in Yellowstone, wrote an extensive report on wildlife conditions in the park in 1935, recommending that the Lamar Valley herd be reduced to "100 or less animals." Mills believed there were too many bison in Yellowstone, and that the current population was probably greater than under primitive conditions. The ranching operations seemed to be a loss of "energy, time, and money." Although Yellowstone had helped save America's bison from extinction in the United States, Mills added that the bison "has been saved and there is now no necessity of fearing that the species will disappear." In spite of Mills' much lower recommendations, the Service maintained the population level at close to 800 through the remainder of the 1930s.[77]

Fauna No. 2 also provided statistics on bison losses in recent decades. Since the army began its bison management in 1902, 682 animals had been slaughtered, 279 had been shipped live, and 48 "outlaws and cripples" had been destroyed. In addition, 124 bison had died from disease.[78] In 1935, the year Fauna No. 2 was published, George Wright expressed his displeasure with live shipping, whether of bison or elk, and whether to other national parks or to state or local parks. He believed that such activity involved the "inadvised mixing of related forms and the liberation of certain species in areas unsuited to their requirements," which brought "great and irreparable damage in many instances."[79]

Regardless of such disapproval, live shipping remained a regular activity in the parks, as did slaughtering and occasional destruction of "outlaws." Yellowstone superintendent Edmund Rogers reported in late 1937 that 59 bison, including "some old animals that we wish to take from the herd," were being held for live shipment. The park planned shipments to the Springfield, Massachusetts, zoo; to an individual in Wolf Creek, Montana; and to Prince Ri Gin, in Korea. In addition, plans were made to send bison carcasses to the Wind River Agency in Wyoming for distribution to local Indians. In Wind Cave National Park, where until the mid-1930s the Bureau of Biological Survey had been in charge of wildlife management, efforts were begun to reduce bison and elk to satisfactory numbers. The Service reported the following year that both Wind Cave and Platt national parks were reducing their bison populations and shipping carcasses to nearby Indian tribes.

These live shipments and distributions of carcasses may not have won much political advantage, but the distribution of buffalo robes was at times intended to reap political gain. Recognizing this potential, Director Cammerer wrote to Secretary Ickes in 1936 that disposition of the hides "to friends of the Service and the Department, upon their special request, has been and will be helpful in maintaining a special interest in matters relating to this Department and the Service." Yellowstone superintendent Rogers noted that requests for hides had been received from a number of highly placed individuals, such as Senator Robert F. Wagner of New York and Clyde A. Tolson of the Federal Bureau of Investigation.[80]

Platt and Wind Cave shared another management practice with Yellowstone, as all of these parks set up fenced-in areas for wildlife (particularly bison) to be viewed by the public. Only a few hundred acres in size, Platt had no choice but to build a display area for viewing bison that had originally been shipped in from a nearby wildlife preserve. The Park Service took over wildlife management in Wind Cave with fences already in place, and despite declared intentions to remove the fences, continued to maintain an animal enclosure for public viewing.[81] As for Yellowstone's bison, Director Albright had stated in 1929 his determination to make the animals "more accessible to the visiting public." The problem as he saw it was how to manage the bison population "under nearly natural conditions and at the same time get it near the main highways where it can be easily and safely observed."[82]

Predictably, the biologists opposed confining park wildlife. In 1931 George Wright made his opposition clear to Albright, pointedly reminding the director that the purpose of park wildlife "does not end with their being seen by every tourist," and chiding that people see many such animals "when the circus comes to town." To Wright and his fellow biologists, an animal enclosure had the appearance of a "game farm" and was an inappropriate display of park wildlife to the public.[83]

Wright's position was reflected in Joseph Grinnell's remarks to Director Cammerer in 1933, after Yosemite's fenced-in Tule elk herd (not native to the park) had been returned to its native habitat in California's Owens Valley. Keeping a close watch on Yosemite's wildlife management, Grinnell wrote to Cammerer applauding Superintendent Charles Thomson's decision to remove the elk from the park. In reference to overall national park policy, Grinnell remarked that parks were not places "in which to maintain any sorts of animals in captivity," adding that it was the "free-living native wild animal life that . . . gives such rich opportunity for seeing and studying." He took it for granted that maintenance of free-roaming wild animals was the Service's "general policy."[84]

Grinnell was mistaken, however. Yellowstone's most ambitious effort to display bison came in 1935, only two years after Grinnell's letter to Cammerer, when the park established "Antelope Creek Buffalo Pasture"—an approximately 530-acre tract south of Tower Falls in the northeast section of the park. Located along the park's main tourist road, the pasture accommodated about thirty bison and included a five-acre "show corral" to assure visitors a view of the animals.[85] An important part of the park's wildlife display for several years, the Antelope Creek enclosure would be discontinued in the 1940s by Director Newton B. Drury, sparking a heated controversy over the very policy issues that Grinnell and the wildlife biologists had raised.

Predators

Park Service leadership in the 1930s still harbored antipathy toward large native predators—a serious matter to the wildlife biologists, who wanted them protected. Again, the Service's actions in this regard exposed internal disagreements over policy and underscored difficulties that the biologists faced in seeking to change traditional practices. Already by 1931, when Director Albright announced the policy of limiting predator control to what was absolutely necessary, wolves and cougars had been virtually eradicated from all national parks in the forty-eight states.

The new predator policy had only limited effectiveness. Of the triumvirate of carnivores most targeted for reduction by the Park Service in past decades (wolves, cougars, and coyotes), only the coyote remained in substantial numbers, except that the Alaska parks had populations of wolves. Despite the new predator policies, coyotes continued to be hunted during most of the decade, mainly on an occasional basis; and limited control of wolves was undertaken in the Alaska parks.[86]

Indeed, the 1931 predator policy itself reflected long-standing bias against the coyote. Instead of a flat prohibition, the policy stated that there would be "no widespread campaign" against predators, and that "coyotes and other predators" would be shot only when they endangered other species. Thus, the policy did not totally eliminate predator control; rather, it only restricted control to no "widespread" campaigns. And it specifically identified the coyote as a potential target—the only species so designated. Moreover, at the 1932 superintendents conference, a lengthy discussion of predator policy focused mainly on how to deal with coyotes. The consensus was that coyotes were to be subject to "local control," and that reducing this species would be a matter of each superintendent's discretion. In fact, two biologists in attendance, Joseph Dixon and Harold Bryant, conceded that coyote reduction might at times be necessary.[87]

By far the strongest support for coyote control came from park management circles. Horace Albright wanted to kill coyotes when they did damage to "more useful species." He particularly feared that antelope populations were threatened, and that without the current "intensive" control of coyotes there would soon be no antelope in Yellowstone. Roger Toll, Yellowstone's superintendent, concurred, asserting that a herd of antelope and deer was "more valuable than a herd of coyotes." He stated that it was not predators, but elk, deer, and antelope that were "the type of animal the park was for."[88]

With support from leaders such as Albright and Toll, "wholesale coyote killing" (in the words of a Park Service report) continued in Yellowstone until the fall of 1933. Earlier that year, in Fauna No. 1, George Wright's team of wildlife biologists had declared a more restrictive predator policy than before, which may have been a factor in easing Yellowstone's aggressive coyote control. As stated in Fauna No. 1, predators were to be "special charges" of the Park Service and would be killed only when the prey species was "in immediate danger of extermination," and then only if the predator species itself was not endangered.[89]

In truth, the 1930s did witness a decline in the killing of coyotes. Under the guidance of Sequoia superintendent John White, biologist Harold Bryant, and especially George Wright, the Service began to rely on "increased scientific data rather than ancestral prejudice" to address the predator issue. In November 1934 Director Cammerer issued a prohibition of all predator control unless written authority was obtained from his office. Yet the following year, in Fauna No. 2, Wright and Ben Thompson acknowledged that coyote management was still controversial. They defined Park Service policy as allowing "judicious control of coyotes" to be undertaken in any park with the necessary authorization from Washington.[90]

Ongoing coyote control demonstrated that these predators were still not true "special charges" of the Park Service. Particularly in Yellowstone, pressure to reduce coyote populations continued, although it apparently diminished after 1933. A matter-of-fact report in March 1935 revealed a cavalier attitude toward eliminating coyotes, as one ranger described how he spied a pair of coyotes copulating "just at daylight" near lower Slough Creek, then shot one of the animals dead.[91] By contrast, some Yellowstone staff doubted the wisdom of continued coyote control. Assistant Chief Ranger Frank W. Childs recommended in April 1935 that the park suspend the killing of coyotes for at least two years, with the intention of carefully studying the resulting effect on prey populations. Childs and others recognized the conflict between, on the one hand, efforts to reduce elk populations, and on the other, killing predators that were presumed to reduce the numbers of elk. He suggested that scientific research might prove that ending coyote control permanently would be best for the "general wildlife balance" in the park.[92] Despite such opinions, evidence indicates that by 1937 interest in further coyote reduction had intensified.[93]

Demands for predator reduction in Yellowstone and other parks were based on concern for the protection of the ungulate species, so that they could be both enjoyed in the parks and hunted on adjacent lands. Also, ranchers ranged livestock on nearby lands and wanted protection from predators. Hunters and ranchers urged the Park Service to reduce or entirely remove major carnivores from the parks. In effect, this stance allied them with Albright and those in the Service opposing predators. Others argued for a more cautious approach. In November 1935 Crater Lake superintendent David H. Canfield responded to the Southern Oregon Livestock Association's "sweeping condemnation" of predatory animals in national parks. The association was particularly anxious about coyotes in the vicinity of Lava Beds National Monument, a park under Canfield's supervision. Canfield countered that the wildlife problems of the area would be addressed through scientific research. Subsequent research in Lava Beds supported protection, rather than control, of coyotes.[94]

Although not bold, the Service's official policy for protection of predators motivated sportsmen's associations and other groups to oppose initiatives for new national parks in the Kings Canyon area of California and Olympic Mountains in Washington. As elsewhere, such groups wanted predators eliminated to protect game species. Resentment of Service policies led the California state legislature to petition Congress to force strict predator reduction in the national parks—to no avail.[95] As viewed by Joseph Grinnell, longtime opponent of predator control, this proposal would have been a "calamity" to those "who see in national park administration the last chance of saving [for] the future entire *species* of certain animal groups." Putting predators in an ecological context, Grinnell wrote to Director Cammerer about the need to preserve the "biotic mosaic" of each park, including predators. The Service should maintain the whole "biotic superorganism uninjured—to the benefit of *all* its constituent species and populations."[96]

In striking contrast to Grinnell, Horace Albright remained alarmed about what effects the discontinuance of coyote control would have on the grazing species, particularly antelope. His letters to Cammerer on predators and antelope were plainly worded. In October 1937 Albright deplored the ongoing, as yet inconclusive, studies of the coyote's impact on Yellowstone's antelope population. He advocated "open war" on coyotes for the purpose of studying stomach contents to determine the extent to which coyotes fed on antelope. In fact, he urged reducing the coyote population under almost any pretext, stating that in spite of Park Service policy or the results of the studies of coyote stomachs, he would "continue to kill coyotes on the antelope range for the reason that the coyotes are of no possible advantage in that part of the park, can rarely be seen by tourists . . . while on the other hand there will always be danger of depleting the antelope herd. It must be remembered that one of the animals most interesting to tourists is the antelope." Albright also feared that, if protected, the coyotes would "over-run adjacent country," causing conflict with land managers and owners outside the park.[97]

Even as Albright campaigned against the coyotes, the Park Service planned more research on this predator, and in 1937 Adolph Murie initiated a study of Yellowstone's coyotes. Murie's report, *Ecology of the Coyote in the Yellowstone,* appeared in 1940. It indicated that coyote predation did not appreciably affect prey populations, having only a "negligible" impact on elk populations. Murie noted that in view of the National Park Service's "high purpose" of preserving "selected samples of primitive America," the parks' flora and fauna should be subjected to "minimal disturbance." He concluded that coyote control was "not advisable under present conditions."[98]

Coming from one of the most outspoken Park Service biologists, Murie's conclusions drew severe criticism from within the Service. Indeed, some individuals in top management apparently wanted Murie fired.[99] Moreover, already aware of Murie's findings and the Wildlife Division's opposition to coyote reduction, Albright wrote to Cammerer in January 1939, repeating his disagreement with the biologists. Believing there was nothing to be gained "either in wildlife management or in service to the public" by protecting the coyotes, Albright feared that, if not controlled very strictly, "powerful predators" such as the coyote were certain to menace the "more desirable species of wildlife." Despite the criticism, Murie's findings gained support from Director Cammerer. As Cammerer stated in his 1939 annual report, the coyote was a "natural and desirable component of the primitive biotic picture," not affecting the well-being of any of its prey species and "not requiring any control at present"—words that sound as if they were written by Murie himself.[100]

Cammerer also noted in his 1939 report that Murie had begun long-range studies of the wolves in Mt. McKinley National Park. Public demands for wolf control in McKinley (which resulted from fear that this predator was reducing Dall sheep and other popular wildlife populations) prompted Murie's study, which would extend into the mid-1940s. As with the coyotes in Yellowstone, the Service sought to establish a scientific basis for its treatment of Mt. McKinley's wolves. Again, however, Horace Albright's comments highlighted the differences between the recommendations of the wildlife biologists and traditional Park Service attitudes. In his January 1939 letter to Cammerer, the former director stated that he found it "very difficult" to accept the idea of protecting McKinley's wolf population in the "territory of the beautiful Dall sheep." Albright believed it was a "grave risk" to spend so much time and effort caring for predators, a responsibility that in his opinion "does not or need not fall on the National Park Service at all."[101]

Writing to Cammerer in May 1939, Park Service biologist David Madsen reflected on the state of national park predator management. He noted the ambivalence that still existed and cited Adolph Murie's belief that the Service was troubled with "confused thinking" and did not have a "philosophical point of view" on predators. In part, Madsen attributed this indecisive attitude to a lack of scientific information that affected the thinking of all Service personnel, both managers and biologists. He saw a "need for enlightenment" on the predator issue, to help the Service handle the "crossfire" between the scientists and groups such as sportsmen and livestock owners.[102]

Influenced by the wildlife biologists (who found some support from park management—from Director Cammerer to Yellowstone ranger Childs), the Park Service moved very slowly and erratically during the 1930s toward a scientific understanding of predator and prey populations and the discontinuance of predator control. Murie's work at Yellowstone and Mt. McKinley and the coyote studies at Lava Beds evidenced a willingness in the Service to use scientific research to address specific predator concerns. Nevertheless, as Madsen recognized, a strong ambivalence attended the issue. The scientific perspective was countered by traditional attitudes favoring popular game species over carnivores and by agitation from organizations of livestock owners and sportsmen. Such pressures would persist.

Fish

Continuing the emphasis of the Mather administration, fish management under Albright and Cammerer's leadership was primarily intended to enhance sportfishing. In its management of fish, more than of any other natural resource, the Park Service violated known ecological principles. With extensive hatching and stocking continuing in the national parks, the Service shipped fish eggs to nonpark areas in an effort to improve fishing elsewhere in the country. Thus its manipulation of fish populations and distribution extended far beyond national park boundaries. The Yellowstone Lake Hatchery was particularly active, shipping millions of native and nonnative fish eggs to numerous states and even to some foreign countries.[103]

In 1928, five years before Fauna No. 1 appeared, the Park Service had detailed a biologist from the Bureau of Fisheries to be the Service's specialist in "fish culture" and to coordinate with the bureau in raising fish for stocking national park lakes and streams. The specialist was probably David Madsen, who was converted to permanent Park Service employment in April 1935, assigned to the Wildlife Division. Reviewing fish culture activities in the national parks, Madsen observed that in the past "other agencies" had run national park fish programs, often with very little direction from the Service. However, the Park Service had recently decided to use wildlife rangers to do the planting and had hired Madsen, thereby assuming more control over what species were planted, and where.[104]

So deeply entrenched was the tradition of fishing national park rivers and lakes that the wildlife biologists themselves seemed ambivalent and did not seek to discontinue this activity. In Fauna No. 1 the biologists had commented in a section appropriately entitled "Conflicts with Fish Culture" that fishing in parks was an "important exception to general policy." Granting the long-established fish management practices, they conceded that the benefits to park visitors overruled the "disadvantages which are incidentally incurred" by allowing fishing.[105] Madsen, too, recognized that the Park Service's fish management was "entirely inconsistent" with other wildlife policies, and that "indiscriminate introduction" of nonnative fish had adversely altered the natural conditions of park lakes and streams. Yet as a fish culture specialist he appreciated the popularity of fishing in the parks and stated that the sport should be "maintained and in some instances developed to the highest point possible in the interest of the visiting public."[106]

Nevertheless, the biologists were largely responsible for the slight modifications in the Service's fish policy that did occur in the 1930s. Fauna No. 1 contained recommendations to reduce populations of exotic species already present in the parks and to prevent the invasion of additional exotics. The report also advocated setting aside one watershed in each park to ensure "preservation of the aquatic biota in its undisturbed primitive state." No introduction of fish or fish food would be allowed in any of these watersheds, except what naturally occurred; fishing would be permitted, but only if it did not "deplete the existing stock."[107]

In April 1936 Director Cammerer issued the Park Service's first written fish management policies, almost certainly prepared by Madsen. The continuation of fish culture activities in parks was a given in the new policy; in fact, the document's introduction specifically stated that it was a policy for "fish planting and distribution." Still, the policy favored protection of native species, emphasizing the intent to "prohibit the wider distribution" of exotics within park waters. Exotic species were not to be introduced in waters where only native fish existed; and in waters where exotic and native fish *both* existed, the native species were to be "definitely encouraged."[108]

The new policy left substantial options to park managers, however, thereby reducing chances of significant change from earlier practices. A superintendent was permitted to stock waters previously barren of fish unless he determined that the lake or stream was of "greater value without the presence of fishermen." In waters where exotic species were "best suited to the environment and have proven of higher value for fishing purposes than native species," stocking of exotics could continue if approved by both the park superintendent and the director. Cammerer refined this last point in his 1936 annual report by specifying that native species would be "favored" in waters where such species "are of equal or superior value from the standpoint of fishing."[109]

The new fish management policy thus allowed continued alteration of national park aquatic habitats for the promotion of sportfishing and the enhancement of public enjoyment. The Service continued to plant exotic species in large numbers in waters such as Yellowstone's Madison, Firehole, and Yellowstone rivers in the years following issuance of the permissive 1936 policy. In some locations, as at Mammoth Beaver Ponds in the Yellowstone River drainage, previously fishless lakes were first stocked about the time the policy was declared, and such stocking continued for years afterward.[110] Not even mentioned in the new policy, the shipment of millions of fish eggs (including both native and exotic species) from national parks to nonpark areas continued unchecked throughout this period. Director Cammerer reported in 1937 that twenty million rainbow and Loch Leven trout eggs (both exotic species) were collected near Yellowstone's west boundary, with only one-fifth of them returned to park waters and the rest shipped elsewhere.[111]

Park Service biologist Carl Russell's remarks to the North American Wildlife Federation in March 1937 reflected the continuity in national park fish policy. He asserted that the new policies would mean continued "maintenance of good fishing" and that the Service was "definitely" committed to fishing as a "recreational activity in parks." Similar observations came from other biologists. Victor Cahalane commented in 1939 that the Service deemed fishing to be acceptable because of the "readily replaceable nature of fish resources," and because sportfishing resulted in "recreational benefits far outweighing any possible impairment of natural conditions." Evidencing the ambivalence among the biologists, Cahalane also stated that it was the National Park Service's responsibility to address the contradictions "existing between use of fish resources and of other natural resources within the parks."[112] Nevertheless, the widespread acceptance of angling in national park waters ensured that the contradictions would remain largely unresolved.

Forests

As with fish, the management of national park forests in the 1930s continued established practices. But the forestry policies were strongly challenged by the wildlife biologists. The conflict over forestry practices exposed fundamental differences between the biologists and much of the rest of the Park Service. The failure of the biologists' challenge to forest management demonstrated the weakness of their position within a traditional organization and, conversely, the considerable bureaucratic strength that the foresters were gaining.

National park forestry operations expanded tremendously during the New Deal, receiving far more funds and support from the emergency relief programs than any other natural resource management activity in the parks. The 1933 act creating the Civilian Conservation Corps specifically called for protection of the nation's forests from fires, insects, and disease damage—goals that matched perfectly those of most national park managers. So important did forestry become in the overall work of the CCC that the organization was at times referred to as Roosevelt's Tree Army.[113]

In his 1933 annual report, Horace Albright's comments on the initial work of the CCC foreshadowed the tremendous expansion of national park forestry. The director stated that the newly established CCC crews were accomplishing "work that had been needed greatly for years," but that had been "impossible" under ordinary appropriations.

> Especially has the fire hazard been reduced and the appearance of forest stands greatly improved by cleanup work along many miles of park highways; many areas of unsightly burns have been cleared; miles of fire trails and truck trails have been constructed for the protection of the park forests and excellent work accomplished in insect control and blister-rust control and in other lines of forest protection; improvements have been made in the construction and development of telephone lines, fire lookouts, and guard cabins; and landscaping and erosion control [have] been undertaken.[114]

During the buildup of CCC-funded forestry programs in 1933, Director Cammerer designated John Coffman the Service's "chief forester," in charge of the newly created Division of Forestry, now separate from the Service's educational program.[115] The forestry management policies that Coffman and Ansel Hall had prepared provided guidance for the Park Service throughout the decade. Under these policies the forests were to be "as *completely protected* as possible" against fire, insects, fungi, and "grazing by domestic animals," among other threats. This comprehensive protection was to be extended to "*all* park areas" associated with "brush, grass, or other cover."[116] Backed by new policies and staffed by thousands of CCC enrollees, Coffman's forestry programs became an increasingly important force in national park operations during the New Deal era.

Significantly, although the Park Service had begun building a cadre of wildlife biologists, the bureau did not hire plant biologists or botanists per se. Rather, it hired "foresters," who were deeply influenced by the management practices of the U.S. Forest Service, particularly regarding control of forest fires, insects, and disease. With the foresters maintaining such traditional attitudes, the wildlife biologists were left with few allies to argue the case for ecological management in the parks. Central to the biologists' concerns were the various prefire protection activities—the very kinds of development Albright enthusiastically endorsed. They objected to building fire roads through natural areas and clearing hazardous dead trees and snags that contributed to the fuel buildup and increased the possibility of fire.

Indeed, the wildlife biologists were never in agreement with the forest management policies written by Coffman and Hall. Although forests were not the focus of George Wright's initial wildlife survey, preserving natural habitat, including plants, was recognized as fundamental to successful park management. In contradiction to ongoing Park Service forestry practices, Fauna No. 1 urged that park forests not be manipulated, stating, for instance, that "it is necessary that the trees be left to accumulate dead limbs and rot in the trunks; [and] that the forest floor become littered."[117] Nevertheless, the CCC programs provided funds and manpower for extensive clearing of forest underbrush and dead trees. This work increasingly alarmed the biologists.

Roadside clearing, a widespread practice in national parks, was intended as a fire protection measure but, in the words of a Park Service manual, was equally important as a means "to improve the appearance of the immediate landscape of the main drive" through parks. A conflicting view came from Wright, who wrote Director Cammerer early in 1934 of the need to consider "all sides of the question" regarding clearing of hazardous debris along park roadsides, including the concern for "wild life values." Wright realized that clearing dead limbs and trees affected habitat. He urged that the Service "reconsider" and determine "exactly under what conditions and in what parks roadside clean-up is a benefit and to what extent it should be carried on." He also told Cammerer that the biologists had discussed the issue with park superintendents and rangers, and that there was "anything but unanimity of opinion on the value of this work." Although some superintendents and rangers recognized the impact on natural conditions, others believed cleanup did not help prevent fires.[118] Nevertheless, clearing was widely accepted in the Service and remained a common practice in the parks.

An opinion even stronger than Wright's came from Adolph Murie in the summer of 1935, during an extended debate over whether or not to clear a twelve-square-mile area on Glacier National Park's west slope, just north of McDonald Creek, a forested area damaged in a recent fire. With many of the trees only partially burned, the tract seemed ripe for another fire, which could spread to adjacent, unburned forests. A meeting in the park in July provoked disagreement on the propriety of cutting and removing all of the dead trees, whether standing or fallen. The contentious debate reflected sharp divergence between the wildlife biologists and the foresters on fire protection and overall national park policy.

Following the meeting in Glacier, Murie reported to the Wildlife Division in Washington his intense opposition to the proposed clearing. In a lengthy letter, he wrote that the burned area was still in a natural condition and questioned the desirability of "removing a natural habitat from a national park." Requiring roads for trucks, bulldozers, and other equipment, the clearing operation would cause "gross destruction," which, he believed, would interfere with the normal cycles of forest decay and growth and create instead a "highly artificial appearance of logged-off lands." Removal of the trees would reduce the area's organic material and its soil fertility, and would cause drying of the soil and increased erosion.

Moreover, Murie argued, this large clearing project could be used to justify "almost any kind of landscape manipulation" in the future. "For what purposes," he asked, "do we deem it proper to destroy a natural state?" His answer was that almost no purpose justified such destruction. Murie concluded his argument with an opinion surely unheard of in national park management before the wildlife biologists began their work under George Wright: "To those interested in preserving wilderness, destroying a natural condition in a burn is just as sacrilegious as destroying a green forest. The dead forest which it is proposed to destroy is the forest we should set out to protect."[119]

Murie's remarks were quickly challenged. Lawrence F. Cook, head of John Coffman's forestry operations in the western parks, had also attended the meeting in Glacier. Cook found Murie's report "rather typical" and took a directly opposite position, fearing the long-term loss of green forests. "Nature," he commented, "goes to extremes if left alone." Despite the Service's best protection efforts, "gross destruction" had resulted from the fire. Beyond adequate detection, fire protection depended on "easy access" into the forests and the "reduction of potential fuel" through clearing—both of which would result from the proposed work in Glacier. Cook anticipated a rapid recovery of forest growth, but only if the area were cleared of dead trees so it would not be burned over by another, more damaging fire. Seeking to protect the beauty of the forests, he also recognized that this part of Glacier was intensively used; it was seen, he claimed, "by more travellers than any other in the park." Cook argued that the question was not whether to allow nature to take its course in the national parks, but to what extent the Service "must modify conditions to retain as nearly a natural forest condition as possible for the enjoyment of future generations."[120]

In a separate memorandum to Coffman, written the same day, Cook expressed concern that the Service's foresters had been accused of being "destroyers of the natural." Their construction of truck trails and fire lookouts and their clearing of damaged forests had been criticized not only by the biologists but by some superintendents, rangers, and landscape architects. Cook insisted that the foresters were seeking to preserve the "natural values" of the parks, while also providing for the "greatest use and enjoyment of the parks with the least destruction." He summed up his credo of national park management, and fire protection in particular: "The parks have long since passed the time when nature can be left to itself to take care of the area. Man has already and will continue to affect the natural conditions of the areas, and it is just as much a part of the Service Policy to provide for their enjoyment as it is to preserve the natural conditions. There is no longer any such thing as a balance of nature in our parks—man has modified it. We must carry on a policy of compensatory management of the areas." He added that "forest protection" is a "very necessary part of this management." Without protection the Service faced the destruction of "any semblance of biological balance, and scenic or recreational values, as well as the forests with which we are charged." Certainly Cook's views prevailed within the Service, and CCC crews cleared a vast area of the McDonald Creek drainage (such that even today, as a veteran Glacier biologist put it, negative effects "are still very evident on the land").[121]

In truth, the Park Service's biologists and foresters all claimed they were seeking to preserve "natural values," which would allow for the "greatest use and enjoyment of the parks with the least destruction." But the two groups had fundamentally different perceptions of what constituted "natural values" and what constituted "destruction" in national parks. Adolph Murie opposed the extensive alterations that resulted from the Service's fire protection methods employed before, during, and after fires. His letter on the proposed clearing in Glacier concluded: "My feeling concerning any of this manipulation is that no national park should bear the artificial imprint of any man's action of this sort. We have been asked to keep things natural; let us try to do so." But Cook's philosophy of national park management reflected the official forestry policies; with funds and manpower coming from the CCC program, the Service continued its intensive protection and suppression activities, rejecting Murie's concepts.[122]

The conflicting approaches to national park management were evidenced in disagreements over other aspects of forestry. Continuing the practices of the Mather era as affirmed in the forest policies, both Albright and Cammerer supported aggressive war against forest insects and disease, regularly calling on the Bureau of Entomology and the Bureau of Plant Industry for expert assistance. In his last annual report (1933), Director Albright noted that "successful campaigns" had been waged against insects in park forests, ending or reducing several major epidemics. The Service, he added, had sought to eradicate infestations of the bark beetle in Yosemite and Crater Lake, and the mountain-pine beetle in Sequoia National Park. Both Glacier and Yellowstone faced insect infestations of such magnitude that studies were being made to determine if control efforts were even practicable. It seemed to Albright that the forests in the national parks were truly under siege from insects, as well as from disease. Among many threats, blister rust was "spreading rapidly," threatening the western parks. "Unless checked," Albright warned, it was "only a matter of time" before blister rust would invade the white pine forests of Glacier and the sugar and white pines of the California parks.[123]

As with fire protection, the CCC provided the Park Service with funds and manpower to wage intensive campaigns against forest insects and disease. Again the wildlife biologists challenged these efforts. George Wright wrote to Director Cammerer in August 1935 favoring use of the New Deal work relief programs, but cautioned against too much "zeal for accomplishment," particularly in insect and disease control. Generally the biologists accepted limited control in and around park development, directing their criticism toward more widespread control efforts. Wright would largely confine control to "heavily utilized areas" most frequented by visitors.

The piñon pine scale infection in Colorado National Monument was, he pointed out, a natural phenomenon that seemed "best to leave undisturbed" outside developed areas. Similarly, reporting on CCC work in Grand Canyon during 1935, Victor Cahalane commented that the Wildlife Division "disapproves of insect control, outside of developed areas," unless a native plant was threatened with extinction.[124]

Much more critical, Adolph Murie lashed out at Park Service insect and disease control efforts after a 1935 visit to Mount Rainier. He acknowledged to Wright that "possibly some effort" was necessary to save "certain outstanding forests." But he opposed extensive control, emphasizing that in its forest management the Service should not "play nursemaid more than is essential." Especially alarming were efforts to kill off native beetle populations and to control the blister rust disease by eradicating *ribes* (native currants and gooseberries, which serve as an alternative host to the blister rust fungus). With both *ribes* and beetles native to the parks, Murie urged leaving them alone and "permitting natural events to take their course. . . . The cure is about as bad as the disease." *Ribes* was, in his words, "just as desirable in the flora as is pine," and he concluded that "justification for destroying a species in an area should be overwhelming before any action is taken."[125]

Predictably, arguments such as Murie's did not sway the foresters. In his letters to Coffman on fire management, Lawrence Cook rebutted the biologists' position, defending the Service's forest disease and insect control policies as an essential part of park management. As with fire suppression, the foresters believed that "some modification," including insect control, "is necessary to preserve for the future the living values of the parks." And, indeed, aggressive forest insect and disease control continued while CCC money and manpower were available. Late in the decade Director Cammerer reported on blister rust control and beetle eradication in a number of parks, noting the support of the Bureau of Entomology and dependence on the CCC program.[126] The termination of the CCC just after World War II began would drastically reduce the resources available to the Park Service for control work—but the policies remained in force, waiting for postwar funding.

The wildlife biologists had found a voice in national park policy and operations, but they frequently clashed with the foresters, who continued their practices despite the biologists' objections. Decades later Lowell Sumner reflected that "even George Wright was unable to make much progress" in establishing ecologically sound forest management.[127] The biologists' criticism of various forest practices had little effect on Service policies, a reflection of the support the foresters enjoyed from Park Service leadership. The policies on forest fires, insects, and disease were aimed at maintaining the beauty of the parks and enhancing public enjoyment, and were much more in line with mainstream national park thinking than were many of the ideas of the wildlife biologists.

In 1940, at the end of Cammerer's directorship and with the biologists' influence in decline, the foresters were truly in the ascendancy. The Park Service's official organizational chart, revised in mid-1941 (a year and a half after Interior secretary Ickes transferred the wildlife biologists to the Bureau of Biological Survey), showed the Branch of Forestry with no less than three divisions: Tree Preservation, Protection and Personnel Training, and Administration and General Forestry.[128] Also, foresters entering the Park Service continued to be influenced by the U.S. Forest Service. Many national park rangers who did not have the specific title of forester had nevertheless been schooled in forestry. In addition, the so-called ranger factory, just beginning at Colorado Agricultural and Mechanical College in the late 1930s (and which would flourish during ensuing decades), trained young men to become rangers under a park management program administered by the forestry school.[129]

Altogether, an alliance was building between the Park Service's foresters and its rangers (they would be combined organizationally in the mid-1950s). The influence of this alliance was bolstered by the fact that the two groups fed directly into leadership positions, in charge of national park policy and operations. With an increasing number of forestry graduates attracted to the Service, forestry evolved into one of the most powerful professions in the organization, attaining full "green blood" membership in the Park Service family. By the end of the decade the foresters' bureaucratic strength began to rival that of the landscape architects and engineers under Thomas Vint and Conrad Wirth.[130] Although not always in full accord, these professions, in alliance with the superintendents and rangers, formed the core of the Service's leadership culture and would dominate national park philosophy and operations for years to come.

Expanding Park Service Programs

During the New Deal the Service sought (as stated in a 1936 internal report) to "enlarge its field of usefulness" through increasing the viability and the social utility of the national park system by expanding the system and making it more accessible to and popular with the public.[131] Extending from Roosevelt's inaugural to the beginning of World War II, the New Deal fostered vast expansion and diversification of Park Service activity and brought dramatic changes in the composition of the national park system. It placed new responsibilities on the Service (especially in the fields of recreation and assistance to state parks), brought different kinds of parks into the system (such as historic sites, reservoirs, national parkways), and accelerated physical development of the parks to provide for public use and enjoyment. Its leadership always possessed of a keen entrepreneurial bent, the Park Service had suddenly entered flush times. The New Deal would fund many programs that bolstered the Service's expertise in recreational tourism, such that it could lay strong claim to national leadership in that field.

Park Service leaders got virtually everything they could have hoped for from the New Deal. Even before Congress passed the act establishing the CCC, Director Albright recognized the legislation's potential. In early March 1933, approximately two weeks prior to the act's passage, Albright wrote to Assistant Director Arthur Demaray that the share of funds allotted to the national parks would depend on the Park Service's preparedness—how much it could demonstrate that it was ready to spend. As recalled by Conrad Wirth, the landscape architect who would ultimately take charge of the Service's many CCC programs, Albright was seeking "to justify a good, sound park program should the funds suddenly become available." The director quickly prepared estimates of $10 million for construction, including roads, trails, and other developments. He asked the park superintendents to assess immediately their ability to take advantage of the new funds, and called for an updating of national park master plans to prepare for the infusion of New Deal money. With Roosevelt's emergency relief programs the Service was, as later recalled by Arno Cammerer, poised to "absorb . . . a large segment of such work and to benefit greatly therefrom."[132]

Albright also contacted state park authorities around the country, advising them that the CCC would become involved with state as well as national parks. Of all CCC activities, assistance to the states in recreational planning and development most expanded the Park Service's operations. Funded by the CCC and given solid encouragement from the very first by the Service's directorate, the state parks assistance program began in 1933 and gained momentum rapidly under the leadership of Conrad Wirth. Wirth was named assistant director for recreational land planning—bureaucratic status that indicated the importance placed on these programs. His principal aide was Herbert Evison, former secretary of the National

Conference on State Parks, the organization that Mather and Albright had helped found in the early 1920s in their efforts to encourage a stronger nationwide park system.[133] Wirth quickly built an impressive, far-reaching program, developing proposals for creating new state parks and overseeing the planning, design, and construction of the facilities necessary for state parks to accommodate public use.

Soon employing thousands of CCC workers in state park projects, the Service constructed roads, trails, cabins, museums, campgrounds, picnic grounds, administrative offices, and other state park facilities—work that replicated the CCC projects Wirth was overseeing in national parks.[134] Through assistance to the states, the Service's expertise in intensive physical development of parks extended far beyond national park boundaries. Also, in both state and national park construction, the Service's architects and landscape architects of the 1930s directed CCC craftsmen toward a harmonious blending of new construction with the surrounding park landscapes. Following the traditions of rustic architecture established earlier in the national parks, CCC laborers created many structures that later generations would praise for their beauty and craftsmanship. Altogether, the focus of CCC development was overwhelmingly in support of public recreational use of parks, thus reinforcing within the Service this aspect of park management.

Added to the Park Service's state programs was a national survey of potential recreational lands that could help meet public needs. The survey came about as a result of Park Service lobbying while it was participating on the National Resources Board, established by Roosevelt in 1934 to study the nation's natural resources and land uses. As recalled in an internal document, the Park Service submitted an "urgent" recommendation to the board that there be a study to determine recreational needs.[135] Late in 1934 the Service completed such a survey, but one that it viewed as only preliminary. It quickly began campaigning to expand the survey and to institutionalize existing cooperation with the states by gaining full congressional sanction for activities that theretofore had been only administratively authorized. The lobbying paid off. The resulting Park, Parkway, and Recreational Area Study Act of 1936 permitted the Service to make a comprehensive national survey of park and recreational programs and to assist states in the planning and design of parks.[136]

This act constituted a decisive political and bureaucratic commitment to the recreational aspects of park management and to all levels of parks, from state and local to national. Using mostly CCC funds, Wirth promptly implemented the act, building on the 1934 preliminary survey to detail the nation's park and recreational needs in a report entitled *A Study of the Park and Recreation Problem of the United States,* published in 1941. A comprehensive document, the study argued for the expansion of recreational facilities throughout the country. Furthermore, in cooperation with the Park Service, forty-six states worked on statewide surveys, with thirty-seven of the reports ultimately completed, and twenty-one published. In addition to these studies, the Service undertook a survey of seashores and major lakeshores in the United States, identifying numerous areas eventually to be included in the national or state park systems and in many cases to be put to intensive recreational use.[137]

The Service's development of parkways for "recreational motoring" furthered its leadership role in national recreational programs. Even before the New Deal began, the George Washington Memorial Parkway, Colonial Parkway (connecting Yorktown and Jamestown, Virginia), and Shenandoah National Park's Skyline Drive were under construction as part of the national park system. Major additions to the parkway program came later in the decade with authorization of the Blue Ridge and Natchez Trace parkways. All of these new scenic highways received massive amounts of New Deal emergency relief funds. They also received staunch support from Park Service leadership, which regarded them as perhaps the most "spectacular new phase of national park planning and development during recent years."[138]

As part of its nationwide recreational work, the Park Service urged authorization of the "recreational demonstration area" program, another type of park planning and development to accommodate intensive use. The Service recognized the potential for acquiring marginal agricultural lands located near urban centers, with the lands to be converted into recreational areas—a concept promoted in 1934 by Wirth while serving as Director Cammerer's representative on a presidential land planning committee. Intended to become state or local parks, the demonstration areas were to be developed for picnicking, hiking, camping, boating, and other similar uses. Having, as Wirth saw it, "unanimous approval and support" from within the Park Service, the program began in 1934, with the Federal Surplus Relief Administration purchasing the lands and the Park Service supervising their conversion into park and recreation areas. Most of the areas, as Cammerer noted in 1936, were meant to serve "organized camp needs of major metropolitan areas." In time, forty-six demonstration areas were established, requiring a substantial Park Service commitment in planning, design, and construction to develop the areas for public use.[139]

Almost all of the recreational demonstration areas were eventually turned over to state or local governments, although some became extensions of existing units of the national park system—for instance at Badlands, Acadia, and Shenandoah. In addition, several demonstration areas were authorized as new units of the park system, including Theodore Roosevelt National Memorial Park in North Dakota and portions of Catoctin Mountain Park in Maryland.[140]

Most of the development that the Park Service supervised in recreational demonstration areas and state parks was undertaken with CCC funds. These monies financed not only the labor (including the enrollees' housing and meals, provided in camps) but also the National Park Service's own professional staff involved in these programs. In addition, major developmental funds came from the Public Works Administration for projects such as electrical and sanitation systems, and road and building construction. Beyond the New Deal's crucial support to state park development, the Park Service recognized the relief programs as "invaluable" to the national parks themselves, making possible the completion of "a wide variety of long-needed construction and improvements."[141]

The Park Service expanded into additional fields during the New Deal era, most notably the management of historic and archeological sites, which theretofore had had no coordinated federal oversight. During the administration of President Herbert Hoover, Director Albright had unsuccessfully sought, by authority of the Antiquities Act of 1906 and other acts, to gain control of historic and prehistoric sites managed by the departments of war and agriculture. Among these sites were Gettysburg, Antietam, and Vicksburg battlefields (managed by the War Department) and archeological areas such as Tonto and Gila Cliff Dwellings national monuments (managed by the U.S. Forest Service of the Department of Agriculture). Immediately on Franklin Roosevelt's taking office, Albright proposed to the new secretary of the interior, Harold Ickes, that the President transfer the numerous historic and prehistoric sites from other departments to National Park Service jurisdiction.

Convinced that the Organic Act provided authority for involvement in historic preservation, Albright also believed the Service could provide the best management of these sites. It already managed Mesa Verde National Park and a number of other prehistoric areas in the Southwest, plus three historic areas in the east: Morristown National Historical Park, and Colonial and George Washington Birthplace national monuments. But Albright also hoped to strengthen the Park Service against its veteran rival, the U.S. Forest Service, by establishing authority in programs alien to the rival bureau. And he wanted to build the Service's political strength in the eastern

United States, where most of the sought-after historic areas (mainly Civil War and Revolutionary War sites) were located, and where very few national park units existed.[142]

This time Albright succeeded. In June 1933 Roosevelt signed two executive orders effecting transfer on August 10 of numerous sites to the national park system, thereby substantially reorganizing the federal government's historic preservation activities. The Service had campaigned for and gained a huge new program, with forty-four historic and prehistoric sites coming into the system along with twelve natural areas. Among the new natural areas were Saguaro and Chiricahua national monuments. The new historic areas included many battlefields, plus public parks and monuments in Washington, D.C., such as the National Mall and the Washington and Lincoln monuments—the Park Service's first major venture into urban park management.[143] Two years later, with the Service's encouragement, Congress passed the Historic Sites Act of 1935, which authorized cooperation with state and local governments in identifying, preserving, and interpreting historic sites.[144] With this act the Park Service increased both its historic preservation responsibilities and its already substantial involvement in state and local surveys and planning.

The reorganizations made early in the Roosevelt era entailed two changes the National Park Service did not want, however. In 1933 it was given responsibility for managing federal buildings in Washington (except for judicial and legislative buildings); along with this, the Park Service suffered a name change: it became the Office of National Parks, Buildings, and Reservations. Management of buildings in Washington added significantly to the demands on the Park Service. Initially, it entailed about fifteen hundred additional employees, a figure that escalated rapidly in the ensuing years. By the mid-1930s the Park Service was in charge of approximately 20.5 million square feet of space in fifty-eight government-owned buildings and ninety rented buildings in and around the District of Columbia and elsewhere—for example, the United States courthouses in Aiken, South Carolina, and New York City.[145] In 1934 the Park Service managed to get its new name (a "much-hated" designation, as Albright recalled it) abolished and the original name restored. Later, in 1939, management of federal buildings was transferred to the Public Buildings Administration.[146]

Finally, additional involvement in recreational programs came when Congress in the early 1930s authorized a National Park Service study of the recreational potential of Lake Mead, the huge new reservoir behind recently completed Boulder Dam on the Arizona-Nevada border. Even before the study was finished, the Service had established CCC camps and begun development along the reservoir's shoreline. Not surprisingly, given the direction the Service was taking in other recreational matters, its study found the potential to be very high, and in October 1936 the Park Service signed an agreement with the Bureau of Reclamation to manage public recreational use on and around Lake Mead.[147]

Ironically, only twenty-three years after a bitter nationwide controversy over the destruction of Yosemite National Park's Hetch Hetchy Valley by construction of a dam and reservoir, the Park Service became a willing participant in the mangement of Boulder Dam (later Lake Mead) National Recreation Area, encompassing what was then the largest reservoir in the world. Philosophical contradictions inherent in the Service's managing a reservoir where the main feature was itself a gigantic impairment to natural conditions were apparent from the start.

In 1932, at the request of Secretary of the Interior Ray Lyman Wilbur, former U.S. Congressman Louis C. Cramton, a longtime supporter of national parks, headed a reconnaissance of the reservoir area, with the study team including national park superintendents from Grand Canyon, Yellowstone, Zion, and Bryce Canyon. Their lengthy report noted the contradictions, observing that conservationists had long fought to protect national parks from "becoming incidental to or subordinate to irrigation and water supply uses." The report warned that heretofore all national parks had involved the "preservation of wonders of nature." Thus, "to deliberately bring into the national park chain and give national park status to such a dam and reservoir would greatly strengthen the hands of those who seek to establish more or less similar reservoirs in existing national parks." The team also warned that designating a reservoir a national park might encourage mining, cattle grazing, and other utilitarian uses of the existing national parks.[148]

However, even these substantial contradictions were readily resolved, to the enhancement of Park Service interests. As with many other park programs initiated during the New Deal era, recreation provided the Service with its principal rationale for entry into the field of reservoir management. Cramton's 1932 report on Lake Mead recommended that the area not be designated a "national park"; rather, the reservoir's *national* importance as a *recreation area* should be declared and that aspect of its management turned over to the National Park Service. The reconnaissance team believed that the Park Service's reservoir recreation work would be "entirely consistent with history and with principle." As justification the report cited the 1916 Organic Act's statement that the Service would manage "such other national parks and reservations of like character as may be hereafter created by Congress."[149]

Thus, by little more than devising the designation of "national recreation area," the Park Service sidestepped the contradictions with its traditionally held purpose of preserving lands unimpaired. It launched a new recreational program centered on huge reservoirs created by inundating western canyons and river valleys. Eventually this program would mushroom for the Park Service, bringing large sums of money and closer ties to the Bureau of Reclamation, particularly during and after World War II. Although within the Service there seems to have been some hesitation about the involvement at Lake Mead—perhaps on the part of Director Cammerer himself—it was nevertheless urged on by Conrad Wirth, spearhead of the Park Service's growth in recreational programs. Wirth, in turn, found support for recreational programs from such individuals as Associate Director Arthur Demaray, and even biologists Wright and Thompson.[150]

The Park Service's recreational programs did in fact draw on the talents of George Wright, who as head of the Wildlife Division represented the strongest potential resistance in the Service to its development-oriented park management. In 1934, recognizing Wright's considerable administrative skills, Director Cammerer appointed him to head the preliminary survey of the nation's recreational needs, which the Service had urged the National Resources Board to authorize. The survey team also included Conrad Wirth and the Park Service's chief forester, John Coffman. Wright wrote to Joseph Grinnell, his mentor at the University of California, that he found the recreational field to be "quite alien." Nevertheless, he supported the Service's rapidly expanding recreational programs. Shortly before his death in early 1936, Wright stated in a paper entitled "Wildlife in National Parks" that it was logical to place "responsibility for recreational resources" under the Service. Moreover, he had earlier given his blessing to the Park Service's involvement with reservoirs.[151]

The chief proponent of preserving natural conditions in the parks, Wright apparently saw the Service's varied recreational efforts as a means of relieving harmful pressure on the traditional national parks. Consistent with the major focus of his career, he wrote to Sequoia superintendent John White in 1935 about his concern that the national parks themselves not "supply mass outdoor recreation"—a prospect that would place a "destructive burden" on the parks. To Wright, adopting the policy of "giving all of the people everything they want within the parks . . . would involve sacrificing the Service's highest ideals."[152]

Overall, the National Park Service responded eagerly to the variety of New Deal opportunities in national recreational planning and development, as well as in the expansion of historical programs. Regardless of the taint of bureaucratic aggrandizement, the Service pursued very seriously—and very idealistically—the development of national, state, and local parks. Its assistance to the nation's park systems and its nationwide surveys and planning laid the foundation for expanding recreational opportunities throughout the country—a contribution that later generations would find easy to forget or take for granted.

It is important to point out that although Conrad Wirth showed little interest in scientific resource management and allowed the biology programs to decline during the last half of the 1930s while he was in charge of CCC funding and staffing, he was nevertheless the Park Service's chief advocate for the creation and development of recreational open spaces, whether as national, state, or local parks. His extensive surveys and planning for new parks during the New Deal (and later during his "Mission 66" program) would ultimately bear fruit with the establishment of dozens of new parks for the public's enjoyment and for the preservation of fragments of the American landscape—a legacy of inestimable value.

New Deal Impacts on the Park Service

The variety of programs taken on during the New Deal impacted the Service and the national parks in significant ways. Prior to 1933 the Park Service administered a system consisting mostly of large natural areas in the West, along with a few archeological sites in the Southwest and historic sites in the East. During the New Deal the Service's expansionist tendencies led it into enormous new responsibilities in recreation and historic site management. Especially with CCC funds, it extended its activities and influence far beyond national park boundaries, becoming involved in complex planning, intensive development, and preservation work with state and local governments from coast to coast. By the mid-1930s, after all of the Service's CCC operations had been consolidated under Conrad Wirth, some observers were claiming that, given the size of the programs under Wirth, there were in fact two National Park Services: the "regular" Park Service, and "Connie Wirth's Park Service."[153]

The Service's official organizational chart, revised no fewer than eight times during the 1930s, reflected the bureau's growing diversification and professional specialization. The sequence of charts showed an increase from three Washington branches and four "field" professional offices (of landscape architects and engineers, among others) in 1928, to a complex organizational maze of ten "branches" (or their equivalent) and four newly created "regional offices" on the 1938 chart. (The regional offices had been established in 1937, largely at Wirth's instigation, to correspond with the regional organization used by the CCC.) On the 1938 chart, specifically identified functions relating to the Service's growth and expansion during the 1930s included management of historic sites, archeological sites, memorials, parkway rights-of-way, and District of Columbia parks and buildings. In addition, under Assistant Director Wirth's Branch of Recreation, Land Planning, and State Cooperation were the Land Planning Division, the Development Division, and the U.S. Travel Division—the last created in early 1937 to stimulate travel to the national parks.[154]

Additional changes for the Park Service were detailed in a 1936 internal report, which noted that in the previous three years Service expenditures had increased "about fourfold and its personnel about eight [fold]." From 1930 to 1933, total appropriations had amounted to $11,104,000 annually. Over the next three years, total appropriations averaged $51,824,000 annually—a remarkable increase. Similarly, personnel figures rose from a monthly average of 2,022 employees in 1932 to 17,598 in 1936, with about three-fifths of the 1936 employees paid from CCC funds. (In Washington alone, management of the federal buildings and the public parks for which the Service was responsible required about 5,000 employees by 1936.) The overall figures included money and personnel for managing the fifty-six historical and archeological parks brought in by Roosevelt's 1933 reorganization, plus staffing for a number of newly created parks.[155]

The various New Deal emergency relief programs that the Service had so successfully tapped funded most of these staff increases. The 1936 internal report revealed that between July 1, 1933, and June 30, 1936, the Service's emergency relief funds totaled $116,724,000, far greater than the $38,748,000 in regular Park Service appropriations. As stated in the same report, the "biggest single factor" in expansion of the Service's operations was supervision of recreational planning and development. The report indicated that, in state parks, up to 91,000 enrollees living in 457 camps had been directed by as many as 5,499 Park Service employees. The relief programs had not only helped bring the national parks "to new levels of physical development," as the 1936 report put it, but had also fostered "new and important fields of activity" for the bureau—the many and varied Park Service programs of the 1930s.[156]

Within the national parks themselves through 1936, the Service managed as many as 117 CCC camps with 23,400 enrollees, and employed as many as 2,405 "national park landscape architects, engineers, foresters, and other technicians."[157] This last figure alone exceeded the total of Park Service employees in 1932, prior to the beginning of Roosevelt's emergency relief programs, and was a reflection of the heavy emphasis the New Deal placed on forestry and recreational development in the national parks. Much later, in 1951, Chief Landscape Architect William G. Carnes estimated that the Service in the 1930s had employed as many as 400 landscape architects at one time. By comparison, the Service employed a maximum of 27 biologists in the mid-1930s—a tiny fraction of those employed in recreational development. Of the biologists, 23 were funded by CCC money and the remaining four were paid through the Service's regular appropriations.[158]

The total funds and positions accounted for by the Park Service during this period attested to the New Deal's interest in recreational development of national and state parks, and also to its emphasis on large resource surveys and national planning. With these programs the Service's foresters, architects, landscape architects, and engineers increased their influence. And by the mid-1930s, the Park Service claimed that its "preeminence" in the recreational field had reached "new heights," with its mission expanded to aiding the conservation of "parklands everywhere."[159] Although certainly meaningful, the emergence of a scientific perspective in national park management seems diminished, even overwhelmed, by the Park Service's extraordinary expansion and development during the 1930s.

It is significant that when Cammerer's health forced him to step down in 1940 to become regional director in the Park Service's Richmond, Virginia, office, one of Secretary Ickes' top choices to succeed Cammerer was none other than Robert Moses, the "czar" of New York's park, parkway, and recreational development. Ickes thought that the New Yorker would provide "vigorous administration"—in sharp contrast to his disregard for Cammerer's abilities. The secretary's interest in Moses, conveyed to President Roosevelt, certainly suggests a perception of the National Park Service as much more of a recreational tourism organization than one committed to scientific and ecologically attuned land management. Moreover, it was Roosevelt's personal animosity toward Moses, rather than any concerns that his aggressive developmental tendencies might overwhelm the national parks, that seems to have led to the President's rejection of Moses as a possible Park Service director.[160]

The varied programs assumed by the National Park Service during the 1930s did in fact draw criticism. Alarmed over the bureau's developmental bent, Newton Drury, head of the Save the Redwoods League and destined to succeed Cammerer as Park Service director, commented scornfully that the Service was becoming a "Super-Department of Recreation" and a "glorified playground commission." Because of these tendencies, organizations such as the Redwoods League, The Wilderness Society, and the National Parks Association believed that the U.S. Forest Service might manage the Kings Canyon area of the Sierra (one of the principal national park proposals during the late 1930s) better than would the Park Service. Such concerns contributed to a delay of congressional authorization of Kings Canyon National Park until 1940 and inspired strong wording in the enabling legislation to protect the new park's wilderness qualities. Aversion to Park Service emphasis on recreational tourism development also caused the Redwoods League to oppose establishment of a national park in the redwoods area of northern California.[161] This opposition helped cause decades of delay, with serious consequences for preservation of the redwoods.

Particularly stinging criticism of changes taking place during the New Deal came from the National Parks Association, which, since its founding in 1919 with Stephen Mather's patronage, had been the public's chief advocate for maintaining high national park standards. The association feared that the traditional large natural parks were threatened by too much development, and that the Park Service was distracted by an overload of new and diverse responsibilities. In a conservative reaction to the sprawl of New Deal programs, the association argued that the National Park Service was run by its "State Park group financed by emergency funds," and that with the new types of parks, the public was increasingly confused about what a true national park was. To the association, the "real impetus" behind the expansion and development of the system was the "recently conceived idea that the Park Service is the only federal agency fitted to administer recreation on federally owned or controlled lands. Some persons even go so far as to assert that its proper function is to stimulate and direct recreational travel throughout the country."[162]

In the spring of 1936, the National Parks Association recommended "purification" as a corrective measure. It urged establishment of a "National Primeval Park System," which would contain only the large natural parks and be managed independently of historic or recreation areas, or of state park assistance programs. The intent of this proposal was to save the "old time" big natural parks from "submergence" in the "welter of miscellaneous reservations" being created. Furthermore, the association proposed limiting future additions to the primeval park system to those areas that had not been seriously impacted by lumbering, mining, settlement, or other adverse human activities. Only the most pristine areas were to be included.[163]

During the 1930s the National Parks Association's highly restrictive approach seems to have had little impact on the Park Service or on the growth of the system. It was, in fact, criticized by individuals within the Service, from Cammerer to George Wright. Cammerer and his staff disliked the primeval parks proposal, believing it would divide the system into first-class and second-class areas. Writing in the *American Planning and Civic Annual* in 1938, former director Horace Albright, one of the principal proponents of Park Service expansion, attacked the restrictive standards as being so "rigid" that they would "disqualify all of the remaining superlative scenery in the United States." Albright rightfully pointed out that parks like Glacier, Grand Canyon, and Yosemite, which had been grazed, mined, or settled before establishment, would not have become national parks had such standards been used in the past. He claimed that those who wanted only "unmodified territory" in the parks were actually allied with "other national-park objectors to prevent any more areas from being incorporated into the system."[164]

In a scathing letter to the National Parks Association, Interior secretary Ickes concurred with Albright. Ickes wrote that opposition to legislation that would include cutover areas in the proposed Olympic National Park or allow recreation development downriver from the proposed Kings Canyon National Park "dovetailed perfectly with the opposition of commercial opponents." He charged the Parks Association with being a "stooge" for lumber companies that also opposed the parks. George Wright's disagreement with the association was much more tempered. In a speech to the American Planning and Civic Association shortly before his death, Wright stated that he no longer feared that the system would be loaded with "inferior" parks, a position placing him in disagreement with the Parks Association. He believed that, in any event, the Service itself could adequately defend against "intrusion of trash areas." More important, the failure to act on truly exceptional park proposals would be much more calamitous than allowing substandard areas to "slip in."[165]

It must be noted that criticism by the National Parks Association and others did not focus on any perceived need for ecologically oriented management of natural resources. Both Newton Drury's assertion that the Service was becoming a Super-Department of Recreation and the National Parks Association's proposal for a primeval park system stemmed from apprehension over excessive development and the kinds of parks being created. Once an area was placed under the Service's administration, the specifics of its natural resource management—the treatment of elk, fish, forests, and the like—seem to have been of not much concern. By implication, where no development problems existed the parks were satisfactorily managed. Expressed largely in terms of opposition to various kinds of development, the critics' desire to protect the parks went against the tide of Park Service recreational growth and expansion under the New Deal. In the end, this criticism had little effect.

Declining Influence of the Wildlife Biologists

Corresponding with George Wright in the spring of 1935 on the need for highly qualified scientists in the parks, Joseph Grinnell stated "quite precisely" his high aspirations for the Park Service's biological programs. Grinnell believed that the country's "supreme 'hope' for pure, uncontaminated wildlife conservation" was the National Park Service, "under its Wildlife Division."[166] A year later, the division had reached its maximum of twenty-seven biologists—but Wright, its founder and chief, was dead. It is difficult to trace all the reasons for the decline of the wildlife programs that followed, but the loss of Wright's leadership surely contributed.

Much later, Lowell Sumner recalled that among the biologists only Wright had the special ability to "placate and win over" those in the Park Service who increasingly believed "that biologists were impractical, were unaware that 'parks are for people,' and were a hindrance to large scale plans for park development." Wright had been able to exert a "reassuring influence at the top, [keeping] hostility to the ecological approach . . . muted." Writing to Grinnell in the fall of 1936, Ben Thompson noted the frequently adversarial role of the biologists, with their negative "I protest" attitudes, which Wright had diverted and diplomatically finessed into "positive acts of conservation." Thompson stated that Wright had succeeded in establishing a division to "protect wildlife in the parks and make the Service conscious of those values." But the "immediate job" after Wright's death had been to keep the wildlife biologists from "being swallowed . . . by another unit of the Service."[167]

Thompson's remarks suggested the vulnerability of the Wildlife Division. By August 1938, while forestry, landscape architecture, planning, and other programs flourished within the Park Service, the number of biologists had dwindled to ten, with six positions funded by the CCC and still only four funded from regular appropriations. The overall total dropped to nine by 1939, as the transfer of the biologists to the Bureau of Biological Survey approached.[168] The transfer came not through any Park Service intention, but as the result of a broader scheme: the compromises made when President Roosevelt rejected Secretary Ickes' attempt to transform the Interior Department into a "Department of Conservation." Ickes had eagerly sought, but failed, to have the Forest Service moved from the Department of Agriculture to his proposed new conservation department. Instead, the Biological Survey was placed in Interior. Soon after (and apparently without Park Service protest), he moved all of the Interior Department's wildlife research functions into the Biological Survey, transferring the Park Service's biologists to the survey's newly created Office of National Park Wildlife. Although biologists located in the parks retained their duty stations, they had become part of another bureau.[169]

Like the national park system, the Biological Survey's wildlife refuge system had expanded greatly during the 1930s. The refuges in effect served as "game farms," which, along with aggressive predator control, constituted the survey's chief efforts to ensure an abundance of game for hunters. Thus, the survey's management practices differed critically from those advocated by the biologists who transferred from the Park Service. In June 1939, about six months before the transfer, Ben Thompson wrote to E. Raymond Hall, acting head of the Museum of Vertebrate Zoology after Grinnell's death, asserting that the survey had "never liked the existence of the NPS wildlife division."[170] Thompson did not explain the cause of the dislike, but differences in management philosophies and policies, plus growth of the Park Service's own biological expertise under George Wright (which very likely diminished the Biological Survey's involvement in national park programs), may have caused tension between the survey and the Wildlife Division.

Aware of the policy differences, Park Service director Cammerer and the Biological Survey's chief, Ira N. Gabrielson, signed an agreement in late 1939 whereby the national parks would be managed under their "specific, distinctive principles" by continuing the Service's established wildlife management policies. The agreement spelled out the policies, using most of the recommendations included in Fauna No. 1. Nevertheless, Lowell Sumner later recalled that the transfer weakened the biologists' influence in the Service. To whatever degree the scientists had been considered part of the Park Service "family and programs," Sumner wrote, "such feelings were diluted by this involuntary transfer to another agency."[171] Although the biologists would return to the Service after World War II, almost another two decades would pass before scientific resource management in the national parks would experience even a small resurgence.

Viewed within the context of the New Deal, the National Park Service's declining interest in ecological management becomes comprehensible. The New Deal impacted the Park Service fundamentally by emphasizing—and, especially, by funding—the recreational aspects of the Service's original mandate. The Park Service, which under Stephen Mather had stressed development of the national parks for public access and enjoyment, used the recreational and public use aspects of its mandate as a springboard, justifying involvement in ever-expanding programs during the 1930s. The emergency relief funds appropriated by Congress during the Roosevelt administration enlarged the breadth and scope of Park Service programs to a degree undreamed of in Mather's time. Under such circumstances the Service continued to respond to its traditional utilitarian impulses, influenced by what its leadership wanted and by its perception of what Congress and the public intended the national park system and the Service itself to be.

It is significant that during the 1930s no public organizations demanded scientifically based management of the parks' natural resources. Pressure from the Boone and Crockett Club, the American Society of Mammalogists, and other organizations that helped bring about the 1931 predator control policies seems to have been focused on that issue alone. It also appears to have subsided following promulgation of the predator policies. The National Parks Association's urging that the parks not be overdeveloped probably constituted the chief criticism faced by Park Service management during the decade.

Even the Service's first official natural resource management policies did not move national park management far from its utilitarian base. The forestry and fish management policies allowed continued manipulation of natural resources, largely as a means to ensure public enjoyment of the parks. The policy on predatory animals issued by Albright in 1931 contained sufficient qualifications to permit continued reduction. Even easing up on reduction met with resistance, including that of Albright himself, who feared the parks' popular game species were being threatened by predators. In addition, the Service's commitment to strict preservation through the research reserve program was never realized. Virtually alone among national park policy statements, Fauna No. 1's wildlife management recommendations, with the expressed intent of preserving "flora and fauna in the primitive state," encouraged an ecological orientation in the Park Service. Still, the ecological attitudes that did emerge were inspired by the wildlife biologists, who failed to gain a commanding voice in national park management.

Unlike the perspectives of the landscape architects or foresters, the wildlife biologists' vision of national park management was truly revolutionary, penetrating beyond the scenic facades of the parks to comprehend the significance of the complex natural world. It was a vision that challenged the status quo in the National Park Service. But without a vocal public constituency specifically concerned about natural resource issues, the wildlife biologists were alone in their efforts. They were insurgents in a tradition-bound realm; for what support they did get, the biologists had to rely on shifting alliances within the Park Service, depending on the issue at hand.

In this regard the 1930s would differ markedly from the 1960s and 1970s, when influential environmental organizations and increasing public concern about ecological matters would bring strong outside pressure on national park management. The failure of the Park Service to pursue options presented by the wildlife biologists in the 1930s left the Service still largely ignorant of its natural resources and unaware of the ecological consequences of park development and use.

Greatly enhancing the influence of professions such as landscape architecture and forestry, the Service's growth and expansion led also to the ascendancy of landscape architect Conrad Wirth as a major voice in national park affairs. After waiting in the wings during the administrations of Newton Drury and Arthur Demaray, Wirth would become director in late 1951. With the Service still primarily interested in the visual, aesthetic aspects of nature, those few wildlife biologists involved in national park affairs under Wirth would face an onslaught of invasive development in the parks not unlike what the New Deal had produced—only more expansive.

Top: A U.S. Army machine-gun platoon drills in front of the Northern Pacific Railroad Company's Mammoth Hot Springs Hotel, in Yellowstone National Park, circa 1890s. Beginning in the late nineteenth century, the army managed several national parks, overseeing road, trail, and hotel construction and initiating policies for the protection and management of natural resources. *Bottom:* National park rangers, circa early 1930s. Created in 1916, the National Park Service adopted army-style uniforms. It also continued park management practices inherited from the army era; the legislation creating the Park Service brought no substantive changes to natural resource management policies. (National Park Service Historic Photograph Collections, Harpers Ferry Center.)

Stephen T. Mather, first director of the National Park Service, aboard the Northern Pacific Railroad's deluxe North Coast Limited, mid-1920s. The photograph suggests the close ties between the early national park movement and the railroad companies. Principal lobbyists for many of the parks, the railroads also helped develop them for tourism. (National Park Service Historic Photograph Collections, Harpers Ferry Center; courtesy National Archives.)

A park ranger, his faithful horse, and a fashionable lady promote the Grand Canyon for the Atchison, Topeka and Santa Fe Railroad Company, 1910. Together with the Fred Harvey Company, the Santa Fe built tourist accommodations along the south rim, where views of the canyon are most accessible. (National Park Service Historic Photograph Collections, Harpers Ferry Center; courtesy Burlington Northern Santa Fe Corporation.)

Top: Glacier National Park's Many Glacier Hotel. An example of chalet-style rustic architecture intended to harmonize with national park scenery, the Many Glacier Hotel was completed by the Great Northern Railway Company in 1915. Promoting Glacier as the "American Alps," the company also built a series of small chalets throughout the park. (National Park Service Historic Photograph Collections, Harpers Ferry Center, George A. Grant, photographer.) *Bottom:* The Norris Museum, at Yellowstone's Norris Geyser Basin. Rockefeller family philanthropy bestowed many benefits on the national parks, including the Norris Museum—a classic example of rustic architecture. Featuring heavy log and stone construction, and commonly seen in the large national parks, rustic architecture became known as "parkitecture." (National Park Service Historic Photograph Collections, Harpers Ferry Center, Dorr G. Yeager, photographer.)

The National Park Service "family" gathers at Mesa Verde for the 1925 superintendents conference. In a successful effort to create a cohesive organization, Director Stephen Mather held numerous conferences such as this. Mather stands on the ladder; to his right is Mesa Verde superintendent and respected archeologist Jesse L. Nusbaum; sitting, hat in hand, on the far right is Rocky Mountain National Park superintendent Roger W. Toll, later promoted to the Yellowstone superintendency. (National Park Service Historic Photograph Collections, Harpers Ferry Center, J. V. Lloyd, photographer.)

Top: The first conference of chief naturalists, held in 1929 on the University of California campus in Berkeley. The naturalists are seated on the steps of Hilgard Hall, where the emerging national park professions of education, forestry, and wildlife biology had offices in their early years. *Bottom:* An evening talk at Old Faithful in Yellowstone National Park, 1931. Through museum exhibits, guided hikes, and evening talks, the naturalists sought to enlighten the public about a park's natural history—but they were not responsible for setting the policies by which natural resources were managed. (National Park Service Historic Photograph Collections, Harpers Ferry Center.)

A coyote caught in a steel trap in Yellowstone National Park, 1929. Staring at the camera, this animal probably soon looked down the barrel of a rifle, one victim among thousands in the widespread effort to rid the national parks of certain native predator species. As the largest national park, and with an abundance of large mammals, Yellowstone led in the formulation of wildlife management policies for the parks. (Yellowstone National Park.)

Top: Bison being stampeded in Yellowstone, circa mid-1920s. Staged to thrill park visitors, these stampedes were discontinued in the 1920s. (National Park Service Historic Photograph Collections, Harpers Ferry Center.) *Bottom:* Park herders prepare to castrate a Yellowstone bison, 1928. Castration, corralling, and winter feeding were among early techniques used at the park's Buffalo Ranch to manage the Lamar Valley bison herd. (Yellowstone National Park.)

National Park Service biologist George M. Wright, in Yosemite, mid-1920s. In 1929 Wright would use his personal fortune to launch the Service's first professional wildlife research. Of all major Park Service programs, wildlife biology was the only one initiated with private funds—evidence of its low priority in national park management. Wright's accidental death in February 1936 weakened the biology programs during the heyday of park development and construction under the New Deal. (National Park Service Historic Photograph Collections, Harpers Ferry Center, Carl P. Russell photographer.)

Biologist Adolph Murie in Mt. McKinley National Park, mid-1920s. Murie exhibits the grit and determination he would need to challenge traditional natural resource practices and promote ecologically informed park management throughout his long Park Service career. An outspoken, articulate critic, he had little success changing managerial perspectives. (National Park Service Historic Photograph Collections, Harpers Ferry Center.)

Top: The managerial perspective: landscape architect and National Park Service Assistant Director Conrad L. Wirth, 1933. Throughout his career, including more than a dozen years as Park Service director (1951–64), Wirth successfully championed development of the parks for tourism, establishment of different types of national parks, and expansion of state park programs. A highly effective bureaucrat, he had little concern for ecological matters. (National Park Service Historic Photograph Collections, Harpers Ferry Center, George A. Grant, photographer.) *Bottom:* Road construction along the Blue Ridge Parkway, mid-1930s. The congressional mandate that the National Park Service keep the parks "unimpaired for the enjoyment of future generations" was never intended to exclude park development—only to control it. (National Park Service Historic Photograph Collections, Harpers Ferry Center.)

Top: A National Park Service team surveys the Lake Mead area for potential recreational uses, early 1930s. The team recommended a new type of park—a "national recreation area"—along the shore of the reservoir being created behind Hoover Dam. Thus, not long after a reservoir had inundated Yosemite National Park's beautiful Hetch Hetchy Valley, the Park Service launched its national recreation area program, in cooperation with reservoir development elsewhere in the West. (National Park Service Historic Photograph Collections, Harpers Ferry Center, Roger W. Toll, photographer.) *Bottom:* The military in Yosemite National Park during World War II. The armed services used a number of national parks for wartime training exercises, convalescence, and recreation. World War II brought about a drastic decline in public use of the parks and a reduction of funding that resulted in deterioration of park facilities. The national parks were left unprepared for the surge of postwar tourism. (National Park Service Historic Photograph Collections, Harpers Ferry Center.)

Top: The headquarters and main visitor center at Grand Teton National Park, 1959. The Park Service's huge Mission 66 construction and development program (1956–66) abandoned the romanticism of rustic architecture. The program's modernistic construction was severely criticized as being intrusive and ill suited for national park settings. (National Park Service Historic Photograph Collections, Harpers Ferry Center, Jack E. Boucher, photographer.) *Bottom:* The Mission 66 Committee, 1956. Smiling and confident, the committee members looked forward to a billion dollars of national park development under Mission 66—but this program would propel the Park Service into conflict with the new environmentalism. By the early 1960s, Howard R. Stagner (far left) would push for investigations into the Park Service's neglected natural science programs. The reports would usher in a new era of national park history. (National Park Service Historic Photograph Collections, Harpers Ferry Center.)

Power in the national parks—natural and bureaucratic: a Yellowstone grizzly and National Park Service director George B. Hartzog, Jr. In the late 1960s, during Hartzog's administration, Yellowstone's grizzly bear management controversy would spark national attention. (National Park Service Historic Photograph Collections, Harpers Ferry Center, William S. Keller, grizzly bear photographer.)

Top: Park staff setting fire to a meadow in Shenandoah National Park, 1978. Reversing policies initiated in the nineteenth century, the Service began controlled, or "prescribed," burning in selected forests and grasslands, intended to simulate natural fire regimes and to reduce fuel loads and prevent unnaturally intense fires. The controversial program remained only partially implemented owing to shortages of funds and staff. (Julie M. Langdon, photographer, National Park Service.) *Bottom:* Hazardous materials at Padre Island National Seashore, 1992. One of the myriad "external threats" to national park areas, hazardous debris from Gulf of Mexico shipping and petroleum development threatens ecology and human health at Padre Island. (Robert J. Krumenaker, photographer, National Park Service.)

Gray wolves return to Yellowstone National Park, 1995. Watched by local schoolchildren, a caravan carrying the beginnings of a new wolf population for Yellowstone passes through the park's famed entrance arch, built in the early twentieth century, when park managers were eradicating wolves. (Yellowstone National Park.)

CHAPTER 5

The War and Postwar Years, 1940–1963

Sometimes I find, Horace, and I am sure you will agree with this, that you can get too scientific on these things and cause a lot of harm.—CONRAD L. WIRTH TO HORACE M. ALBRIGHT, 1956

If the Service is to protect and preserve, it must know *what* it is protecting, and what it must protect against. It is the function of research to get at the truth, to develop the fund of knowledge necessary for intelligent and effective management.—NATIONAL PARK SERVICE, 1961

After removal of the wildlife biologists to the Fish and Wildlife Service in 1940, nearly a quarter of a century would pass before any meaningful attempt to revitalize the National Park Service's biological science programs. By the early 1960s, the Service would come under public criticism for its weak, floundering scientific efforts, described in one report as "fragmented," without direction, and lacking "continuity, coordination, and depth."[1] Moreover, the Park Service would find its management increasingly challenged by conservation groups, its leadership in national recreation programs seriously weakened, and its control over the parks' backcountry threatened by restrictions in the proposed wilderness legislation, which was gaining support in Congress. These challenges would be mounted in the midst of "Mission 66," the Park Service's billion-dollar program to improve park facilities, increase staffing, and plan for future expansion of the system—a highly touted effort to enhance recreational tourism in the parks, and so named because it was to conclude in 1966, the Service's fiftieth-anniversary year.

From 1940 through 1963, national park management was dominated by two directors: Newton Drury, who succeeded Arno Cammerer in August 1940 and resigned in March 1951; and Conrad Wirth, who served from December 1951 to January 1964. (Between Drury and Wirth fell the brief, eight-month directorship of Arthur Demaray, whose preretirement appointment was in recognition of his lengthy and competent service as associate director.) The Drury-Wirth era was marked by the strikingly different personalities and philosophies of these two leaders: Drury, the conservative who criticized Park Service expansion and development in the 1930s, then provided more than a decade of cautious leadership; and Wirth, the highly effective entrepreneurial promoter and developer of the national park system.

Newton Drury was the first director not to have had prior experience in national park management. His principal conservation work had been as executive director of the Save the Redwoods League from late 1919 until he resigned in 1940 to become Park Service director.[2] As the Redwoods League had focused on California issues, Drury's contacts were mainly in that state; he lacked experience in dealing with Congress and the Washington bureaucracy. His personal conservatism was reflected in his loyalty to the Republican Party (which appears not to have been an obstacle to his appointment in the Roosevelt administration) and in his leadership of the Service. Drury believed that the national parks should be limited primarily to the nation's premier scenic landscapes. He took a slow, deliberate approach to improving administrative and tourism facilities in the parks. And he was not enthusiastic about involvement with reservoir recreation management or with the national recreation assistance programs begun during the New Deal.[3] Clearly, Drury did not fit the mold of the previous directors, who enthusiastically boosted park development and expansion of Service programs.

Wartime and Postwar Pressures

Drury's conservative management fit the times. World War II and the postwar years brought drastic reductions in money, manpower, and park development, and a halt to expansion of the national park system. By August 1940, when Drury assumed the directorship, the New Deal programs were already diminishing, yielding to preparations for war and support for the nation's allies. America's entry into the war in December 1941 led to a reduction of more than fifty percent in the Park Service's basic operating budget and the termination in 1942 of the Civilian Conservation Corps, which the Service had used to great benefit.

Personnel cuts were severe. The Service's staffing budget was reduced, and many employees joined the armed forces or went to work in war-related agencies and could not be replaced. Just before Pearl Harbor, the Park Service had 5,963 permanent full-time employees. This number dropped to 4,510 by June 30, 1942 (the end of the fiscal year), and plunged to 1,974 by the end of the following June. By June 30 of 1944, the number stabilized at 1,573, about a quarter of the total of prewar employees. These cuts affected individual parks. Sequoia, for example, had lost more than half of its administrative, ranger, and maintenance staff within six months after the attack on Pearl Harbor. With a wartime economy, including gas rationing and rubber shortages, the number of park visitors also plummeted, from a high of 19.3 million in fiscal year 1940 to a wartime low of 7.4 million in fiscal 1943.[4]

The Service maintained skeletal staffs in its Washington and regional offices; and, by negotiating with the Bureau of the Budget and with congressional appropriations committees, Drury was able to keep professional engineers and landscape architects in key offices. The Park Service's ability to function effectively was further diminished in August 1942 when its headquarters was moved to Chicago to make office space in Washington available for critical wartime use. Although the transfer restricted his Washington contacts, Drury elected to move with the headquarters to Chicago. Wirth went with him, leaving Associate Director Arthur Demaray as the Service's principal representative in the nation's capital.[5]

During World War II the National Park Service was, in Drury's words, reduced to a "protection and maintenance basis." He later elaborated that, overall, the Service had three primary wartime goals: maintaining a "reasonably well-rounded" organization that could be expanded to meet postwar needs, keeping the parks and monuments "intact," and preventing a "breakdown of the national park concept."[6] Indeed, beyond the severe budget and personnel cuts, the war put unusual demands on the park system. Military rest camps were established in parks such as Grand Canyon, Sequoia, and Carlsbad Caverns, while other parks, including Yosemite and Lava Beds, provided hospitalization and rehabilitation facilities. In many instances, the Service converted abandoned CCC camps to such uses. The military held overnight bivouacs in a number of parks and conducted maneuvers in Mt. McKinley and Hawaii national parks, among others. Extended training occurred in numerous parks, including Yosemite, Shenandoah, Yellowstone, Isle Royale, and Death Valley. Defense installations were located in Acadia, Olympic, Hawaii, and Glacier Bay, while two small historical units of the system, Cabrillo and Fort Pulaski national monuments, were closed to the public and used for coastal defense.[7]

The war put pressure on specific park resources, with limited amounts of extraction allowed. For example, early in the war Secretary of the Interior Ickes authorized the mining of salt in Death Valley and tungsten in Yosemite. Seeking a balance between patriotic support of the war effort and protection of the parks, Drury maintained that permission for natural resource extraction in the parks must be based on "critical necessity" rather than convenience, with the burden of proof resting on the applicant.[8] Such concerns were raised in the two principal resource extraction issues that confronted the Service during World War II: demands to cut timber and to allow cattle grazing, both promoted as patriotic efforts to support the war.

Of the wartime requests to cut forests for timber in a number of parks, the most hotly debated was a proposal to harvest giant Sitka spruce trees in Olympic National Park for use in airplane construction. This proposal came soon after Great Britain and France entered the war in the late summer of 1939. With Secretary Ickes' backing, the Park Service refused the request. Under continued pressure, the Service (in the last year of Arno Cammerer's directorship) recommended that spruce trees be taken from two nearby corridors of land intended for a scenic parkway but not yet part of the national park, thus still vulnerable to resource extraction demands. The following year the size of the proposed parkway corridors was reduced to allow cutting in the excluded areas—a means of evading the issue of taking national park resources. Drury, who became director the following August, stated his opposition to any cutting on the lands remaining in the corridors, except as a "last resort," and where "immediate public necessity" could be shown.[9] But local lumber interests persisted in their demands to cut spruce within the park and in the corridors, arguing that quality timber for airplane construction could be found nowhere else. Drury resisted, and sought information on spruce wood substitutes and the availability of spruce elsewhere, especially in British Columbia. The Canadian government refused to release timber statistics, however, citing wartime confidentiality. Pressured by the Roosevelt administration, the Service backed down. Acknowledging a "distinct sacrifice of parkway features," it allowed cutting inside the corridors.

This concession notwithstanding, local businessmen (who had always supported timber company interests and were backed by the Seattle Chamber of Commerce and the city's newspapers) lobbied to reduce the size of the park and open virtually all areas to cutting for wartime and postwar production.[10] Although vacillating, Drury opposed further cutting for any but specific military needs, and only if there were no other available sources of spruce—a position that the congressional House Subcommittee on Lumber Matters supported during hearings in June 1943. In August the Canadian government eased the situation by releasing information on the availability of British Columbia spruce. This change, plus greater production of Alaska's spruce and increased reliance on aluminum instead of wood in airplane construction, led to the administration's withdrawal of pressure to harvest Olympic's forests.[11]

All the same, local campaigns to shrink the park and cut its forests continued well beyond the war years. In a striking example of timid leadership, Drury, with the Interior Department's concurrence, responded to pressure from timber companies by supporting bills introduced in Congress in 1947 to reduce Olympic National Park by fifty-six thousand acres. Swayed by Olympic's assistant superintendent, Fred J. Overly—a forester and timber company cohort who argued that some of the acreage was already cut over and that much of it was remote and difficult to administer—the Service declared that it sought to "attain a better boundary from the standpoint of administration and protection, following ridges wherever possible."[12] Naively, Drury hoped that the timber interests would be placated and would not seek further reduction of the park.

In fact, the Washington office did not have full knowledge of the lands proposed for removal from Olympic. Included were tracts of heavily forested virgin wilderness, particularly in the Bogachiel and Calawah drainages. The proposal brought an angry reaction from conservationists. Under intense pressure, and with timber interest testimony that the reduction was, as Drury saw it, "only a first step" toward gaining access to the rest of the park, Drury, backed by Julius A. Krug (Ickes' successor as secretary of the interior), eventually reversed his position. This belated opposition, along with President Harry S. Truman's reluctance to give in to the timber companies, helped kill the proposed reduction. Vast tracts of Olympic's forests were saved from commercial harvesting.[13]

Nonetheless, park management continued to encourage the removal of hundreds of individual trees blown down by windstorms—a salvage practice sharply criticized by the wildlife biologists as a disruption of natural processes. Receipts from the sale of windblown timber went mostly toward the purchase of inholdings, which served as the park's main justification for the program. However, especially under Fred Overly's direction, the practice became a means by which local companies were able to remove millions of board feet of park timber, much of it healthy. Overly, who became superintendent in 1951 and who almost certainly enjoyed support from the Service's top foresters, set up salvage contracts with timber companies that allowed the cutting of standing mature trees in addition to any windblown timber.

In time this practice became an embarrassment to the Park Service and to the new director, Conrad Wirth. Visitor abhorrence of the loss of healthy trees began to be pointedly conveyed to the park's "seasonal" (or summer) naturalists, who had daily contact with the public—and who conspired to shame the Service publicly and force it to halt the salvaging. Faced with growing congressional concern and an aroused and angry conservation community, Wirth yielded. When Overly later resumed the practice on a limited basis, the director felt compelled to remove him from the park. Revealing the solidarity among Park Service leaders—and in an action not uncommon in the bureau's history—Wirth assigned Overly to another superintendency (Great Smoky Mountains), rather than disciplining him for systematic destruction of park resources.[14]

Just as lumbermen had sought to gain access to national park resources during wartime, ranchers renewed pressure to increase livestock grazing in the parks, claiming the need to provide beef in support of the war effort. Hoping eventually to eliminate grazing from national parks, the Service responded by applying very restrictive grazing criteria. Critical wartime need had to be shown, and postwar needs constituted no justification for grazing increases. Drury argued for protecting all national parks and their "spectacular features" from grazing. But under the force of wartime necessity, grazing could be permitted "in the areas of lesser importance" or in areas where the damage would not be "irreparable." The Park Service also calculated that its employees and those of other Interior Department bureaus could reduce their beef consumption by approximately one-third to compensate for not allowing grazing in the parks—but there is no indication that the Service or the department seriously pursued this idea.[15]

In early 1943, responding to persistent demands from livestock growers' associations and from the War Production Board, Secretary Ickes approved a livestock grazing "formula" submitted by Drury. The formula imposed a ceiling of twenty-eight percent increase in cattle grazing in the parks and eleven percent in sheep grazing. Although affirming the goal of ultimately eliminating all cattle and sheep from the parks, the formula established land classifications for grazing under wartime emergency conditions. The classifications varied from total prohibition of grazing in the most protected park areas, to allowing increases in livestock numbers and range acreage in other park lands, especially recreation areas.[16]

Permits granted under this arrangement resulted in a wartime grazing increase of only about half the percentages set by Drury's formula—a result of Park Service resistance and of support from outside the bureau. For example, efforts by drought-plagued California livestock growers in the spring of 1944 to gain access to grasslands in nearby national parks failed after being opposed by the Service and evaluated by a special committee of representatives from the Sierra Club, California Conservation Council,

Western Association of Outdoor Clubs, and U.S. Forest Service. The demands by livestock ranchers came even after range surveys indicated that national park grasslands in California would support no more than six thousand beef cattle, less than one-half of one percent of the state's 1.4 million head. Recognizing that damage to vegetation would result from livestock grazing, the Service viewed the proposal not as an attempt to support the war, but merely as a means to use national park resources to benefit the local cattle industry. The special committee found insufficient justification for approving the grazing applications. Renewed pressure to allow grazing in the Sierra national parks during postwar drought periods suggests that the urge to increase ranchers' profits was as much a factor as wartime need.[17]

Natural Resource Issues under Drury and Wirth

During and after World War II, the bulk of Park Service day-to-day natural resource management continued to involve field activities of national park rangers—mainly "protection" work, including stocking streams and lakes; reducing elk, deer, and bison populations; and fighting forest fires, insects, and disease.[18] Few changes occurred in the natural resource policies written mainly in the 1930s. Among the wildlife practices, bison and bear management and care of Mt. McKinley's wolf and sheep populations were particularly controversial and reflected the continuing schism between managing for visitor enjoyment and adherence to the ecological principles espoused in Fauna No. 1, still the Service's official wildlife policy. More preservation-minded than his predecessors, Drury frequently supported Fauna No. 1's policies.

Believing that wildlife threatened to overgraze certain park areas, the biologists continued to support the concept of population reduction, chiefly of bison and elk in Yellowstone, deer in Zion, and both deer and elk in Rocky Mountain.[19] Such reductions were undertaken by Park Service rangers, at times with help from state game and fish personnel. The Service continued to oppose permitting sporthunters to participate in the reductions, maintaining that it would be "the first step toward destruction" of the national parks. It would violate the sanctity of the parks and blur their distinction from public lands that were subject to resource exploitation.[20]

Although the Park Service monitored grasslands in an attempt to calculate optimum large-mammal populations, concerns arose that the reduction program needed closer analysis. In 1943 ecologist Aldo Leopold proposed to conduct research on Yellowstone's northern elk herd. As

Yellowstone superintendent Edmund B. Rogers understood it, Leopold wished to direct his research toward "conclusions regarding the current management program" and toward establishing "greater confidence in management measures." Rogers rejected Leopold's request, declaring that money for research should not be diverted from the park's "essential work." Instead, in a perfect illustration of disregard for scientific data, Rogers recommended that the Service's top management should meet and issue an "authoritative statement" regarding future elk reduction. He added assuredly that such a statement should "automatically take care of confidence in the program by interested agencies and the general public"—a notion perhaps true at the time, except among biologists and conservationists. Rogers' recommendation apparently was to be carried out with no further research.[21]

In the 1940s the Service began greater population reductions of Yellowstone's Lamar Valley bison herd—an action that provoked heated debate over the herd's value as a public spectacle. Although it had gradually eliminated some of its ranching activity, the park maintained its facilities at the Buffalo Ranch and continued its winter feeding program, keeping the Lamar Valley herd at about 800 head, an artificially high number sustained by the feeding. Believing they were faced with an "over-population problem" in the valley that was putting heavy pressure on the winter range, Yellowstone officials gained Newton Drury's support for reducing the herd. With bison thought to be safe from extinction in the United States, it seemed, as Drury stated, "no longer necessary to sacrifice the range" in order to save the species.[22]

The Park Service shot 180 Lamar Valley bison in early 1942, none in 1943, and planned to kill 400 more beginning in January 1944. Allowing for natural increases, this would bring the population to about 350 animals, the level sought by the park. With the support of its biologists, the Service hoped ultimately to put the Lamar herd "entirely on its own resources" (Drury's words, and a clear echo of Fauna No. 1), with the size of the herd appropriate to the productivity of the winter range.[23]

These plans provoked the wrath of former director Horace Albright, who remained steadfast in his desire to ensure the national parks' visitor appeal. Insisting on a huge, spectacular herd of bison in the Lamar Valley, in October 1943 he sent Drury a "pro-forma protest" against further reduction of bison. Albright recalled that during his directorship he had never heard the naturalists complain that a thousand animals would be too much for their range. To him, the range conditions were satisfactory and could support a large number of bison.[24]

The Service's chief naturalist, Carl Russell, an experienced wildlife biologist, advised Drury that "sober consideration" countered Albright's views, and that evidence of poor range conditions was "not hard to find." Russell noted that the elk and bison reduction programs were backed by "a number of competent ecologists," among them Victor Cahalane. The scientists sought to maintain a "natural range condition," not a vast herd of bison. Albright responded that he was "not impressed" with these arguments. He pointed out that when he was Yellowstone's superintendent in the 1920s, the park's "big shows of buffalo" were "talked about in all parts of the country." Seeking to preserve a semblance of the vast herds of earlier times, he urged a rethinking of the reduction policy.[25]

Getting to the core of the issue, Drury responded to Albright, correctly observing that their opposing ideas were derived from "somewhat differing concepts of the purpose and function of national parks in respect to wild life." He stated that in view of the Service's original legislative mandate and the policies "crystallized over the years," the only proper course was to place "all species, including the bison, as rapidly as practicable upon a self-sustaining basis, free from all artificial aids." This policy had been endorsed, Drury asserted, by a "long list" of conservationists and biologists queried by the Service, most of whom readily supported bison reduction.[26]

To bolster his position, Drury quoted at length the supportive comments of Tracy Storer, a wildlife biologist at the University of California and former associate of the late Joseph Grinnell. In response to a Park Service query, Storer had written bluntly (and at odds with views Grinnell had held) that it was "utterly irrational" to protect such species "until 'they eat themselves out of house and home' or fall disastrously to the ravages of disease." Herd size could not build up indefinitely when contained in areas of "fixed size," such as national parks. He stated further that by its very nature, the protection of animals in parks would necessitate "removal of the surpluses that develop from time to time."[27]

Albright countered that national parks were not "biologic units" and thus it was impossible "to avoid some controls." Nevertheless, he urged that the Lamar Valley herd be kept between 800 and 1,000 head, and even proposed that after the war the Service should reinstitute the spectacular bison roundups for the public's enjoyment. The park implemented its reduction policy, however, and Yellowstone's rangers killed 397 of the Lamar

Valley bison in January 1944, leaving a population of about 350. Angered, Albright wrote Drury in February that the reduction was a "serious mistake" and that the herd's "usefulness . . . for the public enjoyment" of Yellowstone had been ignored.[28]

The reduction of bison brought a further decrease in the Buffalo Ranch operation, now mainly limited to winter feeding. Predictably, Albright opposed this decrease, because the ranch had managed the large herds that he wanted maintained, and bison ranching was fascinating to the public. His opposition intensified when he learned later in the year of plans to remove the fencing around pastures at Antelope Creek and Mammoth and terminate the exhibition of bison in these areas. Albright insisted to Drury that display pastures were "absolutely essential."[29]

Drury held firm, replying that the Service should manage bison "as a wild animal in a natural environment, and not as the basis of an 'animal show.' " Yellowstone, he stated, was not to be managed as a "zoological park or game farm." In contradiction to this position, in Jackson Hole National Monument (south of Yellowstone and soon to become part of Grand Teton National Park) the Service yielded to pressure from Laurance S. Rockefeller and allowed a wildlife display area to be established along the Snake River. Privately, Drury hoped this display would not be a success.[30]

Concurrent with the bison management disputes a controversy arose over plans to increase the limited, ongoing killing of wolves in Mt. McKinley National Park to reduce predation on Dall sheep, a wild native species. In 1939, after his investigation of coyotes in Yellowstone, Adolph Murie had begun wolf studies at McKinley, an assignment prompted in part by concern that wolves were responsible for a decline in sheep population. Murie concluded, however, that although sheep had declined in numbers, especially during the early 1930s, they had reached an equilibrium with the park's wolf population. He believed that by culling the weaker animals the wolves were helping to maintain a healthier sheep population.[31]

As a result, the park discontinued its wolf control program, causing an angry reaction from Alaska's territorial legislature and sportsmen's organizations that wanted the wolves eliminated to protect the Dall sheep. Petitioning Congress and the President to allow the killing of McKinley's wolves, the antiwolf faction raised fears that Congress would authorize wolf reduction in all Alaska national parks and monuments. Chief Biologist Victor Cahalane was convinced that many people wanted the wolf totally eliminated from Mt. McKinley—he asserted that when Alaskans spoke of "control" they really meant extermination. Cahalane assured Drury that criticism of the Park Service during the limited wolf control program of the 1930s had arisen because the Service was "not effective in *exterminating* the wolves."[32]

Concerned about the possibility of congressionally mandated wolf control in the park, the Service sent Murie back to McKinley in the summer and fall of 1945 to update his studies. This time Murie determined that the Dall sheep population had continued to decline to the point where its survival in the park was truly threatened. Noting the problems of maintaining natural conditions in parks that were also used for recreational purposes, and also noting the "highstrung articles" in the press against the wolf, Murie recognized the difficulty of preserving wolves even in national parks. Nevertheless, he stated that the "principal fact at hand" was that the sheep population had reached an "all-time low." As a "precautionary measure," he advocated reducing the park's wolf population by no more than ten or fifteen animals. Once the sheep population recovered, Murie added, there would be a "better place for the wolf in the fauna, without endangering a species," and control would no longer be necessary.[33]

It is certain that political pressure had increased for reducing the park's wolf population, but it is unclear to what degree, if any, Murie's recommendations were influenced by this pressure. Given his outspokenness about ecological matters, it seems unlikely that Murie would have recommended predator control of any kind. Indeed, his proposal to kill ten to fifteen wolves was conservative and may very well have been intended as a means of easing pressure while keeping at a minimum the impact on Mt. McKinley's wolf population. In any event, his proposal for limited control of one species to ensure the continued existence of a threatened species was in accord with the Park Service's existing predator control policy—first issued in May 1931, then reaffirmed two years later in Fauna No. 1's recommendations.[34]

With the blessing of Secretary Ickes, the Service made plans to implement Murie's recommendations for limited wolf control. The opposition was not pacified. In December 1945, two months after Murie's reevaluation, a bill was introduced in Congress for strict control of wolves and other predators in Mt. McKinley. Faced with possible legislative reversal of its established policy of limited predator control under the most compelling circumstances, the Park Service gained support from leading scientists and conservation organizations to oppose the bill. Among the scientists was Aldo Leopold, professor of wildlife management at the University of Wisconsin. Recognizing that the bill threatened the Service's ability to make and implement wildlife policy, Leopold informed the House Committee on Public Lands that such a mandated predator reduction was "bad public policy" that would "contradict [the Service's] basic function of preserving the fauna of the National Parks." Leopold argued that Adolph Murie, "widely respected as one of the most competent men in his profession," was far better prepared to deal with this issue than was Congress.[35]

In the spring of 1947 the new secretary of the interior, Julius Krug, added his opposition to the bill, helping to bring the legislative effort to an end. With the park's sheep population on the rise and the wolf population somewhat reduced, the Service in 1952 ended the reduction program.[36] As with bison reduction, the wolf-sheep controversy at Mt. McKinley National Park reflected the influence of Fauna No. 1, which recommended that park wildlife be managed to meet ecological goals, rather than simply to please the public.

Fauna No. 1 also influenced the management of bears. Although population reduction was not an issue with bears, public enjoyment was. And through the decades, bear shows had grown from, in Drury's words, "minor incidents into a well-defined program." More frequent contact with people meant, however, that bears became more accustomed to humans and their food, and more prone to wander into campgrounds or other crowded areas, often vandalizing property and endangering people.[37] Under Drury, the Service terminated the bear shows and sought to end roadside feeding of bears, two activities very popular with park visitors.

Quoting directly from Fauna No. 1, Drury explained the basis of the change in bear policy as being that each species would be allowed to "carry on its struggle for existence unaided" and would avoid becoming "dependent upon man for its support." Wildlife in the parks was to be presented to the public in a "wholly natural" way. Drury saw the bear shows as more appropriate for zoos than for parks, and both the shows and the roadside feeding as potentially harmful to the bears. These attitudes had been endorsed by wildlife biologists such as Joseph Dixon, who in 1940 described the bear shows as "unnatural and unnecessary." In a similar vein, Victor Cahalane wrote Aldo Leopold in May 1942 that feeding the grizzlies in Yellowstone "may eventually prove harmful" to the bears. Acknowledging the lack of research to support this opinion, Cahalane added that the Park Service had "nothing very concrete to bolster our contention that this feeding should be abolished."[38]

As early as 1938, Yellowstone reduced the amount of edible garbage at the Canyon dump in order to divert grizzlies to other dumps not open to the public. Further reduction of public bear feeding came in 1940; and, under orders from Drury, none of the parks held bear shows in the summer of 1942. Very likely the decrease in park staffing after the beginning of World War II (including staff necessary to conduct the shows) forced management into alignment with the biologists' point of view, sealing the fate of the shows. Also, the dramatic decline in park tourism after the war began resulted in less roadside feeding and thus a greater opportunity for the Service to reduce that activity permanently. To support a final decision on bear management, Fish and Wildlife Service biologist Olaus J. Murie began a study of bear ecology in Yellowstone in July 1943.[39]

Predictably, the new bear management policies evoked strong criticism from Horace Albright, who wrote Drury that the Service should give the public a chance to "see bears under safe conditions." As for Murie's bear research, Albright was certain that Murie would "persist in his belief that bears must not be fed." Probably hoping to reverse the new policy, the former director objected to removal of the bleachers used in the shows (just has he had objected to dismantling the Buffalo Ranch structures). He lectured Drury that, with the changes in both bison and bear management, the Service was creating a "world of trouble" in Yellowstone. Turning up the pressure, he noted that the changes were also a matter of concern to Kenneth Chorley, the influential assistant to national park benefactor John D. Rockefeller, Jr.[40]

However, Murie's preliminary conclusions had already recommended against feeding the bears (as Albright expected), and Drury responded to Albright that he was "more convinced than ever" of the correctness of the new policy.[41] Based on Murie's recommendations and management's inclinations, the shows were not revived. To reduce temptation to bears, the Service increased the burning of garbage; yet some dumps not open to the public remained accessible to bears.

Feeding was much more difficult to prohibit along miles of roadsides. With its tremendous appeal, roadside feeding even increased as the number of park visitors rose rapidly after the war. In the summer of 1951, the Service handed out more than a million leaflets to warn visitors about bears, but that same summer Yellowstone alone had thirty-eight injuries from bears, most occurring while visitors fed the animals.[42] Later, in 1959 (and with minimal support from the Park Service), biologists John and Frank Craighead would begin in-depth studies of Yellowstone's grizzly bear population. By the late 1960s, they would become embroiled in a major dispute with the park over bear management.

Although Drury frequently supported the wildlife biologists and their Fauna No. 1 policies, he seems to have backed the foresters' practices with little concern for ecological factors. Like other programs, active management of national park forests declined considerably during the war. Then, with no change in policies—only a renewed determination to implement them—forestry experienced a resurgence during the postwar era. Opposition of the wildlife biologists to forest fire suppression and to reduction of insects and diseases continued to have no effect.[43]

In his postwar annual reports, Drury repeatedly made clear his commitment to traditional forestry practices. With the return of veterans, the Park Service's firefighting capability "improved considerably," in Drury's words, allowing "intense prevention" and "efficient control" of park fires. Smoke jumping had proved effective in fighting fires in Glacier, inspiring the Service to build that program in cooperation with the Forest Service. Increased costs of firefighting (about $200,000 was spent in fiscal 1948, above regular salary costs) and the need to upgrade equipment were partially offset by support from the air rescue services of the U.S. Air Force. The fire program could be even more effective, Drury argued, if the firefighters were not burdened with other duties, and if the parks improved communication systems, maps, and aerial "detection and attack" techniques to combat fires occurring in remote areas.

Spraying and other operations to fight forest insects and diseases were under way in parks such as Acadia, Grand Teton, Yosemite, and Great Smoky Mountains. Aided by the Forest Service and the Bureau of Entomology and Plant Quarantine—and especially by the 1947 Forest Pest Control Act, which funded forest protection operations—both Drury and Wirth advanced these programs.[44] Among the chemicals beginning to be sprayed over extensive areas of some parks was DDT. In Yosemite the Service used it to combat the destructive lodgepole needleminer, even though it noted a decline in fish populations near areas where spraying had occurred. Yellowstone also sprayed DDT and found dead trout and other fish in the treated areas. Despite concerns raised by biologists (Lowell Sumner's warnings date from as early as 1948), DDT continued to be used in the parks through and beyond Conrad Wirth's directorship.[45]

The Park Service's forest management goal remained, as stated in a

1957 informational handbook, to "conserve as nearly as possible the primitive and natural character of the native vegetation"—but it was a goal pursued through strategies long opposed by the wildlife biologists. In a foreshadowing of change, however, a limited fire research program had begun in Everglades in 1951, the first such research in the national parks. Conducted by biologist and "fire aid" William B. Robertson, Jr., the research led in 1958 to the beginning of Park Service experiments in what would become known as "prescribed burning"—allowing selected natural and human-caused fires to burn in ways that simulate natural conditions. Given the Service's traditional mindset, these experiments would influence fire policies only very slowly.[46]

By midcentury a decline in fish populations in a number of parks was attributable not just to DDT but to too much fishing, which brought about, also very slowly, a change in national park fish management. Despite concerns voiced occasionally by the wildlife biologists, the Service had continued to place emphasis on visitor enjoyment of fishing rather than on preservation or restoration of natural conditions in lakes and streams. Anglers faced minimal restrictions on size and creel limits; and as postwar tourism increased, so did the pressure on fish populations. The policy changes, begun mainly during the Wirth administration, were affected by the Service's determination to continue to promote fishing and by the need to cooperate closely with state governments, which in most parks shared jurisdiction over fishing and had licensing authority. Seeking to prevent a serious decline in the quality of fishing, the Service tightened restrictions, more in the large natural parks than in national recreation areas with their huge artificial lakes.

In the mid-1950s Shenandoah and Great Smokies imposed new size and creel limits, set bait restrictions, and initiated catch-and-release programs (euphemistically called "fishing for fun"). Also in the mid-1950s Yellowstone closed its hatcheries and soon ended stocking except for limited planting of native species. It tightened size, creel, and bait restrictions, and in 1960 began a catch-and-release program. Many of the new rules were designed to ensure continued good fishing in the parks. However, the Service also encouraged native fish populations through such policy changes, and by placing certain waters off limits to anglers and resisting the urge to stock the remaining fishless lakes and streams. Similar changes began to take effect in other parks, but because of different state rules, fish regulations varied from park to park. Modified during ensuing decades, fish management in the parks would generally adhere to these basic strategies; still, the Service would continue to allow the taking of native aquatic fauna.[47]

The Status of Wildlife Biology

Although Newton Drury accepted the advice of the wildlife biologists many times, he made no effort to expand the Service's scientific research capability; nor would Wirth, who evidenced only random interest in the wildlife biologists and their policies. The biologists were transferred back to the Park Service from the Fish and Wildlife Service beginning in 1944, when Victor Cahalane, George Wright's successor as head of the Wildlife Division, returned and was stationed in Drury's office. Two years later, the remaining handful of wildlife biologists were reinstated to the ranks of the Park Service (and in October 1947 the Service ended its wartime "exile" in Chicago, moving its headquarters back to Washington).[48]

Under Drury in the middle and late 1940s, the Service prepared several reports that emphasized the need to improve its biological programs, including research—but the recommendations encountered a reluctant Service leadership. In late March of 1944, before Cahalane reentered the Service, Chief Naturalist Carl Russell recommended that after the war Drury should promote park research rather than construction and development, and that the Service should not "attempt to justify an extensive program of post-war construction." He advocated instead a "definite program of post-war studies looking toward full understanding of our responsibilities as *trustees.*" Apparently thinking in terms of a jobs program for research, Russell noted the "lack of organized information" in fields such as history, ethnology, and natural sciences. He claimed that as many as two hundred researchers could be put to work in Chicago, Washington, New York, Berkeley, Cambridge, and other cities.[49]

On March 23, 1944—the very date of Russell's memorandum on research—Dorr G. Yeager, assistant superintendent at Zion National Park, issued a lengthy statement on the deterioration of natural conditions in the park that corroborated Carl Russell's concerns. Yeager identified problems of excessive and poorly located park development, predator control and overpopulation of deer, invasion of exotics, and alteration of riverine systems in the park. Given such problems, he speculated that the park had been so mismanaged that it had become "impossible to maintain a *semi-natural* condition." He believed that in Zion the Park Service might be "forced to admit that the natural condition can never be regained." Yeager recommended approaching these complex problems through scientific investigation. Solutions could be arrived at "only through a carefully planned and executed research program" addressing the Service's responsibilities in biological and geological matters.[50]

In early 1945, perhaps in response to such recommendations, the Service prepared the most comprehensive statement on national park scientific research needs to appear during the Drury era. Instead of a large internal research program, however, the report championed the use of independent researchers and foundations. Noting inadequate funds for in-house research, the report encouraged scientists and university students to use the parks as "field laboratories." Still, the use of outside expertise did not mean that the Park Service should avoid "organizing and prosecuting a vigorous research program when time, funds, and qualified personnel are available." The report called for the Service to create permanent positions (it did not suggest how many) to be filled by technical experts who would oversee the necessary research.

The report further stated that, in addition to guiding the management of flora and fauna, research was needed to support interpretation and development. Park development was to be carried out with a scientific understanding of natural resources to help ensure their preservation. In weighing the relative importance of development and preservation, the report favored preservation, stating that "minor objectives in park development such as might pertain only to Man's convenience . . . must receive secondary consideration when they conflict with the primary objective of preserving the primitive."[51]

There was no substantive response to these calls for improving research. In April 1947 Drury's office issued yet another report, asserting that in light of the importance of preserving natural resources, the Service must "extend and expand its existing research program." It stated that current research efforts were "not altogether satisfactory." Expansion of the program should be accomplished through hiring additional personnel and cooperating with other scientific organizations.[52]

Despite the Service's proclamations, the biologists did not gain additional positions to expand their programs. As Lowell Sumner recalled, only eight biology positions were reestablished after the war. The April 1947 report on research listed even fewer: only six positions, four of them in central offices such as Washington or the regional offices, and two in parks. By either count the number was not sufficient, given the large number of parks to be managed, each with serious wildlife and development problems to be addressed; thus the 1947 report called for at least fourteen additional biologist positions. A 1948 summary of wildlife conditions in the national parks upped this recommendation to sixteen additional positions, half of which should be "bird and mammal men, ecologists, and botanists," the other half to be aquatic biologists.[53] The following year, a similar report noted the variety of wildlife concerns, including moose ecology in Isle Royale; fishery management in Yellowstone, Grand Teton, Glacier, and Rocky Mountain; and management of large and small mammals in parks such as Acadia, Dinosaur, Theodore Roosevelt, and Wind Cave. The same report noted, however, that no progress had been made in securing the necessary biological staff, and it repeated the hiring recommendations made in the 1948 summary.[54]

Documented needs and statements of good intentions notwithstanding, the Park Service made no real increases in its biological program during the Drury administration. Victor Cahalane recalled that Drury was very timid in approaching Congress about the necessity for additional scientific positions. He believed the director was supportive of scientific programs, but only as long as they did not cost anything. It may have also been that the biologists themselves were not persuasive advocates of their programs. But, in truth, management had other priorities. In 1951 Chief Landscape Architect William Carnes reported that the Park Service currently employed "about 140" landscape architects, who were engaged in "planning the development essential to the administration, protection, and public use" of the national parks.[55] This very large commitment of staff reflected the priorities of Drury's last years as director, even before Conrad Wirth would substantially increase planning and development with his Mission 66 program.

Ironically, management's failure to give strong support to the biologists may have been influenced by a concept derived from the wildlife policies established in Fauna No. 1—that under the right circumstances most species in the national parks should become self-sustaining. Species were to be allowed to carry on their struggle for existence "unaided" unless threatened with extinction. Once they were out of danger of extinction, any "artificial aids" provided by the Park Service were to be discontinued. One implication that could be drawn from this policy was that with resources that were not endangered, a more or less custodial oversight would suffice—not requiring an extensive commitment to research or to a large staff of biologists. For example, once bison reduction in Yellowstone had brought the population to the desired level and allowed the range to restore itself, it seemed that the bison would require less management and possibly almost no research. Drury had stated in 1943 that the ultimate goal for Yellowstone's Lamar Valley bison was to put the herd "entirely on its own resources," and the Service was already on the way to discontinuing winter feeding and other operations at Buffalo Ranch.[56]

Similarly, Lowell Sumner told a Park Service conference in October 1950 that the Service was mainly interested in "*watching natural processes unfold*," and that park management consisted "primarily" of "*letting nature alone*." He cited the termination of the bear shows and the efforts to eliminate roadside feeding as examples of current wildlife policies—to return bears to "a normal way of life, based on rustling their own natural food." Sumner noted that, by contrast, other bureaus such as the Forest Service and the Fish and Wildlife Service intensively managed game species, treating many of them as crops to be harvested. Later, as director, Conrad Wirth would clearly signal a hands-off approach by stating that wilderness preservation was not specifically a "program item" for the Service, "because in a sense the less you have to do the better it is being preserved."[57]

Sumner and his fellow biologists were keenly aware, however, that not only did placing species on a "self-sustaining basis" require research, but also that Fauna No. 1 itself called for research to be conducted prior to any "management measure or other interference with biotic relationships." Any significant disturbance of natural conditions required prior knowledge of the resources affected—a policy highly unlikely to be honored with 140 landscape architects and only about a half-dozen wildlife biologists in the parks. Sumner knew that, most fundamentally, Fauna No. 1 had called for "a complete faunal investigation" of all national parks—leaving nature alone did not mean failing to achieve an understanding of the populations and dynamics of species inhabiting the parks.[58]

Having long ago endorsed Fauna No. 1 and still seeking to adhere to some of its tenets, the Service lacked the interest in acquiring what biologist Carl Russell had called a "full understanding of our responsibilities as *trustees*." To do so would have necessitated a substantial buildup of its biological staff; but Drury's unwillingness to act left wildlife biology weak and vulnerable. As Olaus Murie, who had become head of the Wilderness Society, commented to Drury just before Drury resigned from the Park Service early in 1951, the status of the biological programs was "precarious" and the Service had only managed to "hang on to some biologists." To Murie, ecological science had moved up to new levels and the Park Service had undertaken a "high responsibility" in keeping important natural areas unimpaired. Yet with the superintendents' tendency to "oversimplify the task of the research man," as Murie saw it, the biological researcher was frequently seen as little more than a "trouble shooter." The Service's administrators gave the biologists neither "universal approval" nor enthusiastic support.[59]

Beginning in the mid-1950s, Director Wirth's Mission 66 program, ultimately averaging about $100 million per year, would not improve the biologists' status. Although the goals of Mission 66 came to include a strong rhetorical commitment to research—declaring that "guess-work is not good enough for America's national heritage" and that "exact knowledge and understanding based on sound scientific . . . research is essential"—in reality the program included negligible support for biological sciences. Exasperated because biology had been ignored and aware that Mission 66 would not include substantial funding for his programs, Chief Biologist Victor Cahalane resigned from the Park Service in 1955.[60]

Rhetoric aside, Wirth indeed seemed distrustful of science. As Cahalane remembered it, Wirth appeared to care neither about wildlife issues nor about what the biologists were doing, and to believe that scientists were using money that could better be spent drawing park plans. The director's indifference was most evident in his failure to bolster science programs during Mission 66. He made explicit his disregard for science in a letter to Horace Albright in November 1956, expressing the need to "slant a practical eye" toward the issue of overgrazing of Yellowstone's grasslands, a matter of deep concern to the wildlife biologists. In a telling comment, Wirth added: "Sometimes I find, Horace, and I am sure you will agree with this, that you can get too scientific on these things and cause a lot of harm." The director's remarks fell on receptive ears, given Albright's record of opposition to the biologists on numerous wildlife management issues. Albright displayed attitudes similar to Wirth's when he told a 1958 gathering of the National Parks Advisory Board that "there should not be too much emphasis laid on biology." After all, he added, the people were "the ones who are going to enjoy the parks." The former director asserted that "ninety-nine percent" of the people who visit the parks are "not interested in biological research."[61]

With little support for science programs within the Service, outside research received continued emphasis during Mission 66. Indeed, reliance on researchers from universities or other federal bureaus had always figured prominently in park management's thinking. Mather had depended on it almost exclusively, and the Drury administration had called for it repeatedly—a trend that continued under Wirth. Much as the 1918 Lane Letter had done, an April 1958 memorandum to the Washington office and all field offices stressed the need for outside research in cooperation with universities and other bureaus. The Service should seek to "advance programs which will attract qualified scientists to the National Park System for productive research purposes."[62]

Victor Cahalane recalled that during his career the Park Service never got much out of university research, that the research was often too abstract, and that it did not influence wildlife management policies and practices. Furthermore, an internal report in the early 1960s observed that relying on others to do national park research resulted in products "most frequently oriented toward the researcher's interests, and only incidentally toward Service needs and objectives."[63] In truth, by continually emphasizing the use of external scientific research, the Service revealed even more clearly the limitations of its commitment to use its own funds and staffing for such purposes. From Mather's time on, the repeated assertions that the Park Service should rely on research conducted by other institutions were a means to avoid coming to grips with the problem internally, in contrast to the enormous support given to tourism development and related management programs.

In 1958, as Mission 66 approached its halfway mark, the budget for scientific research projects throughout the entire park system, not including biologists' salaries, was only $28,000—a minuscule sum compared to that spent for development and construction. In a letter to Lowell Sumner in December 1958, Olaus Murie stated that Mission 66 had brought about a "period of expediency" in the Service, causing "a confused outlook, in which the biological program suffers." Even with many "splendid people" in the Park Service, Murie believed that the Service's Washington office still did not know "what is taking place in the human mind" with the advances in ecological knowledge.[64]

Yet some Park Service leaders were becoming more aware. In 1960 an internal report by a high-level committee commented that the "research effort" for national parks was so inadequate that the parks' resources were "actually endangered by ignorance." Chaired by biologist Daniel Beard (a former superintendent at Everglades and Olympic, and soon to be a regional director), the committee reported that research seemed "less understood, less appreciated, and less organized than anything else" the Service undertook.[65] The following year, Beard told the superintendents conference that the Park Service had a "surprising lack of understanding of the purpose and needs of research." The Service's tendency to seek research support from universities instead of building its own scientific staff meant that the biologists had to "stand hat in hand in an effort to get foundation support" and to rely on the "peon labor" of graduate students. Similar concerns were expressed to the superintendents by Chief Landscape Architect William Carnes, who stated that there was "little reason to brag about our accomplishments in the research field" and that the Service had not "assumed the leadership" to provide knowledge necessary for park management.[66]

The most influential internal statement on research needs was the inspiration of Howard R. Stagner, chief of the Branch of Natural History and advocate of a strong science program. In 1961 Stagner oversaw preparation of a document entitled "Get the Facts, and Put Them to Work." Released that October, it was sharply critical of the Service's "inadequately financed" research program, which lacked "continuity, coordination, and depth." The report described the parks as "*complex organisms*" that were "*rapidly becoming islands*" surrounded by lands managed for different purposes. It argued that research was necessary for the Service to "know *what* it is protecting, and what it must protect against." The Service "must understand, much more completely than it now does, the natural characteristic of these properties, the nature of the normal processes at work within them, the unnatural forces imposed upon them, and, as well, the relationships of park visitors to the natural environments."[67]

A significant shift from earlier thinking, the insights of both Beard and Stagner reflected a growing concern among conservationists and some Service leaders about the national parks' ecological conditions. Such concern went beyond distress about deteriorating park facilities or the location and appearance of facilities once they were built—the major emphasis of many conservationists through much of the 1950s. "Get the Facts" took a different stance, stressing a "critical" need for scientific knowledge of the national parks and quoting from an international panel of scientists that the parks offered the "principal future hope of preserving some scattered fragments of primeval nature for fundamental scientific research."[68]

"Get the Facts" recommended a long-range research plan with a "logical sequence" of projects, together with adequate funding and staffing of the "highest professional research competence." With this document in hand, Stagner worked to increase funding for the science program. More important, he used the report to heighten Secretary of the Interior Stewart L. Udall's interest in national park science.[69] Udall's response would result in major reports prepared outside the Park Service, focusing even more attention on ecological issues in the parks.

By the Service's own reckoning in the early 1960s, it had almost no scientific research to inform natural resource management or to advise on possible impacts of Mission 66 development. In the rush of Mission 66, and with nearly three decades having passed since the acceptance of Fauna No. 1's recommendations as official policy (including the requirement for "properly conducted investigations" prior to any "management measure or other interference with biotic relationships"), the scientific programs called for in Fauna No. 1 had been rendered virtually impotent.[70]

Reflecting this disregard for science, the wildlife biologists' organizational status remained repressed during the postwar years, culminating in the late 1950s with Wirth's decision to bring the biologists under the rangers and foresters, whose policies they many times deplored. Returning to the Service after the war, the wildlife biologists were placed under the naturalists. In 1947, after a number of wartime vacancies had been filled, the Service employed sixty-one "year-round professional" naturalists who oversaw the park interpretive programs—a staff that dwarfed that of the biologists. Rather than managing natural resources, the naturalists focused on interpreting them, a responsibility usually not of primary concern to wildlife biologists.[71] Also in contrast to the biologists' status, the Park Service foresters continued to enjoy close ties with the rangers, and by the end of Drury's tenure many men with formal college training in forestry occupied key positions such as chief ranger and park superintendent.[72]

The rangers gained strength in 1954, when Wirth established a Washington office for "ranger activities"—the Branch of Conservation and Protection. Indicative of their close alliance with the rangers, the foresters were included in the new office. The head of the branch (in effect the "chief ranger"), Lemuel A. (Lon) Garrison—former superintendent of Big Bend National Park and a rising star in the Service—brought clout to the position. (Garrison's successors would be another former ranger and superintendent, John M. Davis, and then former chief forester Lawrence Cook.) To increase the new unit's influence, Garrison urged that it be upgraded to division status.

In 1957, as the rangers were achieving this new status, they made their bid to gain control of the wildlife biology programs.[73] Wirth responded in October of that year by transferring the biologists to the newly established Division of Ranger Activities. Initially he had intended to place the biologists in a branch separate from the foresters. He changed his mind, however, and ordered a merger of forestry and wildlife biology into one branch under forester Lawrence Cook, who reported to the chief of the ranger division.[74]

The transfer evoked impassioned opposition from former chief biologist Victor Cahalane, who strongly disapproved of the Service's forestry practices. Cahalane wrote to E. Raymond Hall (now with the University of Kansas' Museum of Natural History, and a member of the National Parks Advisory Board), urging that "everything possible" be done to reverse the transfer. In view of the Service's failure to bring its forest policies in line with contemporary ecological principles, the former chief biologist characterized the foresters as a group that "pretends to know everything about ecology but actually has no competence in that field." He cited the foresters' efforts to suppress "as rigorously as possible" natural fires and native insects and diseases, and added that "under the mandated merger [the foresters] will apply the same philosophy to wildlife. Knowing little or nothing about animal ecology, they can work havoc."[75]

Cahalane found a ready listener in Hall, who also disapproved of the Park Service's forest management. Early in 1958 Hall attacked the Service's policy to "practice forestry" that led to disruption of natural succession in park forests. In a telling comment, he observed that the Service persisted in using the term "forestry"—a designation used by bureaus such as the Forest Service, with their focus on the economic benefits of timber production. Why, he asked, when the goal was to preserve—rather than harvest—natural resources, should Park Service foresters not be called "biologists," or "botanists"? After all, wildlife biologists were not known as "game managers." Hall believed that the continued use of the term "forester" contributed to a "fuzziness in policy and in practice as concerns the preservation of natural conditions."[76]

Hall's protest had no effect. The merger held, with the wildlife biologists and foresters remaining in the ranger division and reporting directly to Lawrence Cook, an outspoken advocate of traditional forest practices. Contrary to the views of Cahalane and Hall, the rangers insisted that the "basic principles and procedures" of wildlife management were "identical and parallel" to those of forestry.[77] To assist with wildlife management in the field, the park superintendents soon formally designated fifty-nine park ranger positions as "wildlife rangers," probably filling these positions with men who had long been responsible for such work.[78]

In Washington the wildlife biologists transferred to the ranger division were to be involved in day-to-day field operations. Two wildlife biologists stayed with the naturalist division, recently redesignated the Division of Interpretation. They were responsible for overseeing "all biological research" and recommending policies on wildlife and fish management. The directorate explained that the research and policy biologists were better off in interpretation, where they were removed from day-to-day demands of actual management, and that previously "basic research and investigations" had "suffered" when field management activities distracted the research biologists.[79] However, with only two biologists in research and policy, that aspect of the biology programs remained virtually powerless in the surge of Mission 66 activity.

The Road to Mission 66

Although Newton Drury was not enthusiastic about recreational programs and the development associated with them, they remained viable during his directorship, even growing in certain aspects. As in the 1930s, Conrad Wirth spearheaded the Service's recreational planning efforts in the 1940s. After he became director in 1951, recreational tourism, culminating in the Mission 66 emphasis on intensive public use, would become more than ever the driving force within the National Park Service. It was the antithesis of the scientific approach to park management.

With reduction of Service personnel and termination of the Civilian Conservation Corps, maintenance of roads, trails, and buildings in the national parks had declined drastically during World War II. As anticipated, the end of the war brought a sudden upswing in the number of park visitors. Yellowstone superintendent Edmund Rogers reported that in the first three months following victory over Germany in the spring of 1945, visits to the park were up 56.4 percent. Immediately after the Japanese surrender in August, the number of visits "practically doubled" and continued to increase during the remaining weeks of the travel season. Overall, the number of visitors to the national park system jumped from 11.7 million in 1945 to 25.5 million in 1947. With poorly maintained park facilities, the Service, as Sequoia superintendent John White described it, felt more than ever like engineers "compelled to dam a stream in flood without opportunity to divert the flood waters."[80]

Advance planning for development of the national parks to meet the needs of tourism in the postwar era had begun shortly after Pearl Harbor, with a "Plans on the Shelf" program overseen, not surprisingly, by Conrad Wirth. Although Drury promised in his 1945 annual report that after the war the Service would "do what we were doing before the war, but do it better," in fact he did not effectively promote postwar improvement of park facilities.[81]

Yet Drury did attempt to get development funds, for instance in 1947 citing such problems as "poorly equipped and crowded" campgrounds, "pitifully inadequate" utilities, and hotel and tourist accommodations "vastly in need of enlargement and modernization." That year he estimated that annual appropriations of $45 million were needed over a span of seven years (an overall total of more than $300 million) to take care of physical facilities. Two years later he raised the multiyear request to half a billion dollars. But his efforts brought few results. Concerned about huge war debts, only in 1947 did Congress grant the Service a substantial budget increase—which was quickly followed by a return to minimal funding.[82]

Conservative by temperament, Drury, in his postwar funding quests, was probably inhibited by his longtime opposition to increased development of the national parks. His timid leadership, along with the restrained circumstances of the times, meant that he was not able to obtain sufficient support from Congress to launch an overhaul of national park facilities.[83] Even so, his funding efforts triggered fears of too much park development. In the spring of 1948 the Sierra Club advised the Service of its apprehension about the proposed park development program. Drury promised the club that the projects would not intrude on backcountry areas. In a revealing statement, he added that "perhaps an even more important point" was that the chances of the budget request being approved were "decidedly slim"—in effect, not to worry about it, the Park Service was not going to get the money anyway. Similarly, Drury once assured a group of eight Sierra Club leaders that the Service was unlikely to impair the parks; perhaps exhibiting his innermost attitudes toward park development, he told them, "We have no money; we can do no harm."[84]

Drury took a conservative stance with another program that had arisen and that again revealed the Service's affinity for intensive recreational development. Although funds for improvement of national park facilities remained limited, funds became available for studies of recreation potential at proposed reservoir sites in river basins of the West. The Park Service, having undertaken such a study for Lake Mead in the 1930s and accepted responsibility to manage the new Boulder Dam National Recreation Area, soon expanded its involvement with river basin development. It undertook recreational planning for the reservoir to be created behind Grand Coulee Dam on the Columbia River, and in 1941 agreed to survey the recreational potential of reservoirs planned by the Bureau of Reclamation for the Colorado River Basin.

The Service also cooperated with the bureau on plans for river basins in Texas and California, and in 1943 it began surveying areas for possible recreational use along that part of the Alaskan Highway within United

States territory. Another big opportunity came in 1944, when Congress authorized flood control in the Missouri River Basin and the Service agreed with the U.S. Army Corps of Engineers to conduct recreational surveys of the prospective reservoirs in that basin. Beyond recreation, the surveys included extensive archeological investigation and salvage of artifacts from reservoir sites. Providing funds to help keep Park Service staff on board, and overseen by Conrad Wirth's recreation and land planning office, these programs grew substantially during Drury's directorship.[85]

As had happened with Boulder Dam National Recreation Area, the Park Service moved beyond the initial surveys and planning toward actual management of reservoir recreation areas with marinas and campgrounds and attendant facilities. Drury was uneasy with these responsibilities, believing them inappropriate for the Park Service. They ran counter to his belief that the Service's essential mission was to maintain large natural areas in minimally altered condition while also accommodating the public. Early in 1945 H. W. Bashore, commissioner of the Bureau of Reclamation, requested from Secretary Ickes a clarification of departmental policy on which bureau should assume recreation responsibility at certain reservoirs. Bashore's request related most immediately to Millerton and Shasta reservoirs in California, but it also raised fundamental questions regarding the Park Service's true purpose.[86] In an exchange of letters that became known as the "Black Magic–Ivory Tower" correspondence, Drury and the Interior Department laid out opposing views on the wisdom of expanding Service commitment to reservoir recreation management.

Drury believed that a departmental policy on the emerging field of reservoir recreation would have an "important bearing on the future operations of the National Park Service," and he appealed to Secretary Ickes. Seeking to limit involvement with reservoirs, Drury argued that there was "no black magic" in the management of such areas, and that they did not have to be the Service's responsibility. Noting also that it was cumbersome for two bureaus to divide management of reservoirs (as with flood control and public recreation), Drury then raised his chief concern: the potentially negative effects on the Service and on the national parks themselves. Additional involvement with reservoir recreation would, he predicted, "dissipate our energies and divert them from the performance of our primary functions." It would make the national park system vulnerable by diluting the standards and policies of park management that had evolved over time.

The director noted that Service policies against consumptive uses in the national parks (such as grazing, mining, and timber harvesting) were already disputed. These policies would become even more vulnerable if

Congress and the public could no longer distinguish "true national park areas" from multiple-use areas. Thus the Service should "keep clear of such equivocal arrangements" as reservoirs, and "local or mass recreation" should not be a primary concern. Although recognizing the Park Service's legal responsibility to assist with state and federal recreational planning (stemming from the Park, Parkway, and Recreational Area Study Act of 1936, which the Service itself had promoted), Drury nevertheless wanted to avoid becoming the nation's recreation overlord.[87]

But with Ickes' concurrence, his assistant secretary Michael W. Straus (who would soon succeed Bashore as commissioner of the Bureau of Reclamation) rejected Drury's recommendations as an "open abdication" of a serious responsibility and an attempt to retreat to an "ivory tower," away from the conflicts of recreational management. He argued that both "law and custom" (the Park Service's congressional mandates and its past efforts in recreation) made it the "best equipped" bureau to assume the duties in question. Straus noted the diversification of the national park system (a result of Park Service expansion efforts in the 1930s). He wrote that, beyond maintaining the "purity . . . of natural phenomena," the Service already managed reservoirs such as Lake Mead, and Jackson Lake (in Jackson Hole National Monument, soon to be incorporated into Grand Teton National Park), and historic areas like Independence Hall, the Statue of Liberty, and numerous sites in Washington, D.C. Straus saw Drury's opposition as "narrow-visioned" and urged an agreement with the Bureau of Reclamation for the Service to operate the recreational facilities at Millerton and Shasta reservoirs.[88]

Having once accused the Park Service of becoming a "Super Department of Recreation," Drury seemed to have hoped that he was now in a position to restrict the Service's recreational programs. (Indeed, he was currently overseeing the removal of most of the New Deal–created recreational demonstration areas from Park Service custody, as had been originally planned.) But Drury lost the policy debate with Straus. The Service was assigned to manage recreational facilities at Shasta and Millerton reservoirs in California, and at Lake Texoma on the Red River between Texas and Oklahoma.

Although willing to manage large recreation areas such as Lake Mead and Grand Coulee, which could be construed to be of significance to the nation as a whole, Drury continued to oppose involvement with smaller reservoirs.[89] Ultimately, the Park Service was able to divest itself of recreational management at some lesser sites, beginning with a 1948 agreement for the Forest Service to assume the responsibilities at Shasta Reservoir.

The following year, faced with difficulties caused by "unsatisfactory division of authority" between the Park Service and the Corps of Engineers (as Drury had anticipated), the Service was permitted to transfer all of its management responsibilities at Lake Texoma to the corps.[90]

The recreation programs promoted by Conrad Wirth beginning in the 1930s and expanded during Drury's administration had entrapped Drury in a situation he could not fully reverse. The Park Service's ties to river basin studies and reservoir management put it, as Drury stated it, in an "equivocal" policy and philosophical position. Although committed to protecting the parks' scenic landscapes from intrusions such as dams, the Service, through its reservoir work, lent support to the inundation of scenic canyons and valleys throughout the West. Moreover, Drury's fears that reservoir recreation commitments would make the national park system more vulnerable foreshadowed the troubles that arose when the expansive dam-building programs of the Bureau of Reclamation and the Army Corps of Engineers began to threaten established units of the park system.

Drury, in fact, contributed to these troubles. Park Service involvement with the proposed Echo Park dam project, which was intended to inundate a large portion of Dinosaur National Monument and would become the most controversial of all postwar dam initiatives, had begun in 1941, the year after Drury became director. At that time the Service agreed with the Bureau of Reclamation to undertake the recreational survey for prospective reservoirs on the upper Colorado River. Drury hoped the Park Service might gain meaningful influence in the extensive planning under way for the Colorado basin; moreover, the bureau provided funds for the survey—surely an enticement for the Service. Included in the agreement was an understanding that, because the proposed Echo Park reservoir would drown a large portion of Dinosaur, consideration would be given to redesignating the flooded monument a national recreation area.[91]

Drury's initial willingness to support this plan seems itself equivocal and contradictory to his reluctance to manage reservoirs. The vast Echo Park area had been added to the original (and very small) Dinosaur National Monument only in 1938, and the Service lacked real familiarity with the recently added park lands that were proposed for inundation. Therefore, it had little appreciation of the area's scenic qualities—a key consideration for leaders like Drury—and the Park Service became a willing participant in the dam proposal.

The Service failed to take a position against the dam until the late

1940s, when the area's scenic beauty became more appreciated and it appeared that the dam might indeed be built. This delay nearly led to the flooding of a large part of the national monument. Furthermore, by the time the vacillating Drury began to oppose the dam, other water control proposals were threatening major national parks such as Grand Canyon, Kings Canyon, Glacier, and Mammoth Cave.[92]

Having been a cooperative endeavor with the dam builders, the Service's reservoir work had become, as Drury saw it, "a two-edge sword," with the recreational potential of artificial lakes being used as one pretext to gain approval for dams and reservoirs that would intrude on existing national parks and monuments. The preeminent example of the loss of a park's spectacular, natural landscape had come with the creation of a reservoir in Yosemite's Hetch Hetchy Valley, and the specter of Hetch Hetchy disturbed Drury. There, as he put it, "something commonplace was substituted for something great and fine"—a situation he began to fear could be repeated. In June 1948 he instructed his regional director in San Francisco to be "very cautious" and avoid giving the dam builders "ammunition that will be used against the basic cause in which we are primarily engaged."[93]

In hearings held early in April 1950, the Park Service objected to the Echo Park dam proposal—a position contrary to that taken by Secretary of the Interior Oscar Chapman, who favored the dam. Later that year, in what turned out to be his last annual report, Drury stated that in recent years the Army Corps of Engineers and the Bureau of Reclamation had promoted projects that would "destroy or impair the beauty and interest" of the national parks. He called for balanced, long-range planning to address natural resource issues, including nature preservation. Acknowledging that the Park Service had worked "wholeheartedly and conscientiously" with the dam builders to promote reservoir recreation, Director Drury nevertheless argued that an "artificial body of water" in a park never makes a "satisfactory substitute for a natural scene"—a policy he had ignored in his earlier agreement to allow a reservoir in Dinosaur National Monument.[94]

Drury's lack of enthusiasm for reclamation projects and his late-blooming opposition to the Echo Park dam helped precipitate his sudden resignation early in 1951. Secretary Chapman, who still favored the dam, was severely criticized for forcing a respected conservationist out of office. Reflecting on his difficulty with the secretary, Drury later recalled that, as a dedicated Republican who had survived a decade of Democratic administrations, he had been like a "cat in a strange garret." He noted also the hostility of the Bureau of Reclamation, which, he claimed, was the "dominant bureau in the Department of Interior," and which "more or less colored" Chapman's views.[95]

Arthur Demaray, longtime member of the Park Service directorate, succeeded Drury. Having already indicated he would retire soon, Demaray remained director for only eight months, until December 1951, when Conrad Wirth assumed the office.[96] From the time of Drury's resignation, the Park Service, under orders from Chapman, accepted a diminished role in the Echo Park conflict—a role that surely did not enhance the Service's image within the growing conservation movement. Director Wirth often supplied information to and worked behind the scenes with opponents of the dam. Nevertheless, the hard-fought battle against the dam—ultimately successful in the mid-1950s—was waged mainly by conservation groups such as the Sierra Club and the Wilderness Society, which used their increasing strength to fight reclamation projects.[97]

Even though Wirth opposed the Echo Park dam, his support for other national recreation areas contrasted markedly with Drury's ambivalence. In a 1952 address to the American Planning and Civic Association, the new director expressed pride in the Service's accomplishments in this area, boasting that its many years of experience made it best equipped to undertake planning for reservoir recreation: "We feel that these activities are closely related to other responsibilities of ours, and that it is just common sense that we should undertake them."

Wirth claimed that involvement with river basin development programs put the Service in a better position to defend the national parks against possible intrusions of dams and reservoirs. Like Drury, he favored the most impressive reservoirs, especially those behind Boulder and Grand Coulee dams.[98] His interest in the larger reservoirs would help lead the Park Service in 1958 to agree to take charge of recreation at another huge reservoir: Lake Powell, expected to flood nearly two hundred miles of southern Utah canyon country upstream from the massive Glen Canyon Dam, due to be completed in the 1960s. Just over two years after the defeat of the Echo Park dam, the Service signed on to help manage a reservoir that would drown some of the most spectacular sandstone canyons in North America. The sacrifice of this area to a new reservoir was part of the price of the compromise that had prevented construction of the Echo Park dam.[99]

Much of the land to be covered by Lake Powell had once been proposed for part of the national park system as a large, essentially natural area. Thus, at different times, the Park Service had been willing to manage

Echo Park and the Glen Canyon area either as national parks or as reservoir recreation areas. This readiness to administer certain public lands under whichever management policy was arrived at by the political system was revealed in Wirth's 1952 address to the American Civic and Planning Association. The newly appointed director noted that many people considered Hells Canyon, along the Snake River on the Idaho-Oregon border, to be of "national park or monument calibre." But with the Bureau of Reclamation already planning to dam the canyon, Wirth suggested that it could therefore become a national recreation area under Park Service management.[100] Committed to recreation programs and to managing large natural areas, the Service revealed its opportunistic tendencies when an attractive prospect of either type arose.

As one of the chief proponents of the New Deal diversification of Park Service programs, Wirth did not suffer the equivocation that Drury experienced when contemplating the possible effects of reservoir management on the attitudes and priorities of the Service. With an emphasis on physical recreation much more than on the contemplative enjoyment of natural scenery, recreation areas involved substantially different management approaches, perhaps most notably the allowance of public hunting. These areas also emphasized water sports, which necessitated development of marinas and beaches, beyond the tourist accommodations and administrative facilities typically found in national parks.[101] They nurtured the Park Service's already-ingrained affinity for recreational tourism. And in this regard, Wirth soon focused his considerable bureaucratic skills on a huge new program designed to improve the capability of the national parks themselves to receive the hordes of tourists arriving in the 1950s.

Mission 66

When Douglas McKay, Oscar Chapman's successor as secretary of the interior, announced in November 1955 that he was withdrawing support for the Echo Park dam, he also announced that a narrow, rocky road in Dinosaur National Monument would be improved to provide public access to some of the area's most splendid scenery. This little-used park, which had just been saved from inundation, would be made accessible for greater public use.[102] As Park Service leaders had long argued and as the Echo Park confrontation indicated, preservation could not easily stand on its own in the public forum. Especially in the years before the 1964 Wilderness Act, preservation efforts that were not accompanied by development for public use were vulnerable and likely to fail. Tourism development was important in and of itself, but it also provided utilitarian grounds for preservation. It served as a defense against massive intrusions such as dams and reservoirs and as a means of keeping visitors in designated areas, thereby protecting undeveloped backcountry. National park development was locked with preservation in a state of perpetual tension—both supportive and antagonistic.

Since its founding in 1916, the Park Service had relied on two fundamental justifications in its drive to develop the parks for public use. To begin with, Stephen Mather had urged tourism development in order to attract people to the parks to generate public and congressional support and to ensure the parks' survival. His immediate successors, Albright and Cammerer, had continued this rationale for encouraging public use. But by the 1950s the situation had changed. Except in remote areas like Echo Park, the public had descended on the national parks, and development was justified not only as a means of accommodating visitors, but also of controlling record-setting crowds. From this new perspective, Wirth argued as urgently as had Mather that development would save the parks. By the 1950s the public was (in a phrase that Wirth claimed had been coined by the Service) "loving the parks to death." National park development would control where the public went and prevent misuse through what Yellowstone superintendent Lon Garrison termed the "paradox of protection by development."[103]

This idea became a fundamental principle of Wirth's Mission 66 program: in effect, if visitors were going to use certain areas, prepare for this by improving roads, trails, and park facilities that would limit the impact to specified areas. Public use would be contained, leaving alone the undeveloped areas of the parks. As Wirth stated in his annual report of 1956, park development was "based upon the assumption" that "when facilities are adequate in number, and properly designed and located, large numbers of visitors can be handled readily and without damage to the areas. Good development saves the landscape from ruin, protecting it for its intended recreational and inspirational values."[104]

Shortly after becoming director in late 1951, Wirth claimed continuity with the Drury administration, stating that National Park Service policies were "not expressions of the personal viewpoint of individual directors." But, in truth, he was overlooking not only the philosophical differences between himself and Drury, but also their substantial difference in ability to promote and finesse programs to a successful conclusion.[105] To his advantage, Wirth assumed the directorship more than six years after World War II, when programs designed to facilitate automobile travel (such as the

Interstate Highway System and Mission 66) encountered a more favorable political climate. Wirth also benefited from a heightened public awareness of crowded conditions and deteriorating facilities in the national parks, an awareness that resulted in part from the Service's own publicity efforts.

These efforts apparently involved behind-the-scenes encouragement for the prominent historian and journalist Bernard DeVoto to write an exposé of conditions in the parks. A member of the prestigious Advisory Board on National Parks, Historic Sites, Buildings and Monuments, with which Wirth worked closely, DeVoto got the inspiration to write the article during a 1953 board meeting. Entitled "Let's Close the National Parks" and appearing in *Harper's Magazine* in October 1953, the article blasted Congress for ignoring the parks and leaving the Service like an "impoverished stepchild," or like the widow who "scrapes and patches and ekes out," using "desperate expedients" in an effort to succeed. Citing the deplorable condition of roads, campgrounds, buildings, and other facilities, DeVoto complained that the parks were woefully understaffed, many of them operating with the same number of personnel they had had two decades before, when far fewer people visited the parks. Moreover, park personnel often lived in shameful housing—"either antiques or shacks," some houses like "a leaky and rat-ridden crate." He claimed that "true slum districts" existed in parks such as Yellowstone, Rocky Mountain, and Yosemite. Attempting to shock the public in order to gain greater support, DeVoto recommended temporarily closing many of the most popular parks and reducing the system to a size Congress would adequately fund.[106]

DeVoto's widely read article was pretty much on the mark. The following year, 1954, while the Service continued without substantial relief from Congress, 47.8 million visitors entered the parks. This number set a new record for the tenth straight year and was more than twice the number recorded in 1941, the last big vacation year before World War II. In February 1955, just over three years after becoming director, Wirth conceived the idea of a giant program that would affect the entire park system and benefit congressional districts throughout the country.[107]

In his autobiography, Wirth recalled realizing that efforts to acquire major, long-range funding for national park construction and development should be modeled on the strategies used by agencies involved in massive development projects—such as the Army Corps of Engineers, Bureau of Public Roads, and Bureau of Reclamation. Because their dam and highway projects were so large, those bureaus were able to get huge multiyear funding packages approved by Congress. By comparison, the smaller projects of the National Park Service were more vulnerable and easily cut from the administration's annual budget. To strengthen his bid for large-scale funding, Wirth believed he should propose one "all-inclusive, long-term program" for the parks. He also sensed that if "all the congressmen knew that the parks in their states were part of the [Mission 66] package and would be similarly taken care of within a given time, it seemed that once the overall program got started it would be hard to stop."[108]

Wirth quickly formed committees in the Washington office to plan Mission 66. As the planning became more intense and spread throughout the park system, the director secured an opportunity to present the program to President Dwight D. Eisenhower at a January 1956 Cabinet meeting. In addition to Wirth's successful politicking with Congress, his meeting with the President and the cabinet gained Eisenhower's firm support for a ten-year program, to start immediately. The director formally announced Mission 66 at a banquet held in Washington on February 8, 1956. This festive occasion was sponsored by the Park Service, the Department of the Interior, and—significantly, in view of later criticism of Mission 66—the American Automobile Association, one of Stephen Mather's allies in founding the Park Service in 1916.[109]

In initiating this massive program, Wirth instructed Park Service personnel to "disregard precedents," think imaginatively, and be aware that existing park facilities were based on "stage coach economy and travel patterns." Lon Garrison, first chairman of the Mission 66 Steering Committee, recalled that the committee was instructed to "dream up a contemporary National Park Service," in effect, and to prepare the parks for an estimated 80 million visitors by 1966.[110] As conceived, Mission 66 included not only extensive construction and development, but also significant staff increases (especially for interpretation, maintenance, and protection); an ambitious program to acquire inholdings; and a nationwide recreational survey to assist all levels of government in improving public park and recreational facilities. The Service also included as a broad, yet "paramount" goal of Mission 66 the preservation of national park wilderness areas.[111] Despite Wirth's resolve to "disregard precedents," Mission 66 reflected Park Service trends dating from Mather's time on, especially during the New Deal, the last flush times, when the Service developed the parks, increased staffing, and planned for recreation on a nationwide basis.

Without question, Mission 66's primary focus was the improvement of physical facilities in all parks. Having begun planning as far back as the early 1940s to meet postwar development needs, by the mid-1950s the

Service was, in Lon Garrison's words, "ready for Mission 66," and the park development files were "full of goodies!" Indeed, Mission 66 would encompass hundreds of projects, among them 1,570 miles of rehabilitated roads; 1,197 miles of new roads (mostly in new park areas); 936 miles of new or rehabilitated trails; 1,502 new and 330 rehabilitated parking areas to accommodate nearly 50,000 additional vehicles; 575 new campgrounds; 535 new water systems; 271 new power systems; 521 new sewer systems; 218 new utility buildings; 221 new administrative buildings; 1,239 new employee housing units; 458 reconstructed or rehabilitated historic buildings; and 114 new visitor centers.[112]

Much of the Mission 66 work was based on revised and updated master planning led by the landscape architects, who, as William Carnes, Wirth's chief landscape architect, put it, played the "paramount role" in this effort. The most influential of all national park documents, the master plans determined where and how much a park would be developed. Carnes advocated that national park professionals take "the humble approach," with subdued designs that would not dominate nature. But for intensively used areas, he noted that master plans were particularly complex—"actually a matter of town or community planning." Wirth believed that without the plans it would have been "impossible to organize a sound program." He named Carnes to head the Mission 66 Committee, the actual working group (it reported to the higher-level steering committee) that supervised the updating of master plans to guide each park through Mission 66.[113]

Mission 66 evidenced the power that the construction and development professions had attained within the Service, epitomized by the influence of the landscape architects. Since, from the very first, Park Service directors had enjoyed wide latitude to build their bureaucratic organization as they saw fit, their perception of the mandated purpose and function of the National Park Service was reflected in the organization and staffing that evolved under their direction. In the early 1950s, just prior to Mission 66, William Carnes claimed that there were more landscape architects in the Park Service than any other profession, and that the Service was the "largest single user of landscape architects in the country—possibly in the world."

Most landscape architects were in the parks, regional offices, and special field offices; a few were in Washington where Carnes was stationed. Numerous national parks had their own landscape architects to provide the superintendent with information and advice. Carnes and those who did not report to park superintendents or regional directors were under Thomas Vint, the widely respected, longtime Service architect and landscape architect. In 1954 Vint had opened new central offices—the eastern and western offices of design and construction, which, with ever-enlarging staffs of landscape architects, engineers, and architects, would shoulder much of the Mission 66 work. By the time of Vint's retirement in 1961 (at midcourse for Mission 66), his design and construction operations included a staff of more than four hundred permanent employees.[114]

In addition to the landscape architects' professional work, their influence was pervasive in other ways. Carnes noted that, beyond those directly involved in field projects, several had become superintendents, four were assistant regional directors, one was a regional director. Another, Conrad Wirth, had become director. And although not themselves landscape architects, most previous directors had worked "closely and understandingly" with the profession, to the extent that they had been honored as "Corresponding Members" of the American Society of Landscape Architects.[115] Thus, in the 1950s, when the superintendents and rangers gained a power base in the Washington office with branch and then division status (and with leadership by former superintendents such as Lon Garrison from Big Bend and Eivind Scoyen from Sequoia and Kings Canyon), they formed with the design and construction professions a cohesive leadership clique to move Mission 66 forward under Director Wirth.[116]

Despite the evident need to improve the parks' physical facilities, Mission 66 encountered severe criticism, far more than previous national park development and construction had faced. With its ambitious size and scope, Wirth's program was confronted by the rising power of the conservation movement, whose leaders could take their case directly to the public and to highly placed politicians, widely broadcasting disapproval of national park management. And in a pre–*Silent Spring* confrontation, development itself was the central issue, not ecological impacts per se, such as destruction of habitat. Concerns included the inappropriateness of the location and the appearance of visitor centers and other tourist facilities, the amount of road construction, the design of roads, and whether highways should wind gently through park scenery or provide for high-speed traffic.

To many, the major objection to Mission 66 was that it tended to modernize and urbanize the national parks. In Everglades, for instance, the dirt road to Flamingo, forty miles from the park entrance, was paved early in Mission 66, thus opening the heart of the park to heavy tourist traffic. As described by Devereux Butcher, a longtime critic of national park management, the small cluster of structures at Flamingo became like a "fishing-yachting resort of the kind that is a dime a dozen in Florida"—including a sixty-room motel, a large restaurant, a marina with accommodations for large boats, marine equipment sales, rentals for outboard and inboard boats (including houseboats), and sightseeing operations for daily tours of the park's Florida Bay. This development not only resulted in the dredging of part of Florida Bay to provide access for larger boats, but also required regular transportation of supplies and equipment by truck along the park's newly improved road, in addition to increased visitor traffic.[117]

Other modern developments such as Grand Teton's Colter Bay Village and Yellowstone's Canyon Village raised the ire of conservationists. Butcher denounced the appearance of Colter Bay's laundromat, cafeteria, boat-docking facilities, parking lots, "de luxe trailer park," and 150 cabins, and he depicted Canyon Village's new overnight accommodations as "dozens of box-like cabins" (an apt description). In Great Smoky Mountains National Park, the "sky-post"—a swirling, modernistic observation tower atop Clingman's Dome, the highest point in the park (and indeed in the state of Tennessee)—became another target of Mission 66 critics.[118] An article in *National Parks Magazine* declared that inappropriate Mission 66 development made the parks seem "urbanized." It claimed that "engineering has become more important than preservation," creating wide, modern roads similar to those found in state highway systems and visitor centers that looked like medium-sized airport terminals.[119]

One of the first Mission 66 visitor centers to be built in strikingly modern design and in a large natural park was completed in 1958 at the quarry site in Dinosaur National Monument.[120] Including an expansive glassed-in area where ongoing paleontological work could be observed by visitors, the new building (like the road being improved into the Echo Park area) was intended to attract tourists to little-known Dinosaur, providing insurance against future threats to use the park's lands for other purposes, such as reservoirs.

National park architecture during the post–World War II era was indeed influenced by modernism rather than by the romanticism of earlier rustic construction. Many Park Service architects had been trained after World War II and were imbued with modern design tastes, while some of the engineers had gained experience with the military during World War II or the Korean War, when design was of necessity strictly utilitarian. In addition, as the Bureau of Reclamation's dam-building operations declined during the 1950s, the Park Service hired a number of engineers from the bureau. Subsequently involved with national park roads and buildings, they presumably had little knowledge of landscape and architectural aesthetics.

Because modern structures required little if any traditional craftsmanship, they were much less labor intensive and cheaper to build. They were favored by a budget-conscious Service, which could get a greater amount of construction with Mission 66 dollars. Also, long-term maintenance for modern, standardized structures was less costly than for rustic log buildings.[121] The modernism of Mission 66 seemed particularly jarring when compared to the log-and-stone rustic architecture of earlier park structures designed to harmonize with surrounding landscapes. In many ways the rustic structures recalled the frontier and the days of Teddy Roosevelt—as if to suggest a primitive America that the parks themselves represented, rather than the urban America symbolized by the standardized designs of Mission 66.

In 1957, already sensitive to negative public comment on Mission 66, Wirth made an ambitious attempt to disarm the critics with publication of a large-format color brochure, *The National Park Wilderness.* The Echo Park confrontation had served as a catalyst to rejuvenate the wilderness movement; and, hoping to allay concerns about park development and to be viewed as part of the movement, the Service used the brochure to portray Mission 66 as a wilderness preservation program. The brochure began by asserting that "clearly" it was the will of the American public that all of the fundamental laws guiding management and development of the national parks were intended to "preserve wilderness values." The remainder of the lavishly illustrated booklet was devoted mostly to emphasizing the importance of wilderness in national parks while justifying development.

The brochure stressed the need to preserve wilderness while preparing the parks to "serve better their increasing millions of visitors." To the rhetorical questions of whether Mission 66 would "impair the quality or reduce the area of park wilderness" and whether wilderness preservation meant abandoning traditional national park hospitality by limiting the number of visitors, eliminating lodges and campgrounds, or "other radical changes," the answer came that the Park Service sought a "sane and practical middle ground." The brochure stated that this would entail "no compromise whatsoever" with the parks' traditional and basic purpose. Indeed, compromise was seen as unnecessary because of a fundamental compatibility between wilderness and development. The brochure identified different kinds of wilderness, including what it called accessible wilderness, available within a ten-minute walk from many park roads, or where visitors could "see, sense, and react to wilderness, often without leaving the roadside." It claimed that wilderness in the parks was being adequately preserved, and that under Mission 66, development could be used "as a means of better preservation." The more a national park was used, "the less vulnerable are its lands to threats of commercial exploitation." Preparing parks "for as full a measure of recreational, educational, inspirational use as they can safely withstand" would establish a "defense against adverse use [and] . . . safeguard park integrity."[122]

Curiously, one of the most striking examples for this kind of argument was Echo Park. During the fight over the dam, conservation groups themselves had encouraged increased recreational use of the canyons and rivers threatened with inundation for the specific intent of calling attention to these areas in order to preserve them. Between 1950 and 1954, the number of people river-rafting each year in Dinosaur National Monument increased eighteen hundred percent, from fifty to more than nine hundred.[123] The annual total would continue to rise. Moreover, as promised in 1955 by Secretary McKay, the narrow, rocky road into the Echo Park area was improved (with Mission 66 funds) as a means of safeguarding Echo Park from possible future destruction. Strong utilitarian pressure to build dams had brought about a strong utilitarian response—a push for sufficient tourism use to justify preservation. In a similar effort, Mission 66 funded completion of Olympic National Park's Hurricane Ridge Road and its attendant facilities, specifically with the intent of increasing public use in order to block persistent attempts by lumbermen to open the heart of the park to timber cutting.[124] With the memory of Hetch Hetchy ever present, accommodating and encouraging traditional national park recreational use seemed an effective means of opposing far more extensive destruction of a park's natural conditions through dams and reservoirs or logging.

Yet Wirth's 1957 wilderness publication failed to pacify the more outspoken critics. Both the brochure and Mission 66 were denounced in early 1958 in a *National Parks Magazine* article by David R. Brower, Sierra Club activist and executive director. Viewing the brochure as a "very effective piece of promotion," Brower argued that it blurred the distinction between easily accessible areas and true wilderness country by stressing a kind of "roadside wilderness," accessible to automobile tourists—and thus compatible with Mission 66 development.[125]

Olaus J. Murie, by then president of the Wilderness Society, agreed with Brower. He wrote to Wirth that although the brochure and other publicity for Mission 66 contained "very high-minded statements," in fact the brochure represented a "certain advertising technique" to promote Mission 66. Murie believed that inspiration for some of the roads being built in the parks arose not from public pressure for new highways, but from plans generated by the "Service itself." He had written to Wirth earlier that criticism of Mission 66 was being expressed by people around the country. Some believed that the bulldozer was the appropriate symbol for Mission 66, and one individual had asserted that the Park Service needed a "Mission 76 to undo the harm done in Mission 66."[126]

Copies of Murie's letters were sent to a number of conservation organizations and leaders, doing the Service, in Wirth's opinion, a "considerable amount of damage." In a six-page response to Murie (also mailed to numerous conservationists), Wirth claimed that Americans were the "most outdoor recreation-minded of any nation in the world"—that they were "as a mass, the world's greatest travelers," and the Park Service should respond to their demands. Moreover, extensive advertising and improved state and federal highways leading to the parks were attracting millions of visitors. The director believed that the "only thing left for [the Park Service] to do is to handle the resulting traffic to the best of our ability." The Service was bringing the parks up to a standard where it could "care for and guide the people who are going to arrive at our gates."[127]

Mission 66 also faced criticism from the Sierra Club, which became particularly agitated over rehabilitation of Yosemite's Tioga Pass Road, running east-west across the park. Leading club members reacted angrily to plans to widen the road where it passed along the shores of Tenaya Lake, one of the scenic gems of the park's high country. The Service planned to blast away even more of the massive gray granite, with its remarkable examples of glacial polish, which before the original road was built had swept down to the lake's edge. At an on-site meeting with David Brower and photographer Ansel Adams, Wirth explained the engineering and economics behind the Park Service's plans, but failed outright to sway either man. Aware that the Sierra Club had long before approved the original routing of the road along the lake, Wirth asked why the current opposition was so strong. Brower responded bluntly that it was now a "different Sierra Club."[128]

Owing partly to Brower's influence and the fight over Echo Park, the Sierra Club was becoming a more aggressive, activist organization, willing to criticize public land managers more openly rather than rely on gentlemanly negotiations, as in the past.[129] This confrontational strategy was reflected in the tone of Ansel Adams' articles protesting the destruction of the glacial polish along Tenaya Lake's shoreline. Writing in *National Parks Magazine* in the fall of 1958, the influential Sierra Club member noted the "slow but irresistible tide" of roads, buildings, and other development that had changed Yosemite. There he believed the "urgencies of bureaucratic functions have blinded those who should see most clearly. The illusion of service-through-development has triumphed over the reality of protection-through-humility." Adams argued further that there were no true guidelines for managing national parks—there was "no *adequate* definition of what is proper in a national park entered in the laws of the land, comprehended by all, and enforced with determination." In a simultaneous article in the *Sierra Club Bulletin,* he angrily denounced the "bulldozers of bureaucracy" and urged that a "vital restatement" of the 1916 National Park Service Act be undertaken to establish definitive guidelines for park management.[130]

Conrad Wirth remarked in his autobiography that in instances such as the Tenaya Lake dispute, conservation organizations had been "looking for a fight" and needed to have a "good cause for raising money." However, the depths of Adams' feelings about the Park Service and its Mission 66 development by the late 1950s were apparent in his personal correspondence as well, as when he told Sierra Club colleagues that the Service must be "thoroughly deflated and thoroughly re-organized. Heads must roll. . . . Everyone is so hypnotized by the MISSION 66 propaganda that the lurking tragic dangers are not apparent." He believed Tenaya Lake to be "infinitely more important than the Park Service!" He wanted a "strong Park Service," but not one that was "both Strong and Bad."[131]

As illustrated at Tioga Pass, Echo Park, and Everglades, Mission 66 brought about improvement of the national park road system. The twenty-seven hundred miles of new or improved roads resulting from the ten-year program included paving, widening, and straightening of many narrow dirt fire roads built in the 1920s and 1930s. Although intended primarily as "motor nature trails," the improved roads in many instances made access to park backcountry easier for the increasing numbers of hikers, at the very time when wilderness advocates sought greater protection for backcountry. With virtually no sociological research on visitors' use of the parks, Mission 66 did not anticipate how that use would begin to change by the 1960s. The unexpected impact of greater access to park backcountry provided additional ammunition for critics of Mission 66.[132]

Rather than just looking for a fight, the Sierra Club and other organizations had become deeply troubled over the Service's developmental tendencies under Wirth. Their opposition to improving roads through or near national park backcountry would continue throughout Mission 66, for instance with the efforts in the 1960s to prevent excessive modernization of Mt. McKinley's main road. There the extent of improvement was ultimately decreased from what had been proposed.[133] Overall, since conservationists viewed many aspects of Wirth's program as poorly planned development, they had little faith in his argument that Mission 66 advanced park preservation.

Changes in Wilderness and Recreation Programs

Although it promoted Mission 66 as a wilderness preservation program, the Park Service refused to give genuine support to the wilderness bill, intended to set aside vast tracts of the public domain to remain largely unaltered by human activity. The bill had been under consideration by Congress since it was introduced at about the time the Echo Park dam was defeated.[134] It lacked close ties to, or dependence on, corporate recreational tourism, which had been a strong influence in national park management from Yellowstone's earliest times, providing constant pressure to develop the parks. In the quest to leave certain public lands essentially unimpaired, the wilderness bill represented the antithesis of developmental programs such as Mission 66—and it got a cool reception from Park Service leadership.

Earlier, in 1949, Director Newton Drury had encouraged wilderness advocates Howard Zahniser and David Brower to draft a bill that would, as Brower remembered Drury's words, set aside wilderness "inviolate by congressional mandate rather than by administrative decision." But by the late 1950s, in the throes of Mission 66, Park Service leadership had changed its position. To many, the wilderness bill seemed "redundant," as Lon Garrison recalled. National park wilderness areas were, he believed, adequately protected under the Service's 1916 Organic Act—they were wilderness "by original legislative intent." Claiming that the Service was already managing its backcountry "according to wilderness precepts," he stated that "most of us thought that we did not need new specialized legislation." Yet Garrison recognized that the conservationists "did not trust the strength" of the Service's administrative designation of wilderness backcountry areas.[135]

In his criticism of Wirth's national park wilderness brochure, David Brower stated that the Service might have actually intended to "demonstrate that the wilderness bill was superfluous." Brower believed that the brochure's effort to confuse real wilderness with roadside wilderness helped create a lack of clarity which suggested that additional legislative protection of truly wild areas was unnecessary. He noted also that in March 1957 the Service had urged the Advisory Board on National Parks to oppose the bill, and had spoken out against it during congressional hearings in June of that year. In a conciliatory comment, Brower added that the Park

Service "matches with devotion the grandeur of the primeval lands it guards. . . . These men are our friends and we theirs." Yet, he urged, the Service must turn toward true wilderness preservation.[136]

Not at all placated by this gesture of cordiality, Wirth deeply resented Brower's comments and rebuked the National Parks Association for publishing them. In a February 1958 letter to Bruce M. Kilgore, editor of the association's magazine, Wirth stated that he could not "imagine a more unfortunate outburst coming at a more unfortunate time than this one." He described Brower as a "bitter and impatient man" who saw the brochure as "underhand propaganda" in the wilderness campaign.[137] A year later, Wirth wrote to his top staff that continued criticism had made it "increasingly apparent that a greater effort must be made . . . to present the Mission 66 program to the public in its true light." Among other endeavors, he wanted the Service to "strive for public understanding" of the idea that national park development in fact comprised "zones of civilization in a wilderness setting," and that park roads were "corridors through the wilderness linking these zones."[138] These comments reflected earlier remarks the director had made to the Fifth Biennial Wilderness Conference on Wild Lands in Our Civilization, when he described the new Mission 66 road into Mt. McKinley's remote Wonder Lake as "a wilderness road, to bring people into the wilderness, as John Muir advocated."[139]

Wirth firmly believed in the compatibility of wilderness and development. And, as part of the effort to prevent ever-increasing crowds from overwhelming the parks, the Service emphasized park zoning, with master plans demarcating backcountry from areas planned for intensive use and for road corridors. Such "controlled pattern developments" encouraged visitors to stay within specifically designated areas.[140]

In actuality, the planning and zoning process determined backcountry (or wilderness) largely by default: rather than such areas being selected for protection because of special significance, they were the areas left undeveloped by park planners. Forester Lawrence Cook observed in 1961 that the Service considered that "much of the area removed from mechanized transportation" could be "classed as wilderness." But Cook also acknowledged that the Service had not given "much serious consideration" to the effect of development on the "undeveloped remainder" of the parks—a concern that Lowell Sumner had raised about road construction in the 1930s and one that effectively cast some doubt on the 1957 wilderness brochure's extolling of wilderness that was easily accessible from roads.

Moreover, Cook stated that one of the "important problems" was to determine "how far ahead we should project our thinking as to zoning. The

Master Plans do not now limit this except to 'the foreseeable future.' The [Service] should come up with some long-range answers."[141] Not unlike the vulnerability of Andrews Bald research reserve in Great Smoky Mountains, the zoning of park wilderness areas through the master planning process was subject to administrative change any time beyond "the foreseeable future"—perhaps one reason why the conservationists, in Lon Garrison's opinion, "did not trust the strength" of the Park Service's administrative designation of wilderness areas and sought legislation to create permanent wilderness.

In his autobiography Wirth asserted that the Park Service supported the wilderness bill to the extent that "the basic standards already established for [the Service] by Congress would prevail in the national parks"—that is, if the bill would not override the bureau's original congressional mandate and its traditional implementation of that mandate.[142] In contrast, though, Brower's assertion that the Park Service opposed the wilderness bill was in accord with Lon Garrison's remark on the lack of trust in the Service. In its comments on the bill, the Park Service had even claimed that such legislation could weaken protection of wilderness in the parks by reducing national park lands to a "low common denominator," putting them on a par with, for instance, lands managed by the Forest Service.[140]

Howard Stagner, an early member of the Mission 66 Committee (and the true author of the wilderness brochure), recalled that the Park Service was "very cold" toward the wilderness legislation. To the Service it was a kind of "turf situation"—a desire to maintain full control of the national parks' backcountry without additional, burdensome regulations. Stagner also remembered, however, that by 1964, when Congress passed the Wilderness Act, the Service had become "somewhat neutral."[144] Although many of the Park Service's rank and file enthusiastically supported the wilderness bill, the bureau's leadership seems to have drifted from outright opposition to reluctant neutrality.

Of all federal bureaus, the Park Service operated under a mandate that was by far the most closely allied with the goals expressed in the Wilderness Act. Logically, then, the Service might have been expected to seize this opportunity to advance the principle of preserving huge tracts of public lands in a wilderness, or unimpaired, condition, whether or not in national parks. But Stagner rightly identified a key problem: that the Service wanted no interference in its management of backcountry. The Park Service chose to be territorial rather than commit to the principle of greater wilderness preservation. In truth, its deepest commitment was to another principle: to ensure public enjoyment of the parks.

President Lyndon B. Johnson signed the wilderness legislation into law on September 3, 1964, nine months after Wirth left office. Despite claims that Mission 66 was a wilderness preservation effort, the final wording of the act implied misgivings about the Service's treatment of the parks. The Wilderness Act's statement of purpose—"to assure that [Americans do not] occupy and modify all areas within the United States . . . leaving *no lands* designated for preservation and protection in their natural condition" (emphasis added)—suggested that protection was necessary beyond that which the Service was giving the national parks. It suggested the distrust that Lon Garrison had identified and that Wirth had acknowledged when he stated in 1958 that some people felt the "Service is the enemy" and "cannot be trusted to preserve the parks."[145] Borrowing somewhat from the 1916 Organic Act's mandate that the parks were to be left "unimpaired for the enjoyment of future generations," the Wilderness Act was intended to prohibit the very kinds of alterations of natural conditions then being wrought by Mission 66. Wilderness areas were to be managed "for the use and enjoyment of the American people in such manner as will leave them *unimpaired for future use as wilderness*" (emphasis added).[146] A key word in both acts, "unimpaired" was much more narrowly defined in the Wilderness Act, which tied the concept specifically to wilderness conditions.

With Mission 66 under attack by conservationists and the Park Service reluctant to support wilderness legislation, the Service also found its nationwide recreation assistance programs threatened when, in 1958, Congress and President Eisenhower established the Outdoor Recreation Resources Review Commission. Focusing on programs the Service had been associated with for a quarter of a century, the commission was mandated to study all aspects of public recreation, including federal, state, and local programs, and areas such as lakeshores, seashores, urban parks, and wilderness. Wanting the lead role, Wirth sought to have the Park Service conduct the study, and when that failed he encouraged Horace Albright, a member of the commission, to seek the chairmanship. Albright declined, however, because of his advancing age, and longtime national parks supporter Laurance S. Rockefeller was named chairman. Even with Rockefeller in charge, Park Service involvement was minimal and Wirth felt shut out. The Service may have been preempted by its perennial rival, the U.S. Forest Service, and opposed by conservation organizations such as the Sierra Club, Wilderness Society, and National Parks Association. The association asserted that the Park Service in its earlier surveys had deemphasized wilderness in favor of recreation—a factor that surely prejudiced conservation organizations against Service participation.[147]

Especially during its early years, Mission 66 involved substantial planning for expansion of national recreational opportunities, and Wirth credited these efforts with inspiring the new recreation study. Building on work begun even before the advent of Mission 66, Park Service teams completed by the early 1960s a number of surveys of areas suitable for public recreational use, including sites along the Atlantic, Pacific, and Gulf coasts, as well as the Great Lakes and the Ozark rivers.[148] Certainly Wirth's personal efforts, beginning with his promotion of recreational surveys in the 1930s and 1940s and continuing with Mission 66, constituted an impressive contribution to the development of public recreation areas and provided groundwork for the commission's study.

Nevertheless, Wirth was invited to attend only one of the study meetings, and then only after he had complained about the anticipated proposal to establish a new bureau to take over recreation programs. As he recalled in his autobiography (in words that reveal his pique at being excluded), Wirth considered a new bureau unnecessary—it was "our responsibility" (the Service's) to run such studies.[149] But the commission's final report called for a sweeping program to address the nation's recreation needs, including, as Wirth had feared, a new bureau completely separate from the National Park Service to oversee this activity. Surely to Wirth's deep dismay, Horace Albright supported creation of the new bureau, believing, as he later stated, that the increasing recreational responsibilities would "impose too great a burden" on the Service. With the formal establishment of the Bureau of Outdoor Recreation in April 1962, the Park Service's role in national outdoor recreation programs, long cherished by Wirth, was drastically reduced. It was now limited mainly to managing the national park system itself.[150]

The Public Hunting Crisis and a New Look at National Parks

As the Park Service faced the loss of its recreation programs, it was drawn into a heated debate over whether or not to allow another type of recreational activity in the parks—sporthunting. The most publicized controversy yet to arise over park wildlife, the debate would precipitate major reassessments not just of wildlife policy, but of the Service's overall natural resource management policies.

Elk research conducted in the 1950s by biologist Walter Kittams had indicated that even though population control had been under way since the mid-1930s, elk still overgrazed Yellowstone's northern range. Relying on Kittams' recommendations, the Service planned to make a special effort to reduce the herd's population from ten thousand to five thousand head.[151] This planned reduction, unprecedented in size and, as before, to be conducted by park rangers, prompted demands from sportsmen's organizations and from state game and fish commissions to allow hunters to participate in the kill. Their demands soon expanded to permit big-game hunting in all national parks.

Reduction programs of lesser magnitude than Yellowstone's were being carried out in a number of parks, for instance in Zion, Rocky Mountain, Sequoia, Yosemite, Grand Canyon, Acadia, and Grand Teton. In all but Grand Teton, the reductions were the responsibility of park rangers, sometimes assisted by state game and fish personnel. Yet some precedent for public hunting in national parks had been established with the 1950 act adding Jackson Hole National Monument to Grand Teton National Park.

Located just south of Yellowstone, Jackson Hole lies along the migration route of Yellowstone's southern elk herd. Early in the century, biologists came to believe that human settlement had interfered with the migration and with the animals' winter range. To prevent mass starvation of elk unable to reach their winter range, Congress in 1912 appropriated funds to purchase an area (the National Elk Refuge) near the town of Jackson, and the Biological Survey initiated a winter feeding program for the herd. The program continues today. But in 1950, when Grand Teton National Park was expanded to include Jackson Hole, Congress provided, as a concession in the bitter struggle over park expansion, that "qualified and experienced hunters licensed by the State of Wyoming" could be allowed to hunt in the park when the Service and the state determined it necessary for "proper management and protection" of the elk herd. This condition Director Newton Drury reluctantly accepted to secure the addition of Jackson Hole to the national park.[152]

In Grand Teton, elk reduction was to include sporthunting, with hunters to be "deputized as park rangers"—a means of avoiding the appearance of ordinary recreational hunting in the parks. But, as Park Service biologist Robert H. Bendt reported in the early 1960s, public participation in the park's elk reduction effort did not turn out to be "as great as anticipated." Throughout the decade following initiation of Grand Teton's reduction program in 1951, only about fifty percent of the approved hunting permits were used; and an average of only twenty-seven percent of the hunters managed to kill an elk—an overall success rate of about one in eight.[153] Despite such low success in Grand Teton's hunt, when Yellowstone announced plans to increase reduction of its northern herd in 1962, sportsmen's organizations and state game and fish commissioners used this opportunity to seek the opening of all national parks for big-game hunting.

There had always been interest in sporthunting in national parks. But in the early 1960s, pressure from hunting advocates became, in the words of Grand Teton superintendent Harthon L. ("Spud") Bill, "stronger and more disturbing" than before. Still, in his January 1961 recommendation to his regional director on the hunting issue Bill equivocated. He stated that "basically" he did not favor sporthunting in national parks and added that the Service should consider an appeal to the public before "capitulation to the proponents of hunting." Yet he also asserted that the Service had been "maneuvered into a difficult position by the rigid no hunting provision where we have had situations which might have been alleviated by public hunting." Bill advised that the Park Service must be flexible. What worked in one park might not work in another, and the Service must "be in a position to determine when public hunting is necessary."[154]

The Grand Teton superintendent's position was reflected in Wirth's important policy analysis of the issue in February 1961. In a letter to Anthony Wayne Smith, the National Parks Association's executive secretary, Wirth wrote that the Service's ongoing elk reductions were not in the category of mere recreational hunting, but were the result of a "forced situation": with damage from overgrazing, reduction was the only means available for adhering to the Service's mandate to leave the parks unimpaired. He noted that the 1916 act establishing the Park Service allowed "destruction" of animal and plant life that was "detrimental to the use of the parks" and, further, that Fauna No. 1 had recommended keeping native animal populations in line with range carrying capacity. Moreover, Wirth feared that the great public interest in sporthunting would prevent any more national parks from being created unless hunting was allowed. He noted that biologist Lowell Sumner had already recommended that a portion of the proposed Great Basin National Park, in Nevada, be designated a deer management area, with the public to participate in a reduction program overseen by the secretary of the interior—a plan similar to that for Grand Teton.[155]

Wirth then focused on perhaps the heart of the issue, recalling that since the mid-1930s rangers had been carrying out reduction programs, a "disagreeable and time consuming" task for which there was seldom enough manpower or funding. Now, after a thorough review, he believed that public hunting could be conducted in a "controlled and limited" manner to assist the rangers. "On the basis of practical results," public hunting was likely "the most effective [method] to follow in some cases." He emphasized that it was merely a "method" for reducing surplus populations, not a policy that condoned public sporthunting in the national parks. Concluding his analysis, Wirth stated that the Service "must be progressive and we must be realistic" in park management and use "modern knowledge and techniques to further the basic aims of the Service." He asked, "Why, then, should we not permit [public participation in reductions]?"[156]

Publishing Wirth's letter along with Smith's reply in its monthly magazine, the National Parks Association strongly objected to the director's position and threatened to sue the Service to prevent public participation in the reductions. The association conceded that reductions were "in all probability" necessary, yet believed they should be carried out mainly by the "paid staff of the Service." In a slam at Mission 66, the association cited Wirth's expressed concern for funding and manpower in wildlife management and habitat protection, then pointed out that construction funding was far more than that allocated to "management, protection, interpretation, and research" combined—an "imbalance" that needed readjusting. Furthermore, allowing hunting in new parks such as the one proposed for Great Basin could result in a dangerous relaxation of restrictions against public hunting in the older, established national parks.[157]

The association's views were shared by other conservation groups, such as the Wilderness Society, which agreed that reduction was "necessary" yet opposed all forms of sporthunting in the parks. Howard Zahniser, the society's executive secretary, stated this opinion to Olaus Murie a month after Wirth had made his statement. Later that year, the Audubon Society concurred with Zahniser's position.[158]

Opposition to Wirth's policy statement also came from within the Park Service—from none other than Yellowstone superintendent Lon Garrison, probably the Service's most influential park manager. In a memorandum of March 24, 1961, with the notation "NOT FOR PUBLIC RELEASE!" across the top of the first page, Garrison wrote that with the director's public hunting statement, one of the "keystones of the National Park Service suddenly crumbles and the cause of pure park conservation . . . loses much of its vitality." Garrison argued that public hunting would damage the park, with hunters illegally killing other big-game species and generally wreaking havoc in the vicinity of hunting camps. Also, Yellowstone would in essence be in competition with the Grand Teton hunt, where so far hunter participation was less than expected. With the low success rate of those that did take part, Garrison estimated that to reduce Yellowstone's northern herd by five thousand head, twenty thousand hunters would have to participate. In direct contradiction to Wirth's public hunting proposal, he stated that for Yellowstone "*this is not the answer here.*"[159]

Raising a fundamental point, Garrison wrote to his regional director, Howard W. Baker, questioning if in all this controversy the Service had sufficient information to justify the targeted five thousand population figure. He then answered his own question: "Of course not." Garrison judged that the Service might even have "misled the public" into believing that the park had detailed knowledge of the northern herd, when in fact much more information was needed. The ecological relationships among bison, elk, pronghorn, bighorn, deer, and beaver were, he stated, "subtle and not very well known," and Yellowstone's wildlife should not be managed "on the basis of hypothesis and sheer guesswork."

Garrison called for a "long-term study, perhaps five years," into the "ecology of Northern Yellowstone big game species." He summarized the seriousness of the situation by arguing that if the Service did a "good job based on professional research" there would be "no valid criticism." But the stakes were high. In Garrison's opinion, "a mediocre job based on uncertain knowledge spells failure and will provoke a continuing storm of criticism that will jeopardize far more than elk management at Yellowstone. We simply cannot risk failure." The Yellowstone superintendent's recommendations were forwarded to Wirth by Regional Director Baker, with a cover memorandum opposing public hunting in the national parks.[160]

On September 14, probably after extensive deliberations, Wirth issued a statement on "Wildlife Conservation and Management," a major policy reversal that brought him in line with Garrison's recommendations. He declared that public hunting was "neither the appropriate nor the practical way" to carry out the Service's wildlife management objectives—it was "irreconcilable" with national park purposes. Given the tremendously complex ecology of the parks, "competent and adequate ecological research" was necessary. In the "long view," Wirth declared that "management of the natural environments must be based on complete and exact knowledge of all factors involved, and be guided by a program of continuous appraisal of wildlife and other natural conditions."[161] Despite the strength of Wirth's pronouncement, it would prove altogether as rhetorical as the initial Mission 66 commitment that national park management would not be built on guesswork but on scientific knowledge.

The director apparently issued this statement without clearance from Secretary Udall's office, which sparked friction between the two officials. Wirth argued that his statement merely reflected a long-standing national park policy. But hunters' associations and state conservation officials in the West reacted angrily to his new position. Wirth's reversal back to the no-public-hunting policy created a "crisis in public relations," as Udall's assistant secretary, John A. Carver, later recalled.[162]

Park Service rangers carried out Yellowstone's massive elk reduction in the winter, killing more than forty-five hundred of the northern herd. Smaller numbers of elk were shot in Rocky Mountain and Glacier, and deer in parks such as Acadia, Sequoia, and Grand Canyon. In 1962 opponents of the policy introduced a bill in the U.S. Senate requiring the Park Service to consult with state officials on the need for reductions, and authorizing the secretary of the interior to use hunters in reduction programs. Attacked by conservation organizations, the bill failed to pass.[163]

With the Service beset by critics, in April 1962 Secretary Udall called for thorough studies to be conducted on its science and resource management. The studies would address concerns expressed long ago in Fauna No. 1 and by wildlife biologists such as Lowell Sumner and Adolph Murie—concerns now echoed by conservation organizations and by high-level Park Service managers, including Daniel Beard and Howard Stagner. In one request Udall asked the National Academy of Sciences to undertake a review of the "natural history research needs and opportunities" in the national parks. In another he called for a "blue ribbon" committee of highly respected wildlife specialists to study the Service's wildlife management policies and practices. The National Academy selected William J. Robbins, a prominent biologist with the National Science Foundation, to chair its study. Secretary Udall personally persuaded A. Starker Leopold, professor of biology at the University of California at Berkeley and son of the late ecologist Aldo Leopold, to head the wildlife management review.[164]

In response to the "crisis in public relations," and coming nearly half a century after establishment of the National Park Service, prestigious committees from outside the Service were to undertake in-depth reviews of research and wildlife management policy. Never before had this happened. Originating within the Service, Fauna No. 1 had lacked the clout that could be derived from reviews by prominent scientists and wildlife specialists brought together by a secretary of the interior. Imposed by Udall's office, these reviews by influential outside experts were awaited by a large and increasingly vocal conservation community.

Looking forward to the reports, Lowell Sumner wrote to his friend and professional colleague Starker Leopold in May 1962 expressing belief that the upcoming studies were probably the "biggest and most hopeful development since George Wright's death cut short the evolution of the original Wildlife Division" in the mid-1930s.[165] And indeed the wildlife management study (referred to as the Leopold Report) and the National Academy's research review, both presented in 1963, would call for a potent infusion of science into national park management. They would constitute an important restatement of the Service's basic goals in managing natural areas of the national park system.

At a 1968 meeting of Park Service scientists, Lowell Sumner took a long look back and asked the question, "Why, among NPS activities, did biology alone fail to recover" after the end of World War II? One of the Service's most experienced scientists, Sumner believed that "the heart of the matter" had been the Park Service's "reluctance to acknowledge the ecological importance of the parks."[166] Indeed, the period during and after World War II was marked by two phases of national park management, both of which witnessed steady resistance to meaningful improvements in the Service's scientific capabilities and its knowledge of the parks' natural resources.

Newton Drury had overseen a period of minimal growth and development during the war and early postwar years. His cautious outlook and preservationist leanings meant that he was timid in advocating park development for tourism and opposed to extensive involvement in reservoir recreation. Commenting on Drury's conservatism, former chief biologist Victor Cahalane recalled that the director was a "state's righter" who believed the federal government should have a very limited role in managing park lands. Drury differed markedly from Mather, Albright, Cammerer, and Wirth, whose aggrandizing ways contributed much to the expansion and development of the national park system.[167] In resource management issues Drury often supported the wildlife biologists, who pressed for decisions based more on ecological considerations than on the desire to satisfy park visitors. Yet he made no determined effort to enhance the biologists' authority in the Service.

Under Conrad Wirth, the next phase of national park management featured Mission 66—the kind of long-range, expansive program Drury never earnestly pursued. National park facilities had badly deteriorated and the parks were strained to the limit by public use that had more than quadrupled between the end of World War II and the mid-1950s. Even the vehement Park Service critic Ansel Adams once admitted that Mission 66 was an "excellent program of providing 'necessities' in terms of expected travel increases." From a "bureaucratic viewpoint," he added, it was "one of the better undertakings of recent years," its purpose "undoubtedly well intended."

Despite widespread criticism of the program, Wirth came to believe by 1961 that the Park Service had not "planned big enough." He later wrote that "instead of having the urgency behind us, we were facing a new dimension—an action program was required that would dwarf the first five years of Mission 66." In fact, Mission 66 funding during its last half-decade amounted to considerably more than during the first five years, the grand total reaching just over one billion dollars by the end of the program.[168]

Many current and retired Park Service employees view Mission 66 as a kind of Golden Age of the national parks—an exciting time of growth, expansion, and development of the system. Mission 66 was the culmination of the vision of Stephen Mather and Horace Albright, who had sought to develop the parks and make them accessible for the benefit and enjoyment of the people. The program was a high point of what might be termed the "landscape architecture approach" to national park management, when, under landscape architect Wirth, development of the parks for recreational tourism dominated national park affairs and went largely unfettered by natural resource concerns. With huge sums of money, Congress backed Wirth's policies—in effect confirming the Service's long-held belief that the basic purpose of the national parks was public enjoyment, rather than scientifically based preservation of natural resources. Appropriation of a billion dollars for park development demonstrated that Mission 66 was what Congress and the people wanted for the parks.

Yet this Golden Age brought changes to National Park Service programs that did not please Wirth and his associates. After building its reputation and leadership in park management and recreational tourism, the Service witnessed its control of national recreation planning given over to the newly created Bureau of Outdoor Recreation, and its administrative discretion over the parks' backcountry threatened by the wilderness bill. Moreover, the Park Service was at odds with the rising tide of conservation. Wirth's comment in early 1958 that some of the conservationists believed that the "Service is the enemy" and "cannot be trusted to preserve the parks" reflected his apprehension that the bureau was losing ground with that important and vocal part of its constituency.[169]

At first, the criticism regarding the deteriorated condition of park facilities that helped bring about Mission 66 was aimed largely at a parsimonious Congress, rather than at the Park Service. Once Mission 66 began, however, the Service itself came under intense criticism from conservationists who argued that the program's construction and development were too extensive, too modern, and too intrusive. Finally, by the time Mission 66 passed midcourse, critics increasingly aimed at another target: Park Service failure to build a science program and to consider the ecological impact of park development. The focus had shifted from, for instance, Bernard DeVoto's early 1950s article identifying a crisis in terms of deteriorating national park facilities; to widespread concerns about modern, inappropriate development under Mission 66; then to "Get the Facts, and Put Them to Work," defining crucial park needs in ecological and scientific terms.

Intended to commemorate the Service's fiftieth anniversary, Mission 66 marked a major transition in national park history. The era that brought to culmination the Mather and Albright vision of developing parks for public use and enjoyment would also witness the resurgence of George Wright's vision to protect, or even restore, the integrity of the parks' natural resources—a vision shared by those wildlife biologists who continued after Wright. Once the studies requested by Secretary Udall from the Leopold Committee and the National Academy of Sciences were released in 1963, the Park Service truly would enter a new era, in which park management would be judged far more on ecological criteria. Yet this era began at the height of national park development under Mission 66 and would confront a half-century of Park Service tradition emphasizing recreational tourism.

CHAPTER 6

Science and the Struggle for Bureaucratic Power: The Leopold Era, 1963–1981

[In the National Park Service] there were clear parallels with the struggle for survival in the natural world. . . . The struggle was not as violent and predatory as in the animal world. It was . . . more like the competition among plants. Certain branches and divisions had a favorable place in the sun. They overshadowed the others and got the major share of the funds, as well as major representation on policymaking and planning committees.—LOWELL SUMNER, 1968

Yellowstone was not created to preserve an "ecosystem."—HORACE M. ALBRIGHT TO GEORGE B. HARTZOG, JR., July 1972

Much of the history of the National Park Service from the George Wright era on involved a conflict not between "good" intentions and "bad" intentions, but between two idealistic factions—each well-meaning but committed to different perceptions of the basic purpose of the national parks. One group, by far the stronger and exemplified by Conrad Wirth's career, emphasized recreational tourism and public enjoyment of majestic landscapes, along with preservation of a semblance of wild America. Wirth's understanding of the mandate to leave the parks "unimpaired" was tied to preservation of park scenery. The other group, represented mainly by the wildlife biologists, whose influence had diminished substantially since the 1930s, focused on preserving ecological integrity in the parks, while permitting development for public use in carefully selected areas. In effect, this group defined "unimpaired" in biological and ecological terms—a concept more compatible with that expressed in the 1964 Wilderness Act.

The conflict between these two factions intensified during the environmental era, when park science and ecology received a strong boost from outside the Service, forcing the bureau's tradition-bound leadership to reconsider its policies and make organizational adjustments. Reflecting ecological concerns, the 1963 reports of the Leopold Committee and the National Academy of Sciences on biological management and science in the parks appeared as Mission 66 was approaching its vigorous conclusion, the apex of a half-century of recreational tourism management. During the 1960s, programs tied to tourism in the national parks would proliferate. As had happened in the 1930s, the ongoing development and the new endeavors that the bureau launched monopolized the attention of the Park Service, in part because they represented a continuation of traditional interests with which the Service felt competent and at ease. Within this context of exceptional recreational tourism activity, the reemerging science programs would seek to thrive.

Mission 66 and Parkscape U.S.A.

Conrad Wirth's Mission 66 had revitalized the National Park Service, lifting it out of the doldrums of the postwar years. Much as the New Deal had done, the program poured large amounts of money into the parks, to bring facilities up to standards the Service deemed appropriate. Horace Albright believed Mission 66 to be one of the "noblest conceptions in the whole national park history," ranking in importance "with the creation of the National Park Service itself."[1] Indeed, the emphasis of Mission 66 on park development and use made it more evident than ever that the large parks in the system were subject to a kind of recreational "multiple use." Taken as a group, they accommodated a range of uses, such as downhill skiing, motorboating, sportfishing, hiking, horseback riding, and hunting (in recreation areas and in Grand Teton National Park)—all facilitated by large-scale camping and lodging accommodations.

The extensive surveys for future parks, begun as early as the 1930s and continued under Mission 66, began to pay off in the 1960s, with the growing national interest in setting aside recreational lands. Congress approved a remarkable array of additions to the national park system during the 1960s and 1970s. New parks were created at a more rapid pace than ever before, with many of the areas providing opportunities for intensive recreational uses. From the shoreline surveys alone, twelve parks came into the system between 1961 and 1972, including Cape Cod, Padre Island, and Point Reyes national seashores, along with Pictured Rocks, Indiana Dunes, and Apostle Islands national lakeshores. Together these new parks contained more than 700,000 acres, with 718 miles of shoreline as initially established.[2]

Wirth's expansionist zeal was rivaled by that of his successor, George B.

Hartzog, Jr., who became Park Service director early in 1964 and used many of the surveys conducted under Wirth to bring about the creation of new parks. A politically astute lawyer and Park Service veteran, Hartzog adroitly capitalized on the momentum of President Lyndon Johnson's Great Society to expand the national park system. Support for his expansion efforts continued through the first administration of President Richard M. Nixon. Under Hartzog ten new parks were created in 1964 alone. Other notable years included 1965, with fourteen new parks; 1966 and 1968, with ten each; and 1972, with thirteen. Overall, between 1961 and 1972 (the year Hartzog's directorship ended), a total of eighty-seven units came into the system, constituting nearly 3.7 million acres. Besides the seashore, lakeshore, and recreation areas, numerous small historical parks were established, plus larger natural units like Voyageurs, Guadalupe Mountains, and North Cascades national parks.[3]

Many of these parks were brought in under a new Service agenda: "Parkscape U.S.A." In the mid-1960s, seeking to maintain the momentum created by Mission 66, the Service devised this successor program, which had as its principal focus the continued expansion of the system, rather than construction of roads and facilities, as with Mission 66. In Director Hartzog's words, Parkscape U.S.A. would "complete for our generation a National Park System by 1972," the centennial year of Yellowstone. The tremendous surge in outdoor recreation during this era placed added pressure on national park areas and increased the urgency to create new parks. In 1966, at the close of Mission 66, total annual visits to the park system had reached 133.1 million, up from 61.6 million when the program began in 1956. By 1972, annual visits climbed to 211.6 million.[4] These figures stemmed in part from the increased number of parks over the years, but clearly the Service's responsibilities and workload were growing and the parks were under a greater burden than before.[5]

Among the new parks, the national recreation areas in particular added to the Service's involvement in recreational tourism. Director Hartzog stated in an article in the July 1966 *National Geographic* (a special issue celebrating the Park Service's fiftieth anniversary and the accomplishments of Mission 66) that the national recreation areas "have been so popular that [the Service knows now] that we do not have enough of them." Hartzog suggested that it was in this category that the "greatest expansion of the National Park System" would take place. In the same article, the director extolled the virtues of Lake Powell, the principal feature of recently created Glen Canyon National Recreation Area. Hartzog viewed this new recreation area as representative of the "spirit of Parkscape U.S.A." His statement identifying the Park Service's significant new program with a large water impoundment came less than two years after passage of the Wilderness Act, which reflected a diametrically opposed philosophy of land management.[6]

During Hartzog's tenure and the Parkscape era, eight reservoirs were added to the system as national recreation areas, among them Bighorn Canyon, Lake Chelan, and Curecanti. Each of these new units marked a continuation of the national recreation area concept initiated in the 1930s with Lake Mead, and each reflected the strength of the recreational tourism urge within the Park Service. Yet Lake Powell (acclaimed in the *National Geographic* article for its sparkling waters and its swimming, waterskiing, and motorboating potential) flooded deep, strikingly scenic sandstone canyons of southern Utah. Water impoundment began in 1963, damaging the riverine ecology downstream in Grand Canyon. The dam that helped create a national recreation area began to degrade natural conditions in a national park.[7]

As the environmental debates of the 1960s and 1970s intensified, the National Park Service was substantially compromising itself as an advocate for nature preservation. Indeed, reclamation interests, which were allied with the Service in national recreation area management, proposed about a half-dozen additional water-control projects in northern Arizona that would create a string of reservoirs all or partly within Grand Canyon National Park itself. Such proposals threatened to alter drastically the very heart of this spectacular park, one of the giants of the system. Conservationists eventually succeeded in blocking the proposals, preventing further degradation of Grand Canyon.[8]

Two of the new national recreation areas created during the Parkscape era—Gateway and Golden Gate—were not associated with big western reservoirs; rather, they were justified as providing the crowded New York City and San Francisco metropolitan areas with significant recreational opportunities. Yet they also brought the especially difficult challenges of urban conditions, not unlike those the Service already faced in managing the numerous parks and monuments in Washington, D.C. With the exception of Washington, the Park Service's chief involvement with cities had been management of small historic sites, such as Federal Hall on lower Manhattan Island and Independence Hall in Philadelphia. Having little experience with large recreational open space in or near major urban areas, the Service had to devote considerable attention to developing the skills and staffing necessary to administer Gateway and Golden Gate. Other national recreation areas near large urban areas soon followed: Cuyahoga

Valley (near Cleveland), Chattahoochee River (near Atlanta), and Santa Monica Mountains (near Los Angeles).[9] The demands of these heavily used parks could not help but heighten the bureau's emphasis on planning, developing, and managing for intensive public recreational use.

Reservoir and urban park management drew Park Service attention to law-enforcement issues much more than before, as crowded public-use zones became scenes of increasing crime and accidents. Even the traditional natural parks like Grand Canyon, Yellowstone, and Yosemite began to experience urban kinds of law-enforcement problems, owing largely to crowded conditions. A 1970 Interior Department report on law enforcement compared Grand Canyon Village (an extensive development on the canyon's south rim) to a small city, with an average overnight population of 6,000 people, plus a daily transient population of 12,000. Similarly, figures for Yellowstone's Old Faithful Village were 5,000 overnight plus 10,000 transient. Yosemite Valley topped them all with 15,000 overnight and 18,000 transient, for a daily total of 33,000. Another internal report revealed that "major offenses" (homicide, rape, assault, robbery, and larceny) had more than doubled in the national park system in just a few years, jumping from about 2,300 incidents in 1966 to about 5,900 in 1970.

In the summer of 1970, a riot in Yosemite and a young boy's death in one of Yellowstone's thermal pools brought greater focus on law-enforcement and safety issues. The widely publicized riot by mostly countercultural youth in Yosemite Valley's Stoneman Meadows on the Fourth of July in 1970 emphasized to Park Service leadership that the bureau's law-enforcement capability needed serious attention.[10] The riot created a crisis atmosphere that made Congress more receptive to increases in law-enforcement funding. Russ Olsen, then assistant superintendent in Yosemite, later observed that Hartzog "parlayed" the American public's concern about law enforcement "into big bucks"; and in March 1971 the director announced the establishment of a law-enforcement office in Washington. He also announced a wider deployment of the U.S. Park Police, a Park Service unit previously engaged in policing parks and other federal properties in the District of Columbia and environs. Hartzog planned to increase the Park Police staff by 40 positions (from 371 to 411), the bulk of the new positions to be assigned to the Service's regional offices and to parks most in need of police authority.

In addition, the director began a "comprehensive" law-enforcement training program, to include 225 entry-level rangers and selected management personnel. He anticipated that by the beginning of the 1971 summer travel season, 50 rangers from throughout the national park system would each have completed 540 hours (17 and a half weeks) of police training. Furthermore, an "intensive" eight-week program was to be conducted for supervisory park rangers from the areas most impacted by crime; and a minimum of 100 rangers hired only for the summer season would receive training.[11]

Exacerbating the situation, law-enforcement emphasis conflicted with the antiestablishment attitudes of the times, as evidenced in Yosemite. As longtime Park Service law-enforcement authority William R. Supernaugh recalled, a critical factor was that park rangers did not understand the youth of this era—their concerns for free expression and their challenge to authority. The rangers were "separated in years and point of view" from the youth of the 1960s and 1970s. Still, the Service's expanded law-enforcement effort would become increasingly important in park management, and part of the customary scene in national parks.

In Washington, Hartzog placed the newly created law-enforcement office with the rangers, a move that reflected their long-established responsibility for such work. With the rangers bureaucratically allied with park superintendents (and solidly within the main feeder group for superintendency positions) the law-enforcement programs, or "visitor protection and safety" programs, as they would become known, were virtually assured of continued strong support from Service leadership.[12]

Following the tragic death of nine-year-old Andrew Hecht, who in June 1970 accidentally fell into one of Yellowstone's boiling-hot thermal pools, safety issues also came front and center. Hecht's parents filed a $1 million tort claim against the Park Service, charging that safety precautions around the thermal pools were inadequate. The Hecht case (and effective pressure applied directly on the Park Service by the Hecht family) brought public criticism on the Service for its overall weak safety program, generating a significant new emphasis on safety in the parks. An increased commitment of funds and staffing included a safety specialist in the Washington office, allied with the ranger operations.[13]

In another program expansion, Hartzog diversified and increased the parks' interpretive activities, particularly focusing on environmental education and "living history" presentations, the latter given at historic areas by Park Service employees dressed in period costumes. The motivations behind these two activities were related. Even living history (especially "living farms") contributed to the Great Society's efforts to improve public understanding of environmental matters. Reflecting Director Hartzog's deep personal commitment to the idealistic values of the Great Society (and in all probability his awareness of potential urban-area political support for the Park Service), the new programs focused on inner-city populations, mainly children, with the hope of enriching their everyday lives and perhaps their appreciation of nature and the nonurbanized world.[14]

Hartzog's "Summer in the Parks" program, providing entertainment and recreational opportunities for city dwellers, began in Washington, D.C., in the late 1960s and became a cornerstone of the Service's urban efforts. The director soon added other programs, involving many parks in the system. Among the new endeavors was the National Environmental Education Development program, which provided materials and curricula for environmental studies, kindergarten through high school, especially in schools near units of the national park system. Associated with this, Hartzog's Environmental Study Area program identified special areas in the parks where student groups could conduct field studies. In January 1970 the Park Service reported that it was operating 67 study areas throughout the system, and that 50,000 children were participating in the education program, with increasing involvement expected soon. Even more popular, the living history programs had spread to 114 parks by the mid-1970s.[15]

To assist with these efforts, Hartzog had an interpretive planning and design center built at Harpers Ferry, West Virginia, which opened in March 1970. He located the new facility adjacent to the Stephen T. Mather Training Center—another new Park Service operation (begun during the Wirth era, but officially opened under Hartzog in 1964). The Mather center emphasized interpretive training for Service employees. Also in the early 1970s, Hartzog created a new unit in the Washington office to provide oversight and policy guidance for urban issues: the Office of National Capital and Urban Affairs, which included the divisions of Urban Park Planning and Urban Park Programs.[16]

In addition, in 1971 Hartzog centralized most park development activities by combining the eastern and western offices of design and construction (which Wirth had enlarged to push through Mission 66) into a single office in Colorado, the Denver Service Center. By the early 1960s, at the midpoint of Mission 66, Park Service design and construction offices had employed more than 400 people, including engineers, planners, architects, landscape architects, graphics specialists, and construction representatives, along with administrative support positions. When the Denver Service Center officially opened in 1972, it employed approximately 350 persons, committed in one way or another to national park planning, design, and construction. In addition, the center stationed specialists in the parks to oversee major projects. Reflecting the continuing emphasis on development, service center employment would increase with preparations for the 1976 Bicentennial, which included a huge design and construction program. The center's staffing would peak in 1978 at about 800 employees.[17]

A high point in the growth of Park Service activities during this era came with the demanding responsibilities in Alaska—the extensive planning for new parks, as mandated by the Alaska Native Claims Settlement Act of 1971. From the first, the Alaskan effort was exceptionally political, involving other Interior Department bureaus and the U.S. Forest Service, plus the state government of Alaska and numerous Alaskan native groups. Tight congressional and departmental deadlines, along with intense surveillance from both environmental and private-enterprise groups, added to the pressure, so that the attention of Park Service leadership was continually drawn to this arena. The ensuing 1978 proclamation by President Jimmy Carter of national monuments totaling about 41 million acres in Alaska (mostly creating new units, but also including additions to some older parks) initiated huge additional land management responsibilities for the Park Service. Carter's action was sanctioned two years later when Congress passed the Alaska National Interest Lands Conservation Act, increasing by approximately 2.6 million acres the Alaska lands placed under Park Service administration.[18]

The 1980 legislation also designated about 32.4 million acres of Alaska park lands as wilderness—more than ten times the total acreage so designated in all other parks. Congress thus bestowed on the Service the responsibility for more wilderness acreage than that of any other land-managing bureau—a factor that helped obscure the Service's reluctance to support passage of the 1964 wilderness legislation and, subsequently, its restrained implementation of the act. The antithesis of development, wilderness designation meant that restrictions would be placed on national park backcountry management, thereby protecting the designated areas not only from excessive use by the public, but also from the managerial and developmental impulses of the Park Service itself.

The Park Service met the act's ten-year deadline to evaluate roadless park lands of five thousand acres or more for their wilderness suitability, and submitted to the secretary recommendations for wilderness in forty-nine parks. Further wilderness reviews continued. Realizing that public use of these areas had to be controlled, the Service initiated formal backcountry planning for wilderness and other undeveloped areas. From the beginning of the review process, however, environmental organizations charged that the Service was opting for smaller wilderness designations than it should. They claimed that, using land classifications recommended by the 1962 report of the Outdoor Recreation Resources Review Commission, the Service had devised a purist definition of wilderness to exclude certain undeveloped lands. It also planned what the environmental groups viewed as unnecessarily wide buffer areas between wilderness zones and developed areas (especially roads), rather than extending the zones close to development.[19]

In a particularly striking case, the Service limited its wilderness proposal in Great Smoky Mountains National Park to conform with its plans to construct a second transmountain highway through the park. Opponents, claiming the Park Service was allied with "crowd recreationists," gathered sufficient support to defeat the road plans. The wilderness proposal was increased in size; however, entangled in the politics of road construction, it was never enacted by Congress. Similarly, the Service's expansive plans for developing Cumberland Island National Seashore into a recreation hub to accommodate more than fourteen hundred visitors a day foundered in the face of strong opposition seeking to protect the Georgia sea island's relatively undeveloped natural setting. Instead, access to the park was calibrated to allow a far more limited number of visitors per day, and more than twenty thousand acres of the park were designated as wilderness or "potential wilderness." Overall, Congress frequently disagreed with the Service's more limited proposals and increased them in size before enacting them into legislation. In part because of the opposition of local congressional members and a changing national political climate, several large parks containing huge tracts of de facto wilderness never gained the added protection of the Wilderness Act, among them Yellowstone, Grand Canyon, and Big Bend, in addition to Great Smoky Mountains.[20]

With the creation of several dozen new parks of different types during the 1960s and early 1970s, the growth of the national park system was so rapid that the Service established a task force to make recommendations on how such growth might be controlled. This was apparently the first formal effort of its kind in the bureau's history. The report, submitted in early 1973, stated that indeed "areas of questionable quality" had been included in the system, and that they had "overreached [the Park Service's] capability to manage a System at the desired level of quality." Without naming them, it recommended that some park areas, mainly "in the recreation category," should "not be administered by the National Park Service," and that there should be "no further urban recreation areas added to the System." Yet, instead of reductions, large numbers of new parks (including recreation areas and the Alaska expansions) added tremendously to the burden. Later studies to reduce the size of the system were conducted sporadically during the 1970s and 1980s. These efforts failed to produce results, in part because additions to the system most often were a result of congressional politics and largesse, and proposals for removal would be zealously resisted.[21]

Although Park Service leadership might have wished to get rid of certain parks it considered unworthy, it never ceased to promote overall growth of the system. Just before the 1973 report appeared, the Service issued a long-range National Park System Plan for natural areas, a document intended to guide expansion. The plan presupposed continued expansion, and (inspired by the rising public interest in environmental issues) stated that the national park system should "protect and exhibit the best examples of our great national landscapes, riverscapes and shores and undersea environments; the processes which formed them; the life communities that grow and dwell therein." The plan sorted the nation's natural history into physiographic and biological regions, representation of which would form the basis for a "completed National Park System." Identifying "gaps" in the system, the plan divided the regions into types of areas. For instance, the Great Basin region, centered in Nevada and Utah, contained areas of "mountain systems," "works of volcanism," "hot water phenomena," and "works of glaciers."[22]

Ironically, at the height of the environmental movement the Service contemplated expanding the national park system on the basis of scientific and ecological characteristics, while only grudgingly accepting ecological science as part of park management. The Park Service may have thought of itself as being ecologically aware, but it remained largely uninformed about its biological resources and oblivious to the ecological consequences of park development and use. Its reluctance, dating from the 1930s, to pursue recommendations of the wildlife biologists for scientifically based preservation of natural resources had no doubt allowed a vast multitude of both anticipated and unforeseen changes to the parks' natural conditions—changes that might have been avoided had the Service understood the parks better.

Before passage of the Wilderness Act in 1964, the Park Service was the only federal bureau with a mandate specifically encouraging preservation of natural conditions on public lands; thus it might have been expected to assume a leadership role in the emerging environmental movement. Instead, entangled in its own history and the momentum of its tourism and park development, the Service had to be awakened to ecological management principles by outside critics. Ecological management inherently required far deeper understanding of natural resources than did scenic preservation and tourism management, a factor that brought new pressure on a traditional Park Service. The Service's vacillating response would stand in marked contrast to its energetic support of law enforcement, safety, interpretation, and other matters related to tourism.

The Leopold and National Academy Reports

Appearing in 1963, the Leopold and National Academy reports were threshold documents. As the first studies of their kind—reviews of Park Service natural resource management conducted by experts from outside the bureau—they had much greater effect than would the numerous reports on park management and science that appeared in subsequent years. Indeed, long after its appearance the Leopold Report would be particularly well remembered. Not only did it receive widespread publicity, with reprints in several national publications, but also, as noted in the *Sierra Club Bulletin*, it enunciated ecological principles at an "extremely high political level." Additionally, its resounding nationalistic challenges became a kind of call to arms for the Park Service: that each park should be an "illusion of primitive America," and that the Service should preserve or create the "mood of wild America." These phrases inspired a patriotic, ethnocentric goal—to maintain the landscape remnants of a pioneer past as they were "when first visited by the white man" or when "viewed by the first European visitors." Ignoring Native American perceptions of landscapes and wilderness and the possibility of ecological change resulting from Native American use of lands, this New World imagery suggested a kind of wilderness *pastorale* that had enormous appeal to many in the Park Service.[23]

Reflective of the growing awareness of ecology and the complex interrelatedness of nature, the Leopold Committee responded to Interior secretary Udall's request to analyze specific wildlife management issues by placing the concerns in a broad ecological and philosophical context. It put "in good perspective," as Conrad Wirth commented, "the immediate, as well as the distant view." A. Starker Leopold, chairman of the committee and primary author of the study, acknowledged that the report was "conceptual not statistical," with emphasis on the "philosophy of park management and the ecological principles involved."[24]

The Leopold Report set the stage for serious tension within the Park Service when it stated flatly that the "major policy change" recommended was for the Service to "recognize the enormous complexity of ecologic communities and the diversity of management procedures required to preserve them." Even more, it urged that scientific research "form the basis for all management programs" and that "every phase of management" come under the "full jurisdiction of biologically trained personnel of the Park Service"—extraordinary challenges to a bureau long focused on accommodating tourism.[25]

In August 1963, five months after the Leopold study appeared, the National Academy submitted its report. As Leopold had done with his committee, biologist William J. Robbins both chaired the committee that prepared the report and was the principal author. The agreement by which the Park Service authorized the academy's study clearly reflected ecological concerns. It noted that the parks were "complex natural systems" that "constitute a scientific resource of increasing value to scientists in this country and abroad," and that for proper management they needed a "broad ecological understanding and continuous flow of knowledge."[26]

In a substantially longer document than the Leopold Report, the academy discussed the scientific aspects of managing natural systems, made detailed recommendations for change, and bluntly criticized the Service's failure to support science. It portrayed Park Service scientific research in unflattering terms. The program lacked "continuity, coordination, and depth," and was marked by "expediency rather than by long-term considerations." Further, it "lacked direction" and was "fragmented," "piecemeal," and "anemic," with insufficient funds being requested or appropriated. Overall, the report noted that the Service had little appreciation of research and its potential contributions to park management. To the academy it seemed "inconceivable" that scientific research was not used to ensure preservation of such "unique and valuable" properties as the national parks.[27]

Asserting that the Service had "some confusion and uncertainty" about the purposes of the parks, the report defined the parks as "dynamic biological complexes," which should be considered a "system of interrelated plants, animals, and habitat (an ecosystem) in which evolutionary processes will occur under such control and guidance as seems necessary." Beyond the large, popular mammals normally of concern to the Service, species unknown to the general public (such as blind fish found in park caves, or thermophilic algae and other organisms associated with hot springs) presented national park management with "challenging questions of a fundamental character."[28]

The National Academy made numerous recommendations potentially affecting Park Service organizational structure, personnel, and budget: in these regards, its impact was greater than that of the Leopold Report. Most important, it argued that the Service needed a "permanent, independent, and identifiable" scientific research unit that should have "line responsibility," not "simply an advisory function." To the academy, strength and "independence" were key elements for the program's success. Directing the program should be a "chief scientist," who would supervise natural history research and the research staff, and an assistant director for research in the natural sciences, who would handle the administrative aspects of research and related activities. Both positions should report to the Park Service director, thus avoiding intervening and possibly antagonistic levels of bureaucratic authority. In addition, the Service should assemble a staff of about ten "highly competent" scientists in the Washington office, who would evaluate research needs and thereby determine necessary scientific staffing in the parks.[29]

To further ensure independence from park managers, the report urged that scientists be stationed in parks but answer directly to the chief scientist in Washington. The research program should also be supported by special centers that would be established in or near selected parks. To be commensurate with science funding in other federal land-managing bureaus, Park Service science should receive about ten percent of the bureau's annual budget (at the time it received only a tiny fraction of one percent). Moreover, the report recommended that a scientific advisory committee be created for natural history research, and that, as necessary, each large natural park should have its own advisory committee.[30]

In a foreshadowing of resistance to substantive change to the science programs in the years ahead, Park Service leadership reacted defensively to the National Academy's barbed criticism. Although the Service responded with rhetorical enthusiasm to the report and Director Wirth urged that every employee "should become familiar" with it, in reality the leaders did not care for the document.[31] Howard Stagner, longtime member of the Park Service directorate, later recalled that they even considered suppressing the report, mainly because they did not want the blunt criticism to be made public. Stagner stated that the document's language was such that it could be "very damaging"; thus the Service decided, "Let's distribute it and say we agree with it." The Park Service did authorize release of the study, but, in Wirth's words, it did "not seem necessary . . . to reproduce the full report for general distribution." Rather than formal publication, the National Academy put the document out in typescript, as a soft-bound, in-house report. Perhaps as a result, it seems to have received very little attention in the press and was largely forgotten by Park Service rank and file, other than scientists.[32]

Ironically, the National Academy's study was influenced by the Park Service's own internal, unpublished report, "Get the Facts, and Put Them to Work," prepared in October 1961 by Stagner with input from biologist Lowell Sumner, as part of Stagner's effort to prompt a review by the academy. In some instances the same wording even appeared in the two documents. Both viewed the science programs as "fragmentary" and "piecemeal," and one of the academy's sharpest criticisms—that the Service's science programs lacked "continuity, coordination and depth"—was taken verbatim from "Get the Facts."[33] It is obvious, however, that even though the Park Service would allow certain criticism from within ("Get the Facts" could be absorbed in the bureaucracy and rendered ineffective), its leadership disliked being publicly reproached and sought to limit the impact of the academy's report.

The environmental era raised resource management questions that clearly required scientific data. Regarding the Leopold Report, Conrad Wirth stated that it put the Service's 1916 congressional mandate into "modern language."[34] In fact, written by scientists (mostly biologists), both the Leopold and National Academy reports gave a *scientific* perspective to national park management—a kind of ecological countermanifesto that marked the beginning of renewed efforts to redefine the basic purpose of the national parks. In the short span of a few months in 1963, the Park Service found its natural resource management subjected to far greater scrutiny than ever before and faced recommendations for radical changes in its organization, operations, and policy. Much of National Park Service history since 1963 may be viewed as a continuing struggle by scientists and others in the environmental movement to change the direction of national park management, particularly as it affects natural resources.

The Pursuit of Bureaucratic Power

Following the reports, efforts to infuse science into park management were affected by two underlying factors. Perhaps the more daunting was that, both explicitly and implicitly, the reports called for a redistribution of power within the Park Service. A full and committed response by the Service would have required sizable increases in staffing and funding for natural resource management, including research. Ultimately, bureaucratic leadership would be shared with those advocating scientifically based management—a concept virtually alien to Park Service leaders and field personnel of the early 1960s. At the time of the Leopold and National Academy reports, science was buried in the Service's large and complicated organizational chart, split between two divisions and receiving little of Director Wirth's attention.[35] But the Leopold Report's declaration that all national park management should come under biologically trained personnel suggested that an extraordinary change in the Service's entrenched power structure was necessary. In effect, biologists would have to become full members of the bureau's leadership culture. This possibility—along with the National Academy's recommendations that directorate-level scientific positions be created, that ten highly qualified scientists be placed in the Washington office, and that funding for science be increased to ten percent of the Service's annual budget—constituted the more formidable aspects of the potential redistribution of power.

Compounding the problem of sharing power within the bureau was a second major obstacle: the complexity of ecologically oriented park management. The Leopold Report repeatedly emphasized the challenges inherent in attempting to restore the parks to a semblance of primitive America (efforts akin to what would become known as ecological restoration). "The implications of this seemingly simple aspiration are stupendous," it stated, given the "enormous complexity" of both the ecological communities of the parks and the means required to manage them. The report's summary statement gave notice that restoration of the parks to their primitive condition required "skills and knowledge not now in existence."[36]

Both reports issued in 1963 called for scientifically informed management—with which the Service had only limited experience, even considering the George Wright era of the 1930s. Decades of indifference to science reflected the attitude that neither research nor professional scientific skill was necessary for proper park management. The lack of experience with complex scientific land management probably helped foster the Park Service's naive, rhetorical "can do" response to the reports—even though the bureau at times acknowledged the reports' warnings about the difficulty of ecological management.[37] It is also likely that the Service was reacting positively to the National Academy study for the very reason Howard Stagner later gave—as a means of masking negative feelings toward the report.

Overall, Park Service leadership seems to have underestimated the extent of the challenge it faced: the amount of effort necessary and the degree of change required in its traditional power structure and park operations. Scientists hired by the Service in the 1960s and 1970s were similarly challenged. With little or no experience in a bureaucracy of such size, they confronted the bewildering task of developing an insurgent science program in the face of management's long apathy toward science—an indifference that had not ended with the appearance of the reports in 1963.

Relating to the factor of complexity, the reports' promotion of much stronger science programs implied a slower, more cautious pace than before for many park actions. If research were to become an integral factor in decisionmaking, it would force greater deliberation over both long-range planning and daily operations. As Director Hartzog admonished his superintendents in 1965, they would have to be cautious with their management and development and be "alert to the requirements for studies that arise" out of their park programs. When development was contemplated, each superintendent had to know what effect it would have on the "ecology of the surroundings" and had to be sure that a development site was "not of such scientific value as to justify . . . proposing a different location." Hartzog emphasized that if they did not know the answer to such questions, the superintendents should ensure that research was conducted "before—not after" breaking ground.[38] Such an admonition was, however, largely rhetorical. Although Hartzog was well aware of the need for prior scientific studies, most often that need would be ignored by park management. Easily decreed, it was not easily enforced. The bureau's leadership would provide lip service to the new science initiative, and would prove reluctant to share power and to accept substantive changes in its mode of operations.

In urging that the science programs be headed by a chief scientist under an assistant director who would in turn report to the director, the National Academy sought to ensure that science would be "independent of operational management" in order to promote objectivity in research and recommendations. The concern for independence—a central issue in the years to follow—stemmed from the traditional academic and professional ethic of maintaining the integrity of scientific research and recommendations. The scientists were to become the chief advocates for the new ecological approach to park management, and their views would frequently conflict with the pragmatic, day-to-day interests of park superintendents, who were accustomed to being in charge of all activities in their parks. Commenting on the question of independence, Robert M. Linn, a scientist in the Washington office, asserted in 1967 that scientific research should be separate from management; otherwise, it would always be "suspect," in that management could "dictate a prejudiced result." Much in accord with concerns expressed by the National Academy, he added that scientific research should be "as free as possible to criticize the parent organization."[39]

The organizational changes in the years after the Leopold and National

Academy reports moved very slowly and erratically toward the scientists' goals of independence and freedom to criticize. Even building the basic framework for a credible Park Service science program was bound to be a lengthy process. The Service could not merely hire several dozen scientists and with little forethought send them to the parks to do research and advise the superintendents. Thus its initial response to the reports included hiring a chief scientist, who was to design and implement a science program.

Director Wirth stated soon after the academy's report that the Service's "organizational deficiencies" would be "corrected." He promised to set up an "identifiable unit" (rather than an independent unit) of science, having "close liaison with administration and operating arms of the Service," and having "freedom of communication on professional matters." In 1964 Director Hartzog established the Division of Natural Science Studies and named George Sprugel, Jr.—a highly respected biologist with the National Science Foundation—to be the Service's chief scientist.[40]

Sprugel's division included all research scientists in the Washington office, several of whom were transferred from a branch within the Division of Interpretation. This action gave science a clearer identity and an elevated status (from a branch to a division), at the same time breaking an organizational tie with interpretation that had been in place much of the time since the early 1930s. Sprugel reported to an assistant director for resource studies, a new position that included history and archeology programs. This arrangement, it was asserted, would allow a dialogue among researchers from different disciplines. However, it may have served more to restrict the status and visibility (and certainly the independence) of the science program, in that science was only one division among three reporting to an assistant director.

Sprugel sought to establish his programs through systematic research planning for the major natural parks. He created teams of Park Service scientists and naturalists that, working with experts from outside the Service, studied the parks to determine the particular research requirements of each. Although the scientists wanted short-range research to address "stop-gap" management concerns, they principally planned for long-range studies that would address "every feature and factor represented in each natural area," providing a "basic ecological understanding" for park management. The first "natural science research plan" produced was for Isle Royale National Park, followed by plans for Everglades, Great Smoky Mountains, and Haleakala national parks. By the late 1960s, the research plans were being folded into more comprehensive documents, the "resource management plans." Although intended to provide the ecological component of park master plans, the resource management planning effort languished in the 1970s without aggressive systemwide support.[41]

In a move providing some degree of independence, Sprugel was given direct supervision over all field research scientists, in accord with the academy's recommendation for control of the research assignments and protection of the objectivity of the scientists' findings.[42] Thus researchers were not to be under the authority of the park superintendents. Yet Robert Linn recalled that placing in a park a scientist who was "directed in his activities and paid by someone else" was viewed as a "direct threat to the concept of the Superintendent as the 'captain of the ship'"—a factor that Linn believed led to "major problems" and to "constant pressure" by management to gain control of the science programs.[43]

In the late 1960s, recalling recent efforts to strengthen the science programs, biologist Lowell Sumner noted "clear parallels with the struggle for survival in the natural world." He acknowledged that the struggle within the Park Service was not "as violent and predatory as in the animal world"; rather, it was more like the "competition among plants," in that particular branches and divisions had a "favorable place in the sun." These plants "overshadowed the others and got the major share of the funds, as well as major representation on policymaking and planning committees." Also, in Sumner's opinion, biology had been "dismembered into 2 camps"—wildlife rangers and scientists. Of the two, the wildlife rangers had the more "favorable place in the sun."[44]

With the effort to build scientific programs, the Park Service seems to have made more meaningful attempts to distinguish between "research science" and what was gradually becoming known as "resource management." Resource managers (the wildlife rangers) continued to perform the actual in-the-park treatment of flora, fauna, and other natural elements, including elk reduction, fisheries management, and firefighting. They also assisted the scientists with some data gathering and other routine functions. The wildlife rangers were selected to do natural resource management either because of their experience in such activities or because of related academic training, although some apparently had limited qualifications. Field oriented rather than academic, they assisted the superintendents in making, as well as implementing, decisions on natural resource issues. (In the Washington office, the wildlife ranger division consisted mainly of individuals transferred in from the big natural parks where they had had experience managing large mammals.)[45] By contrast, the scientists had the responsibility for formal research, including preparing research designs, then gathering data and interpreting the findings.

In Wirth's September 1963 response to the National Academy's report, he made it clear that research scientists were to play a limited role in park management. In spite of the academy's admonition that science should not be "simply an advisory function" and that it should have "line responsibility," the Park Service director stated that the researchers' "basic responsibility" was to submit findings and recommendations, and that their authority was only "advisory." The scientists would "not make decisions, or give orders pursuant to putting recommendations into effect." The superintendents, assisted by the wildlife rangers (the resource managers), would give such orders and then implement them. There would be little sharing of bureaucratic authority with the scientists. Moreover, the Service was developing two organizationally separate biological programs—research and resource management—an arrangement that Washington office scientist Robert Linn would later characterize as "biology divided."[46]

In contrast to the research scientists' rather consistent lack of common understanding with superintendents, the wildlife rangers maintained strong ties with park managers. With wildlife and forestry management combined in the Washington ranger office (as in most field offices), the rangers not only had a substantial role in resource management, involving both animals and plants, but also enjoyed much greater bureaucratic influence than did the research scientists. Early in 1964, an internal Park Service study recommended reorganizing the ranger division and renaming it "Resource Management and Visitor Protection."[47] The very title of the new unit (a title also adopted in many parks) helped ensure that the rangers would continue to be in charge of resource management in addition to law enforcement.

While head of the ranger division in the mid-1960s, Spud Bill, former superintendent at Grand Teton, soon to be Hartzog's assistant director and then deputy director, sought to make sure that the rangers kept a tight hold on resource management duties. In a memorandum on "The Role of the Park Ranger in Resources Management," Bill asserted to Director Hartzog that it was "wholly logical" for resource management to be a "park ranger function and basic responsibility." He recalled that, in decades past, specialization of national park work in such fields as interpretation and maintenance had taken these activities from the rangers, which he feared had diminished their status. But, as the Service moved toward more intensive management of natural resources, he was convinced that the ranger should be the principal player. The strength of the resource management program must "spring from and be motivated from the program operation level"—it was the ranger, the "man on the ground," who was familiar with the park and who should remain the "keyman" in resource management. Much later, veteran law-enforcement ranger and resource manager William Supernaugh commented that as the parks' resource management programs grew, their personnel "largely came from law enforcement"—the ranger staff. The wildlife rangers continued, in Supernaugh's words, to be within the rangers' "empire."[48]

As part of the ranger domain, historically allied with park management, the wildlife rangers were a steady source of support to whom the superintendents could turn for advice in the new era of ecology. Much more inclined than the scientists to make decisions without the delays and questions resulting from research, the wildlife rangers tended to act if, in Supernaugh's words, the situation "felt right."[49] Their management style surely suited the perceived needs of superintendents, who were long accustomed to decisive action.

This adherence to tradition was almost certainly encouraged by the wildlife rangers' close association with the Park Service's foresters, whose leaders remained little influenced by current ecological concepts. Well after the Leopold and National Academy reports were issued, the foresters (believing, as biologist Robert Linn later stated, that "all fire is bad and must be put out") continued to push for suppression of fires and elimination of certain insects and diseases. In 1966, in a clear illustration of minimal ecological concern, the acting assistant director in charge of the foresters and wildlife rangers wrote that although the Leopold Report's intention was to "restore and maintain the natural biotic communities" within national parks, these natural communities have "little justification for retention as national parks except as they are utilized by man, i.e., the park visitor." Recalling such attitudes, Lowell Sumner asserted that the "trouble with ecological considerations" in the parks had been that they were "frequently in conflict with some of the programs of other Service units—programs such as native forest insect control, filling in of swamplands to enlarge campgrounds, road and trail building into essentially pristine ecological territory, or suppression of natural fires in parks whose distinctive vegetation was dependent on the continuing role of natural fires."[50]

With the rangers generally representing the traditional perspective on natural resource management and the scientists much more attuned to current ecological thinking, discord between the two groups was sure to arise in the Washington office. In a statement prepared in late 1966 and issued under Deputy Director Bill's signature, Robert Linn (then acting chief scientist) observed that "evolving traditions, reorganizations, realignments, etc." had created the separate wildlife ranger and science functions, with each group having a "traditional dominion or assigned mission." Efforts to strengthen the science programs had rapidly precipitated territorial disputes. In a restatement of his "biology divided" observation, Linn wrote that "unhappy animosities" and "unhappy moments" resulted from "conflicts of 'jurisdiction' and conflicts of opinion." Responding to this problem, in 1967 the Service created a Natural Resources Committee for the purpose of "creating and/or maintaining coordination" between the wildlife rangers and the research scientists.[51]

An internal Park Service statement dated July 1964 (shortly after George Sprugel became chief scientist) noted that although the Service had reorganized to accommodate science, funds for research were "so limited" that even stopgap studies to address "pressing natural history problems" could not be satisfactorily accomplished. Mainly, Park Service scientists had to seek support from outside researchers to "initiate and carry out basic studies" in the parks, and funding had to be "obtained elsewhere" than from the Service. By 1965 the Park Service's own funding for research projects had reached only $105,500—up almost $80,000 from 1963 but still a minuscule amount compared to the hundreds of millions of dollars being spent on park development under Mission 66. The scientific research programs had not, as Director Hartzog observed early in 1965, "achieved the pace" he had hoped for; rather, they had met with "mixed results." Hartzog cautioned his superintendents that research was not a "fringe activity," but a "real and practical requirement" that needed recognition. Nevertheless, research received little budgetary support.[52]

Park Service leadership asserted that a key factor in limiting research funds was that some members of Congress believed that scientific research was not a proper function for the Service. Sprugel recalled that during his tenure as chief scientist (1964–66) a "budget problem" existed, resulting partly from congressional opposition to allowing line-item funding for research.[53] Shortly after assuming office in early 1964, Director Hartzog had been bluntly advised by no less than the chairman of the House Interior Department Appropriations Subcommittee, Michael Kirwan of Ohio, that research was not any business of the Service. This attitude prompted the director to avoid use of the word "research" in budget requests to Congress; he substituted the designation "resource studies," to disguise the program. In addition to the monies appropriated for resource studies, Hartzog used his emergency funds to finance research. Robert Linn recalled the director's use of "hip pocket cash" to ensure that the research programs continued.[54]

In Linn's opinion, the funding question was one of the chief problems to "plague mightily the science program." He noted too that hip-pocket funds "rarely legitimize a program and rarely come in amounts sufficient for major efforts." Linn claimed that the Park Service had "never presented effectively" to Congress an explanation for "why research was necessary to carry out the mandates that Congress gave to the Service," or why the Service should conduct its own scientific research. In truth, the failure of Service leadership from Mather's time on to determinedly pursue scientific research funding reflected a lack of concern for acquiring an in-depth knowledge of the parks' natural conditions.[55] At a conference of national park scientists in the late 1960s, Lowell Sumner concluded that funding and staffing for research were still "*peanuts* in comparison to those of larger and more powerful branches and divisions of the Service." He quoted a former Park Service landscape architect, Al Kuehl, who had frequently commented that if managers "think [biology] is important, they'll *find* the people and money" to do the work.[56]

Along with the question of funding, the Park Service faced the persistent problem of control—in Linn's words, "who should direct the work of the [park] scientists." George Sprugel recalled that even though he had line authority over the scientists, the superintendents exerted strong influence over them, and that superintendents and regional directors often sought to impede efforts to advance science in the parks. Sprugel remembered Hartzog as "friendly" toward the science programs, but viewed the director's top lieutenants as difficult obstructionists, unwilling to tolerate the new scientific approach to park management.[57]

Exasperated with what he saw as insufficient funding and overall weak support for his research programs, Sprugel resigned from the Park Service in September 1966. His departure occurred just over two years after he assumed the position of chief scientist and more than three years after the Leopold and National Academy reports declared the need for strengthened science programs within the Service. But in Sprugel's opinion, many of the Service's leaders had only "paid lip service" to the two reports and were "very hardnosed" in their resistance to science. Representing a new focus in park management, he had felt like an "outsider" with old-line managers, and it seemed as if he had been "thrust down their throats" by Director Hartzog.[58]

Announcing the chief scientist's resignation, a strongly worded article entitled "Science: Sense and Nonsense" appeared in *BioScience* and reflected Sprugel's concerns. Sprugel recalled that the author, reporter Harold Simons, had obtained his information from a highly placed source—no less than Assistant Secretary of the Interior and prominent biologist Stanley A. Cain, who, prior to his appointment to the Interior Department in 1965, had served on both the Leopold and National Academy committees. The article characterized Park Service efforts to respond to the reports of these committees as being "sorry, at best." The Service had "turned its back on scientific advice." Simons stated aptly that three years after the academy's report the same harsh criticism "could be made, along with the same threats, needs, and recommendations" that the academy had identified. Although Sprugel himself had "enlivened the scientific approach" to the parks, "all signs" indicated that the Service had not yet seen the "scientific light."[59]

Despite such a negative outlook, shortly after Sprugel's resignation the status of the science programs rose significantly. The Division of Natural Science Studies, reporting to an assistant director, was redesignated the Office of Natural Science Studies, reporting directly to Hartzog. Moreover, in the spring of 1967 Hartzog enticed Starker Leopold to become Sprugel's successor as chief scientist, thus bringing into the Service the Leopold Report's principal author. Hartzog had tried hard to get Leopold to join the Service, pushing the matter even after an initial refusal. He reached an agreement that while Leopold served as chief scientist he could continue his work at the University of California and be stationed in Berkeley. To ensure smooth operation of the science programs, Hartzog promoted Robert Linn, who had been acting as Sprugel's replacement, to serve as Leopold's deputy.[60]

Hartzog had long admired Leopold, and no doubt recognized the prestige he could bring to the Service and its science programs. In what Hartzog termed a "brilliant address," Leopold had told the superintendents conference gathered at Yosemite in October 1963 (several months after his report had been issued) that the Service—as well as Congress—needed a "complete overhaul" in its attitude toward research. He believed the research necessary to manage the parks "intelligently" was "simply enormous," and that without scientific input there was no way for "ecological management" to take place.[61]

With the chief scientist reporting directly to Hartzog, the Office of Natural Science Studies had at last attained for science the kind of status and independence envisioned in the National Academy Report of 1963—a situation enhanced by Leopold's reputation. And, as Leopold's deputy based in Washington, Robert Linn became "a part of the Director's squad," or "inner circle," as he put it. Heretofore, Linn felt, the scientists had had "no real representation on that august body."[62] At that point, it seemed that little more independence could be expected as long as science remained a function of the Park Service.

However, in the spring of 1967, just before Leopold took office, U.S. Senator Clifford P. Hansen of Wyoming proposed the National Park Service Natural Science Research Act, which would create a fully independent scientific wing of the Service. The bill included establishment of a Commissioner of Natural Science Research, who would have status approximately equal to that of the Park Service director. Senator Hansen claimed that the legislation was drafted "in response to some of the dramatic crises" that the national parks faced, and was "based upon the findings and recommendations" of the National Academy's 1963 report.[63]

In fact, the academy had only recommended that science be "independent of operational management" within the Service. Park Service leaders believed that Hansen's real intent was to wrest decisionmaking power from the director in order to gain control over wildlife management issues, particularly the still-unresolved issue of public hunting of Yellowstone's elk. Acting chief scientist Linn wrote to his former boss, George Sprugel, asserting that Hansen, reacting to concerns about Yellowstone, had "dragged out the old [National] Academy Committee Report" and made a bill out of it, to create a new "agency or super-agency within the National Park Service." Considering the potential power of a Commissioner of Natural Science Research, Linn added that there was "not much room for a war between a Director and a Commissioner." In his response to Linn, Sprugel evidenced his past frustrations with the Service, noting that with regard to the bill, nothing he had seen led him to believe "that tying the hands of the Director when it comes to a research program might not be to the benefit of the Park Service." Predictably, however, the bill promised a situation Hartzog could not tolerate. The Park Service claimed that it was already "organizing as effectively as possible" to have research support for park management, and the proposal was never enacted into law.[64]

Even the independence attained by the science programs under Starker Leopold did not last long. Leopold resigned as chief scientist effective June 1, 1968—exactly one year after his appointment. Having accepted the position on a conditional basis and with personal reservations, he soon realized that he could not satisfactorily address the needs of both the university and the Park Service. On Leopold's recommendation, Hartzog appointed Robert Linn as the new chief scientist. But soon, in 1969, the Office of Natural Science Studies lost its high organizational status when Hartzog removed it from his direct supervision and buried it in a cluster of eight divisions under one associate director.[65]

Thus, within a period of about two years, Park Service science had risen to a prominence it had never before known—then dropped back to a rank-and-file level. Given the timing of these shifts, it seems likely that the elevation of science was tied to Leopold's personal status and influence. Once he was gone, the Service quickly lapsed into customary organizational arrangements. The bureau's traditional leadership culture had reasserted itself and reduced the visibility of science and its role in management.

The status and independence of the science program were further affected when in the fall of 1971 Hartzog suddenly ordered the transfer of Washington office staff scientists to the regional offices, to become "regional chief scientists" reporting to the regional directors. At the same time, supervisory authority over those biologists stationed in parks was taken from the chief scientist in Washington and turned over to the regional directors or park superintendents. In the opinion of Chief Scientist Linn, this move had very likely been spawned by the antagonism of the more traditional managers, who resented their lack of control over the scientists, perceived by some to be engaged in research "hobbies" in the parks. Linn believed that Hartzog had come under "constant pressure" from the superintendents on this matter.[66]

Linn later recalled that he had not been consulted prior to Hartzog's sudden announcement of the reorganization, made at a meeting of regional directors in a hotel near Washington's Dulles Airport. The very fact that this important restructuring of the science programs was pulled off as a surprise move indicates that its proponents intended to catch Linn off guard and force the issue. At a coffee break following the announcement, the director discussed the matter with Linn, offering to reconsider if he had any "real heavy objections." In Linn's words, the "biggest mistake I ever made as chief scientist was not vigorously objecting." Caught in a sudden power play—and no doubt under tremendous pressure from a phalanx of the Service's directorate seeking to preserve a long tradition in which superintendents had virtually complete control in their parks—Linn assented. Told that he would be able to make the selections for the regional chief scientist positions, and assured by at least one regional director that in reality "nothing will change," the chief scientist hoped that the situation would work out satisfactorily.[67] With the regional directors closely allied with the superintendents (indeed, most had themselves come from the superintendency ranks), Hartzog's move greatly increased management's control over science. It was a definitive rejection of the National Academy's recommendation for a program independent of operational management.

Although individuals with high academic credentials had been employed before, the scientists hired in response to the Leopold and National Academy reports constituted the first sizable number of Ph.D.s to come into the Service. Academics with little bureaucratic experience and embodying a challenge to established views of park management, they had difficulty finding common ground with traditional managers. Their research could delay decisionmaking and could present ambiguous conclusions. It could also challenge a superintendent's preferred course of action. In many ways the scientists were caught between their desire to do independent scientific research and their need to participate directly in park decisionmaking. Participation would necessitate working closely with management during, for instance, analysis of policies, preparation of park planning documents, and analysis of the potential impacts of management actions. Such cooperation could rob the scientists of valuable research time and threaten their independence of thought and action.[68]

With their goal of independence (a goal seemingly less important to other research-oriented professions in the Service, such as archeology or history), the research scientists had been implanted in a bureau with long-established modes of operation that rejected the alien concept of an independent scientific voice. The desire for independence may have reflected the degree to which scientists wished to be free to criticize management—but it may also have served to estrange park managers even more. Park Service leadership preferred that science be integrated with management, in the hope that it would be responsive to the managers' immediate needs. With regionalization, the goal of an independent scientific research program (which, as Linn had put it, should be "as free as possible to criticize the parent organization") had all but vanished, buried beneath the Service's management traditions.[69]

Regionalization brought no consistent organizational arrangement for scientists located in the parks. Each reported either to a staff person under the superintendent (usually the chief ranger) or to the regional chief scientist. Such varied reporting arrangements further splintered the science programs. Attempting to retain some authority and cohesion in his programs, Linn secured approval for creation of a Natural Science Coordinating Council, to consist of the chief scientist, his remaining Washington staff, and all of the regional chief scientists. The council was to meet quarterly. Yet with the regional and park scientists under supervisors who had varying degrees of interest in science, even this arrangement faltered. Within about six months of the reorganization, Linn learned that two of the regional chief scientists had been told not to "call Linn." In his view, this marked "the end of a centrally directed science program."[70] He recalled that even "simple inquiries to the field" from his office became treated by superintendents "with distrust and disdain and as an act of trespass." In addition, Linn stated that park research at a "number of locations" was "reduced by the encroachment of 'more meaningful' management activity." Budget allocations for long-range research competed with the superintendents' desire to use those funds for more "instant success" with park projects such as "snowplowing, dangerous tree removal and a variety of visitor services activities."[71]

The frustrations of biologist Ken Baker, stationed at Hawaii Volcanoes National Park, exemplified the new situation for field scientists. Baker wrote to his friend Lowell Sumner that "just prior to reorganization" Chief Scientist Linn had approved his annual operating budget of $5,200, but that "after reorganization I got $0.00. That's what I said, $0.00. So here I sit, without any mileage for my personal vehicle, animal feed, telephone expenses, etc." The Washington office had passed the science funds to the regional offices, but, as Linn indicated, the money was diverted to other purposes. Baker described a similar loss of funds for the Death Valley biologist. With the reorganization, Baker believed that the chief scientist had become "nothing more than a figure-head and that Research Biologists have been tossed to the lions (Superintendents)." Emphasizing the role of the park superintendents more than the regional office, he wrote bitterly that the biologists were "under the control of Superintendents. They control our purse strings and whoever does that controls the Biologist." After Sumner conveyed this information to Starker Leopold, the former chief scientist responded that having the biologists dependent on the superintendents "is not the way I envisioned this program working nor the way it should work."[72]

The remarks of Linn and Baker on funding pointed to a key problem for scientific research: the ability of regional directors, superintendents, and chief rangers to manipulate funds and personnel. Always quite limited, science funding was generally subject to the discretion of park management, and more so after the regionalization. Superintendents could now shift funds with virtual impunity, especially when it involved the related fields of science and natural resource management. When a superintendent turned research funds over to the chief ranger, the ranger could instead use these funds for resource management (under his control) and thus free up his own resource management funds for other purposes, such as law enforcement or safety. As Linn's remarks on snowplowing and tree removal suggested, scientific research funds not infrequently ended up supporting routine park operations.

Similarly, Roland H. Wauer, a leading natural resource management strategist during this era, later observed that the majority of Park Service scientists found themselves "dealing with resource management issues rather than [scientific research]."[73] Despite their desire for independence, scientists were repeatedly drawn from research into resource management. Among other things, the shuffling of funds and personnel for science and for resource management and other ranger work seriously blurred the situation, making it difficult, if not impossible, to track the actual yearly funding or staffing for science.

Nearly two years after regionalization of the science programs, the extent of the loss of central control over the programs was made evident. An August 1973 Washington office memorandum on the "status of the Service-wide natural science program" pointed out that "attempts" to "ascertain the scope of ongoing [science] projects and to determine the actual funding level by Regions" had proved "somewhat unsuccessful." Had the Service's directorate been truly committed to science, surely it would have insisted on close tracking and accountability. But, in Linn's opinion, with the Service's priorities focused on daily park operations, "long-range promises" for science had done "badly in a marketplace of instant success."[74]

Reorganizations during the 1970s brought continued fluctuations in the status of the Washington office science programs. In 1973, under Hartzog's successor, Ronald H. Walker, and following Linn's departure, the chief scientist position rose to approximately the same status it had had during Starker Leopold's tenure, reporting immediately to the director. But under succeeding directors, more shifts occurred. By 1976 the science office had fallen back to division status, again buried in the organization as one of six divisions reporting to an associate director. In 1977 Starker Leopold and Purdue University biologist Durward L. Allen conducted a review of national park science programs. With their report and with an order from Assistant Secretary Robert Herbst that the "scientific program be upgraded and coordinated throughout the Service," the bureau again came under pressure to raise the status of science.[75]

As with the Leopold and National Academy reports, outside authorities once more sought to influence Park Service organization and effect shifts in power. Leopold and Allen recommended that the position of associate director for natural science be established, and that the incumbent operate in effect as the chief scientist. After months of stout resistance by Park Service leaders reluctant to share authority with science, this recommendation took effect. In 1978 Linn's successor as chief scientist, Theodore W. Sudia, was promoted to a new position of associate director for science and technology.[76] Although the title "chief scientist" was abandoned, science for the first time had attained the associate director level—fifteen years after issuance of the Leopold and National Academy reports.

Overall, during the 1970s, the organizational fluctuations of the science programs demonstrated that a kind of intellectual and policy push-and-pull was taking place—between the dictates of the Leopold and National Academy reports on the one hand and the traditional mindset of the Service on the other. As Linn put it, "representation at the top of the bureaucratic pile" was an "important function" for science, and the highest levels of the Park Service "MUST include [scientific] knowledge and concern." But the Service had only very slowly yielded some of its power to scientists. In marked contrast with science, perennially favored functions such as administration, design and construction, and ranger operations had maintained a consistently high organizational status throughout this period, generally at the assistant, associate, or even deputy director levels.[77]

Possibly contributing to the instability of the scientists' status was the revolving-door directorship imposed on the Service during the 1970s. For various political reasons, between December 1972, when Hartzog resigned as director, and May 1980, when Russell E. Dickenson took over, the Park Service had three directors: Ronald Walker, Gary E. Everhardt, and William J. Whalen. These eight years manifested the most rapid leadership changes in Service history, with an average tenure of about two and one-half years, and with each director having his own perception of science and its role in national park management. Robert Linn believed that Hartzog's support for science was much greater than that of his immediate successors, a view seemingly contradicted by the rise of science under Director Walker.[78] Clearly, though, science had had difficulty securing sufficient bureaucratic strength to resist sudden organizational shifts, especially when the directorship itself was susceptible to rapid turnover. Yet in 1978 in the Washington office, science at last gained the associate director level—an organizational status it has maintained.

The period of frequent turnovers in the directorship helped bring about another shift in power within the Park Service's top echelons. As director, George Hartzog had dominated the Service. But because his successors lacked his bureaucratic strength and finesse, the regional directors assumed greater control, in effect filling the vacuum created by Hartzog's departure. And by the late 1970s the number of regions had increased from eight to ten, which augmented the regional directors' overall strength. Richard H. Briceland, who as associate director headed the science programs from 1980 to 1986 and encountered firsthand the authority of the regional directors, described the dispersal of power as "distributed anarchy." He believed it seriously impeded efforts to run an effective Servicewide science program. Similar to the earlier observations of Chief Scientist Linn and Hawaii Volcanoes biologist Ken Baker, Briceland stated that not infrequently he had seen regions or parks divert funds from scientific research to maintenance or other park activities.[79]

Briceland also recalled how the regional directors defeated an effort to elevate science throughout the Service. At a 1986 meeting Director William Penn Mott, Jr. (who had succeeded Russell Dickenson the previous year), proposed that science personnel be placed in high positions of authority, immediately under the regional directors or superintendents, in all regions and major natural parks. This plan, promoted by Briceland and others, had been presented to the regional directors well before the meeting. To some extent it would replicate throughout the system the status that science had finally attained in the Washington office.

Revealing his deference to the strength of the regional directors (and in a democratic management style in startling contrast to that of the often-intimidating Hartzog), Mott submitted to their judgment, asking them to vote on the issue. Unwilling to surrender more authority to the scientists, they voted nine to one against the proposal, believing that it "wasn't needed," Briceland recalled. Support for the proposal came only from Western regional director Howard H. Chapman, who was building science programs in parks such as Sequoia, Yosemite, and Channel Islands. This important vote by the regional directors effectively constrained the scientists' authority in the regions and parks.[80]

Environmental Legislation and Change

In seeking stronger influence in national park management, the scientists were bolstered by the environmental movement and the resulting legislation. Particularly important were the Wilderness Act, the Endangered Species Act, and legislation on specific kinds of issues (such as amendments to the Federal Air Pollution Act and the Water Pollution Control Act).

The National Environmental Policy Act of 1969 had special potential to inject the scientific perspective into park management. This act specifically called for "use of the natural and social sciences" in plans and decisions substantially affecting the environment. Through "environmental impact statements" it also required interdisciplinary analysis of alternatives during planning. To comply with the act, greater scientific knowledge would have to be used in managing public lands, including those under the care of the Service. Nevertheless, as recalled by veteran Park Service manager and analyst John W. Henneberger, the Service in the early 1970s thought it should be exempted from this legislation, believing the parks were already managed properly.[81] To comply with the act, the Park Service created only a few science positions, most of which were stationed in the huge new Denver Service Center, to deal with that office's many complex planning and developmental projects.

Within the parks themselves, new environmental legislation brought unanticipated changes for natural resource management. William Supernaugh recalled that although wildlife rangers had performed both resource management and law enforcement, their resource work became more complex as they sought to help the superintendents comply with new laws and regulations. At the same time, professional law enforcement itself was becoming much more demanding. The complexity of both types of work led to a division of labor, which tended to separate resource management from law enforcement.

The duties of natural resource managers now included a variety of increasingly specialized concerns, such as management of caves, threatened species, nonnative species, fires, and wildlife, in addition to monitoring of air quality, biocide use, and coal, oil, and mineral mining activity (where legal in parks because of prior rights). Their responsibilities also included preparation of resource management plans. Evolving slowly over time, these plans fostered a broader ecological understanding of the parks because they required analyses of historic changes, existing natural conditions, and descriptions of current and anticipated natural resource management needs, including research, for each park.[82] With such duties the resource managers became, as Park Service scientist Bruce M. Kilgore stated in 1978, the "key people in bridging the communications gap between science and management." Kilgore foresaw a continuing professionalization of the resource management staff, in which individuals would have advanced college degrees and "extensive and effective experience" as managers of flora and fauna in the parks.[83]

Historically there was, as Supernaugh put it, a "direct line" between wildlife rangers of the 1960s and latter-day natural resource managers. As resource management became more professionalized and more ecologically oriented, rangers with the education and interest in biological management often chose that field over law enforcement, and took on the increased legislative mandates and ecological problems faced by the parks. (In some parks, mainly those that had no strong wildlife ranger contingent, the park naturalists assumed these responsibilities.) In the 1970s the Service dropped the "wildlife ranger" designation in favor of the more inclusive title of "natural resource management specialist." Some parks created separate divisions for resource management and law enforcement, although, as with other organizational arrangements, there would never be complete consistency throughout the Service. In the Washington office, the formal separation of law-enforcement rangers and resource managers occurred in 1973, when natural resource management got its own division and was placed under a different assistant director.[84]

In addition, scientific resource management in the Service was enhanced by the creation of a number of special research offices. In 1970, during hearings on the proposed North Cascades National Park, U.S. Senator Henry Jackson of Washington prompted the Park Service to cooperate with the University of Washington in conducting a program of scientific studies on the "ecological, environmental, and sociological aspects of park and wild land management." The agreement reached that year established the first Cooperative Park Studies Unit—a university-based scientific research office that became the prototype for similar arrangements across the country.

As the program evolved, Park Service scientists at the studies units would bring the Service's research contracts to the host university, benefiting both professors and graduate students. Many Service scientists became adjunct professors, teaching part time and serving on graduate committees. Advantages to the Service included increased use of university professors and graduate students and increased access to technology (especially computers). The agreements also provided for reduced overhead charges by the university, thereby lowering research costs to the Service. The program got a fast start; by 1973 there were agreements with eighteen universities (some units were established without a Service representative on campus). By 1980 units existed at thirty-five schools, a figure that dropped to twenty-three in 1983, then rose to thirty-one in 1988. Included were such universities as Oregon State, Texas A&M, Idaho, and Hawaii.[85]

These research offices addressed the needs of individual parks, as well as groups of parks that shared similar concerns. For instance, in one of the more successful efforts to deal with broad natural resource questions, the cooperative park studies unit at the University of Massachusetts focused on shoreline stabilization at several parks along the East Coast, among them Cape Hatteras, Fire Island, and Cape Cod national seashores. In cooperation with other universities the unit studied barrier island dynamics, involving continual sand deposition from the forces of wind and water, which often affected park recreational development and nearby urban areas. The research findings, emphasizing the natural processes of constantly shifting island profiles and mass, became the basis of official policy.[86]

Adding to the responsibilities of the study units, legislation of the 1970s on special environmental problems such as air and water pollution increased the need for scientific information in park management. The Service moved very slowly to address air and water concerns, and did not establish an air quality office until the late 1970s. Yet this office soon became one of the largest and best-funded research operations in the Park Service. Similarly, a water resources division emerged in the 1980s, developing substantial expertise in research, resource management, and water rights issues.[87]

Furthermore, in accord with the National Academy's recommendation that "research laboratories or centers" be created in parks "when justified by the nature of the park and the importance of the research," the Park Service established several science "centers," usually associated with individual parks. Building on the model of the Jackson Hole Biological Research Station (opened in the early 1950s in Grand Teton National Park to study and monitor the area's elk population), research centers were created in, for instance, Everglades and Great Smoky Mountains national parks.[88] However, these two new centers came about more from fortuitous circumstances than from any systemwide review and planning by Park Service decisionmakers about which parks or groups of parks needed science centers.

In Everglades, creation of the South Florida Research Center in the mid-1970s resulted mainly from the personal interest and political power of Assistant Secretary of the Interior Nathaniel P. Reed. By the late 1960s the proposed Miami Jetport had threatened the park, catching the Service unprepared and thus compelling it to rush to gather data in hydrology, geology, ornithology, and other fields that would strengthen the park's defense. To many, this effort made clear the need for a strong science program at Everglades. Reed, a south Florida native vitally interested in the welfare of the Everglades, proposed a scientific research center in the park and successfully engineered its establishment and funding. As initially intended, the center was also to serve nearby Biscayne National Monument and Big Cypress National Preserve.[89]

In establishing the center, Reed faced adroit, stubborn resistance from Park Service leaders, who did not relish the competition and interference of a potentially powerful research voice in the park. Moreover, once the center was set up, the Service did not appreciably increase its operating funds, leaving the center weakened by inflation and causing it to terminate support for Biscayne and Big Cypress. To the dismay of scientists, successive Everglades superintendents gradually diverted the center's research funds to resource management. Much of the burden of the latter program therefore fell on the center, likely freeing up the park's own resource management funds for other, often-unrelated ranger operations. Special short-term "project" funds were used to augment research; but in the opinion of the center's second director, Michael Soukup, and the assistant research director, Robert F. Doren, "such erratic funding" did not lead to a "strong stable [research] program." They asserted that the Park Service failed to develop the "organizational, financial, and personnel requirements for a science program to match resource needs" of the Everglades—a park with profoundly complex ecological problems and under tremendous pressure from outside its boundaries.[90]

The Uplands Field Research Laboratory established in 1975 at Great Smoky Mountains National Park also came about in a fortuitous way. In the early 1970s a Cornell University graduate student in biology, Susan P. Bratton, was hired by the regional office to work in Great Smoky Mountains. There she noted the park's serious lack of scientific information. The park also lacked resource management capability. Yet it faced such problems as management of the mountaintop balds, exotic plants and animals (especially the voracious European wild boar), and Cade's Cove—the park's large historic district, where cattle grazed in areas inhabited by rare plants. Bratton believed that, despite regional office interest, the park managers did not truly want a biologist. She recalled that when she arrived in the park, "old guard" management held her suspect and did little to advance science.[91]

This perspective changed with the appointment of a new superintendent, Boyd Evison, who wanted to bolster scientific input. The coincidence of his and Bratton's interest in improving natural resource management (and the continuing support of the regional office) led to establishment of the Uplands Field Research Laboratory in 1975. This center soon grew into a small multidisciplinary operation, with aquatic, bear, and wild boar specialists, among others, and including both researchers and resource managers. Nonetheless, had there not been a superintendent sympathetic to science, who would take advantage of the regional office interest and of Bratton's work, the uplands laboratory might never have become a reality. Bratton herself believed that neither the laboratory nor the South Florida Research Center came about as a result of any "systemwide thought process." In her opinion, there was "absolutely no overall policy for this kind of thing"—personalities influenced Park Service research, and the centers were a result of "personalities and chance."[92]

The short and troubled existence of the National Park Service Science Center, near Bay Saint Louis, Mississippi, provides perhaps the most glaring example of opportunism and the lack of an overall policy or long-range commitment to science centers. Inspiration for the center (formally established in late 1973) came largely from U.S. Senator John Stennis of Mississippi. Stennis wanted to fill space vacated at the National Aeronautics and Space Administration facility near Bay Saint Louis to help boost the local economy while the Apollo space program wound down—reasons obviously unrelated to any concern for national park science. The facility included laboratories equipped with sophisticated computers and additional up-to-date research capability, such as remote sensing. As intended, other bureaus used the technology; for instance, the U.S. Geological Survey placed scientists there, helping to form a cluster of scientific expertise at the facility. The Park Service decided that its own operation at the science center would provide assistance for planning, inventorying resources, and conducting ecological research throughout the national park system.[93]

From the first, the center did not fare well. Although it had some permanent funding (or "base funding"—appropriations mainly obtained by Senator Stennis), the center operated to a considerable degree on short-term project money. Thus, every year it depended on regional offices and parks for sufficient projects to keep operating. This tenuous funding situation helped lead to failure, as parks and regions proved uninterested in using the center's expertise. More focused on accommodating Stennis than on developing an effective science center, Park Service leadership had agreed to establish the office but did not ensure adequate funding.[94]

In 1974, and again in 1975, the Washington office authorized task force studies of the center's operations. The 1975 study pointedly criticized the attitudes of the center's scientists and the effectiveness of their work. Recommending that such problems be corrected, it nevertheless concluded with the statement that "gut reaction has been to abolish the facility." Angrily reacting to such criticism, the center staff, in its 1976 annual report, accused the Service of failing to provide an "approved role, mission and organizational identity," which caused the office to experience a "series of disappointments in trying to implement studies and services." Intended to assist parks systemwide but left without systemwide support, the center had indeed become isolated. Its 1976 report claimed that the center had found it "impossible 'to do business' in some sections of the Service." Based on the task force recommendation, the center was disbanded early in 1977, just over three years after it was established. This contrived, half-hearted effort had come to an end.[95]

In addition to the centers, the National Park Service was slowly building scientific research offices in individual parks. These were similar to the centers at Everglades and Great Smoky Mountains, except that they were more integrated into traditional park organizations and did not usually have identities as discrete as those of the centers. In 1967 Acting Chief Scientist Robert Linn had noted that the Service was moving Glen Cole to Yellowstone to become supervisory research biologist, overseeing biological work in that park as well as in Glacier and Grand Teton. Linn expected to hire a biologist for Grand Teton (Cole's old position) and one for Glacier. He believed that together these would "make a pretty good research nucleus" for that part of the system. He also planned to hire scientists for Hawaii Volcanoes and Grand Canyon national parks, and one to be shared by Saguaro and Organ Pipe Cactus national monuments.[96] Almost immediately, Yellowstone's small science office would be subjected to strong criticism over its recommendations for grizzly bear and elk management—criticism that would persist over many years. After some delay, the park's science program would grow and diversify.

Especially concerned with fire management, Sequoia National Park built up a small research staff in the late 1960s and the 1970s for diverse studies on matters such as grazing impacts, threats to park wilderness, and air quality. Yosemite and a number of other parks would follow suit, including Indiana Dunes National Lakeshore, which developed a science program during the 1980s. By the early 1990s, Indiana Dunes had four research scientist positions plus additional support personnel. Backed by park legislation specifically addressing scientific research, Channel Islands National Park set up an unusually strong natural resource management and science program during the 1980s and early 1990s.[97] Necessitated by special circumstances, science and resource management programs at Redwood National Park operated on an even larger scale. In 1978, after the hard-fought campaign to secure its expansion, Redwood was faced with a massive thirty-thousand-acre rehabilitation project resulting from commercial clear-cutting of trees outside the original park boundary, but on lands included in the expansion. The legislation enlarging the park authorized $33 million for restoration of the cutover lands, which included work on landscapes, vegetation, and streams and required a staff of forest ecologists and other scientists, as well as natural resource managers.[98]

In contrast to the varied, uneven success of scientific research offices, the Denver Service Center, created in 1971 as a Servicewide planning, design, and construction office, gained a commanding position in National Park Service affairs. Even though the service center depended largely on project funds, it became a fully accepted and integral part of the bureau's organization. Soon after its establishment, the service center began to hire scientists, especially to address requirements of environmental legislation.

It did so grudgingly, however. In 1968, three years before the service center's creation, Chief Scientist Robert Linn had noted that the two centers then in existence (the large "eastern" and "western" centers, predecessors to the Denver office) employed two ecologists. The situation improved very little by the time the Denver office began operations. A September 1972 memorandum from Johannes E. N. Jensen, assistant director, service center operations, to Director Hartzog reported that there were currently three full-time positions "authorized for EIS [environmental impact statement] activities," to be supplemented with three permanent, but less-than-full-time, scientists.[99]

In an office of several hundred employees devoted to planning, designing, and constructing national park facilities, allocations of staff and funds to address ecological concerns were meager. Jensen stated that the six scientists would have to prepare an estimated 120 impact statements for various service center projects during the coming fiscal year, and "provide some input" for about 75 additional statements. Even by an "optimistic estimate," these statements would require 1,575 workdays, whereas the six employees could provide only about 1,200 days—a difference that might be made up somewhat by borrowing "other personnel on a part time basis." The assistant director admitted that the service center would be "hard pushed to adequately handle the EIS program."[100]

Biologist William P. Gregg, who assumed the responsibility of building the center's science staff to address the mandates of the National Environmental Policy Act, recalled that the Park Service wanted to do little more than meet minimal regulatory requirements of the law in order to avoid litigation over noncompliance. Gregg believed, however, that the act became the "major factor" in hiring scientists in the Denver office in the 1970s. He stated that "more were hired under that aegis"—including himself—than for any other reason.[101]

Early in his efforts to build a program, Gregg arranged a meeting with the Bureau of Reclamation's legislative compliance chief to discuss how to deal with the law's regulations. Aware that the Park Service might be sued if it did not comply with the required impact statement processes, he sought advice from the bureau because, with its water reclamation projects under attack by environmental groups, it was *already* facing litigation and should be well versed in the pitfalls of the compliance process. With little faith in the Service's willingness to comply with the law, the center's lead scientist sought advice on how to avoid litigation from the very bureau whose development in the West had been a significant factor in inspiring the environmental movement and its legislation.[102]

Another problem stemmed from the fact that the scientists' work schedules were tied to the deadlines of the service center's design and construction operations. When the Park Service first began to address its impact statement responsibilities, the center already had a backlog of completed plans and other documents for which statements were required. Responding to a law that mandated analysis of alternatives during the decisionmaking process, much of the work that the scientists first undertook came after the fact, justifying decisions already made. The scientists also began to assist with preparation of impact statements on newly initiated projects. But the center's rapid production pace caused the scientists to continue in a rubber-stamp situation. They only had time to gain some familiarity with the park resources, synthesize what was known from existing scientific literature, and apply this knowledge to the plans pouring out of the Denver office.[103]

Still, without the influence of the National Environmental Policy Act there would have been far less scientific input. As recalled by R. Gerald Wright, a biologist hired in Denver in the early 1970s, the act "gave science a power it never had before." The scientists gradually moved the service center toward some comprehension of the parks' natural resources and how they might be affected by the projects being implemented. But because they were virtually forced on the service center, the scientists found their work resented by those unaccustomed to interference. Wright remembered that the old-time planners were particularly hostile. Indicative of the assertion that Park Service leaders initially thought the bureau should not be subject to the new environmental law, the planners tended to question the scientists' motives and to view science as, in Wright's words, a "constraint on their freedom" to plan as they saw fit.[104] Such unfettered decisionmaking was, indeed, what the act's environmental impact statement process sought to curb.

Wright believed that a mid-1970s reorganization of the service center further impeded the scientists' efforts. The reorganization broke up the science office that William Gregg had assembled, and placed its members within the center's regional, multidisciplinary "teams," which were usually headed by landscape architects, engineers, or planners. This "regionalization," Wright concluded, caused science to lose its "group identity" and, more important, its "independence to challenge the team leaders" whenever service center proposals might be unduly harmful to the parks' natural resources. The traditional elite professions within the Park Service thus gained greater control and curtailed the emerging influence of science in the center. Wright found the situation in many ways comparable to the chief scientist's loss of his programs when Director Hartzog had transferred most of the scientists to the regions and the superintendents only a few years before.[105] At its Denver Service Center—the office having far and away the greatest assemblage of landscape architects, engineers, and other professionals capable of undertaking projects that could alter natural conditions in the parks—the Park Service operated with very limited ecological insight.

Throughout the 1970s the Service's scientific natural resource management efforts had increased, but the progress was erratic, influenced by "personalities and chance" and by a steady resistance to change. Gradually, with scientists hired into the Washington office, the regions, the cooperative park studies units, and parks such as Everglades, Great Smoky Mountains, Yellowstone, and Sequoia, their personnel numbers had risen. An internal report issued in 1980 declared that the Service had "about 100" scientists—an estimate that probably included research scientists as well as administrators of science programs. Although the Park Service still did not have adequate tracking and accountability for science, the report stated that funding for natural science research had reached $9 million. Some of these funds were used to support resource management or other activity rather than research. Having worked in natural resource management during this era, William Supernaugh later observed that the science program remained "a kind of mystery to many park managers," adding that, "what you don't know about you either distrust or ignore—the situation did not lend itself to success." Much stronger than before, the science programs still faced problems identified in the 1963 National Academy Report: they tended to be "fragmented" and "piecemeal," lacked "continuity, coordination, and depth," and were marked by "expediency rather than by long-term considerations." The academy had believed it "inconceivable" that science was not used to ensure preservation of the national parks' "unique and valuable" properties. But in the more than a decade and a half since the report appeared, the Service had failed to establish a comprehensive, coordinated scientific management program.[106]

Policies—New and Old

In its formal natural resource policy statements and its actual in-the-field management practices, the Service was similarly equivocating. It had entered the environmental era with minimal understanding of the ecology of the parks; and in the 1960s and beyond, with pressure to shift toward more ecologically attuned park management, its changes in natural resource management were often impulsive, politically motivated, and scientifically uninformed.

Although the National Academy study affected the role and status of science within the bureau, the Leopold Report had the greater impact on day-to-day resource management. In addressing the issue that had precipitated Secretary Udall's call for the studies, the Leopold Report advocated continued elk reduction on Yellowstone's northern range, although it stoutly opposed "recreational" public hunting in national parks. If ranger staffs were not sufficient to handle a reduction program, members of the public (who should be specially selected and trained) could assist—but for the "sole purpose of animal removal, not recreational hunting." In contrast, for national recreation areas operated by the Park Service (such as reservoir sites), the report endorsed the Service's existing policy of allowing sporthunting. Citing "precedent and logic" and asserting that national recreation areas were "by definition multiple use in character," it declared that hunting should be permitted "with enthusiasm."[107]

On a broader scale, the Leopold Report urged that "naturalness should prevail" in park management. The Service should encourage native plants and animals, discourage nonnative species, and minimize human intrusions in the parks. Further, the report recommended "controlled use" of fire as a management technique and questioned the Service's extensive use of chemical pesticides to combat forest insects and diseases, declaring that such use could have "unanticipated effects on the biotic community" within a park.[108]

The concern for naturalness by no means precluded active manipulation of resources, as with fire management or reduction of elk populations.

Indeed, although it acknowledged the process of continuous change in nature, in several instances the Leopold Report advocated preserving or re-creating a particular "ecologic scene." In its opening paragraphs it asserted that although "biotic communities change through natural stages of succession," such communities could be manipulated "deliberately" through control of animal and plant populations. It recommended as a "primary goal" for national parks that a "vignette of primitive America" should be reestablished or maintained, to approximate the conditions at the time of first European contact—the desired "ecologic scene."[109]

Elaborating on his report, Starker Leopold advised the October 1963 national park superintendents conference that the "manipulation of ecological situations" was a proper means of preserving "what it is that we have set up to display before the public." He stated that in cases where the Service wanted to "show a natural scene typical of an area, we can build it—if we have to." A visitor to Mt. McKinley National Park should "have the opportunity to see the type of scene that was observed by the pioneers . . . or whoever was the first visitor to that area." He added, "This is the objective of ecologic planning in the parks."[110]

In May 1963, two months after the Leopold Report appeared, Secretary Udall declared it official Park Service policy. Previously, the Service lacked a cohesive policy statement for overall park management. Instead, it had relied on myriad "handbooks" developed over the years for guidance on specific activities such as wildlife management, maintenance, concessions, road and trail management, and master planning. Although Fauna No. 1 still influenced natural resource policy, by the early 1960s there was little direct reference to it.

Shortly after the Leopold Report appeared, Director Hartzog ordered the preparation of new, concise policies. They were to be separated into three categories—natural, recreational, and historical—a division generally based on land classifications proposed by the 1962 Outdoor Recreation Resources Review Commission Report and on the perception that Congress had created three basic types of parks. First published in 1967, with slight revisions in 1968 and 1970, the management policies for natural areas included the Leopold Report reprinted in full.[111]

Yet the new policies revealed ambivalence within the bureau. Displaying a remarkable allegiance to tradition in the face of modern ecological concepts, the policies also included the 1918 Lane Letter, the primary policy statement of Stephen Mather's directorship. A product of its times, the Lane Letter had placed heavy emphasis on recreational tourism in the national parks and had characterized the parks as "national playground[s]."

Oriented toward park development, the letter virtually ignored national park science, merely commenting that research should be conducted by other bureaus. Its resource management recommendations included such practices as fighting forest insects and diseases and allowing cattle grazing in areas not frequented by the public. In a July 1964 policy memorandum drafted by the Park Service and included in the new policy book, Secretary Udall noted that the Lane Letter was "sometimes called the Magna Carta of the National Parks." The secretary declared that its policies are "still applicable for us today, and I reaffirm them. . . . The management and use of natural areas shall be guided by the 1918 directive of Secretary Lane." Nearly half a century old, the 1918 statement was unquestionably at variance with 1960s ecological concepts for preserving natural areas. As a philosophical and policy statement on parks, it contrasted strikingly with the Leopold Report. Yet *both* documents were now official policy, and *both* were reprinted in the 1968 policy book for natural areas.[112]

It is especially noteworthy that Fauna No. 1 was not even mentioned in the new policies. Nor were its recommendations included in the appendix, although they were clearly forerunners of the Leopold Report and far more in tune with it philosophically than was the Lane Letter. The new policies also neglected to mention the 1963 report by the National Academy of Sciences. This omission gives credence to Howard Stagner's recollection that the Service wanted to distance itself from the blunt and, as he put it, potentially "very damaging" criticism in the academy's report.[113] Inclusion of the Lane Letter and exclusion of any mention of the ecologically oriented Fauna No. 1 and National Academy Report rendered the new Park Service policy statement much less forward-looking than it could have been. These factors also reflected the uncertain status of ecological science in the Park Service during the 1960s and 1970s and suggested that the Service was less than fully committed to employing scientific knowledge as a basis for natural resource management decisions.

Moreover, fears arose among environmental groups that, with the three categories of parks, the Service might be more inclined to neglect natural resources in historical parks and cultural resources in natural parks. Rejecting the perception that units of the national park system could be categorized into three basic types and managed accordingly, Congress, in the General Authorities Act of 1970, declared the various types of parks to be part of a single system. The act stated that the parks

> though distinct in character are united through their inter-related purposes and resources into one national park system as cumulative expres-

> sions of a single national heritage; . . . and . . . it is the purpose of this Act to include all such areas in the System.

Subsequently, in revisions of its management policies, the Park Service abandoned the three separate administrative classifications. Congress reaffirmed the single-system principle in the Redwood National Park Expansion Act of 1978.[114]

Natural Regulation and Elk

Like the Leopold Report, the new policies sanctioned both "naturalness" and manipulation in resource management. They also endorsed what was becoming known as "natural regulation" or "natural processes" management, stating that managers would "minimize, give direction to, or control" the "changes in the native environment and scenic landscape resulting from human influences on natural processes of ecological succession." By attempting to "neutralize" human influences, the Service aimed to allow the "natural environment to be maintained essentially by nature."[115]

In contrast, though, the new policies provided for the manipulation of populations, such as the taking of fish, a practice long accepted in national parks. They also permitted—and seem even to have taken for granted—the continued reduction of ungulate populations. Wildlife populations would be controlled "when necessary to maintain the health of the species, the native environment and scenic landscape," or to ensure public safety and health. Included in the appendix to the policy book was a September 1967 memorandum from Director Hartzog that dealt solely with control of ungulate populations. Overall, the new policies stressed manipulation more than natural regulation, but left the Service with the option of taking whichever approach it might deem necessary.[116]

Already, with Yellowstone's northern elk herd, the Park Service had suddenly made what would become one of its most controversial natural regulation decisions. Despite the Leopold Report's recommendations to the contrary, sportsmen's organizations and state game officials had continued to lobby for hunters' participation in national park reduction programs. At the same time, reacting to increased media coverage, the public became uneasy about the killing of elk and other park ungulates. In response, Director Hartzog met in early March 1967 with Secretary Udall and U.S. Senator Gale McGee of Wyoming and agreed to halt the shooting of elk on Yellowstone's northern range, a decision that immediately preceded Senator McGee's public hearings on the issue.

At the hearings Hartzog declared that the "direct kill of elk in the park is stopped." In a separate announcement, he stated that the "most desirable means of controlling elk numbers" was through public hunting on lands adjacent to Yellowstone. Winter migration of elk from the park to neighboring lands, where the animals could be hunted, would be "facilitated whenever possible." The park would also intensify its trapping and shipping of elk to other areas. But Hartzog made it clear that if the new policy did not work, the Service would resume direct reduction. Termination of the long-standing population control program did not take effect until the following winter. The policy change was applied to other parks, particularly Rocky Mountain, and soon resulted in termination of the killing of any native ungulate species for purposes of population control.[117]

The policy decision arrived at by Hartzog, Udall, and McGee came not as a result of scientific findings, but because of political pressure. Adding to the pressure was Senator Clifford Hansen's resolution pending before the Senate Interior Committee to prohibit direct reduction of elk in Yellowstone (apparently retaliation for the fact that the hunting community had not been allowed to participate in the killing). The agreement to end the reduction program thus provided a quick solution to increasingly difficult problems: the angry crossfire of public alarm over shooting elk, the demands of hunters to participate in the reduction, and rising concern in Congress.

Soon, however, the Service justified its new policy on the basis of the natural regulation theory. A September 1967 park information paper outlined a program to "encourage the natural regulation of elk" in Yellowstone. The following December a similar park document, entitled "Natural Control of Elk," declared natural regulation to be the preferred management approach. Rather than by shooting, elk populations would be determined by "winter food [availability], by periodic severe winter weather and native predators." The Park Service asserted that "historical and recent knowledge" indicated that such factors would limit the number of elk. (Park biologists soon came to view predators as a much less important factor than the other two.)[118]

With management primarily focused on the most conspicuous plants and animals, the Park Service had, in effect, always practiced a form of natural regulation of the less obvious species by ignoring them (although many were affected by fire suppression, forest insect and disease control, and other park activity). As the Service's earliest official natural resource policy, Fauna No. 1 declared that "every species" that was not threatened with extinction in a park should be "left to carry on its struggle for existence unaided." Natural regulation theory began to be more fully articulated in the 1950s and 1960s as an alternative to artificial control of ungulate populations. However, despite claims of "historical and recent knowledge," the Park Service had virtually no scientific data on the overall ecological effects of a naturally regulated elk herd on the northern range. Nor, since the Service lacked supporting data and had prepared no formal research plans, did it begin the politically motivated program as a truly scientific experiment. Initiated as a comprehensive program without prior testing, it was more than anything else a political experiment.

One of Yellowstone's own scientists, William J. Barmore, protested to the park's lead biologist, Glen Cole, that the Service was "abruptly 'scrapping' current objectives with no . . . supporting information." He seriously doubted that the park could "come up with a satisfactory explanation of the proposed change [to natural regulation] on the basis of objective information available *at this time.*" In fact, in February 1967, immediately before the policy change that had a reasonable chance of resulting in an increase in elk population, Barmore had presented a professional paper stating that aspen on the northern range were in poor condition from overbrowsing by "excessive numbers of elk" that were blocked from their traditional winter grasslands by development outside the park."[119]

In adopting a natural regulation policy, the Park Service disregarded the urgent call from both the Leopold and National Academy reports for scientifically grounded decisionmaking. It also went against the Leopold Report's recommendation that elk reductions continue (a recommendation made in one instance in the context of the report's discussion of "habitat manipulation"—the exact opposite of natural regulation). Although the report had laid out a series of elk management options, it clearly favored direct reduction of the herd, stating that "direct removal by killing is the most economical and effective way of regulating ungulates within a park." At the March 1967 Senate hearings, Leopold himself reaffirmed the need for direct reduction, stating that he had not changed his mind on the matter. Nevertheless, in a somewhat disingenuous effort to justify its change of elk policy, the Service asserted that it was acting in accord with the Leopold Report. As Hartzog stated in his announcement of the new elk policy, it was "based on the recommendations approved by an advisory board to the Secretary of the Interior" (the Leopold Committee). Similarly, Yellowstone superintendent Jack Anderson held that the Service had "followed the recommendations" of the Leopold Report.[120]

Maintaining close involvement with the national parks, Leopold held to his belief that the natural regulation policy was resulting in overgrazing and deterioration of the northern range—a position shared by other critics. In a June 1983 interview—more than a decade and a half after the policy was initiated—he remarked that because of increased elk browsing Yellowstone's "aspen patches . . . shrink every single year. . . . The aspen are simply just vanishing." Blaming in part the absence of recurring fire in the area, he believed that the "other part [of the problem] is elk chewing the remaining aspens." The elk "simply girdle" the trees, "eat the bark right off, and the aspens die and fall down and disappear."

That same month Leopold discussed the issue more fully in a letter to Sequoia–Kings Canyon superintendent Boyd Evison. Coming only a few weeks before Leopold's sudden death, this may have been his final statement on natural regulation. Worried about the "progressive disappearance of aspen," he stated that when ungulates are "destroying vegetation, they should be reduced in number, by predators if possible, if not, by trapping or shooting." He believed that such management issues "are not resolved simply by 'allowing natural ecosystem processes to operate.' " To Leopold, the national parks were "too small in area to be relegated to the forces of nature that shaped a continent."[121]

Although for many species natural regulation had in essence always operated in national park management, the Park Service had acted almost as if the Leopold Report had "discovered" natural regulation. The Service then applied it to elk management and embraced it, proclaiming it to be sound policy. Yellowstone became perhaps the chief focus of the natural regulation policy—in part because the policy significantly affected management of large mammals of interest to the public, animals that Yellowstone had in much greater numbers than other parks in the contiguous forty-eight states. Also, at about two million acres, the park was large enough to encourage belief that it contained some approximation of a "complete ecosystem," where natural regulation of large mammals might be feasible.

Grizzly Bears

Management of Yellowstone's grizzly bear population—another major move toward reestablishing natural conditions in the parks—sparked an angry dispute and again revealed the Service's ambivalence toward scientifically based management. The grizzly bear controversy began in the late 1960s, when Yellowstone superintendent Jack Anderson, backed by his lead biologist, Glen Cole, decided to close the park's garbage dumps, a reliable source of food for the grizzlies since the 1880s, virtually since tourists began coming to the park. This plan was intended to put the bears on a more natural regimen, where they would not depend on food supplies at the dumps and would disperse across the park seeking natural sources of food. As with the ending of elk reduction, the Service claimed that its new grizzly bear management was in accord with the Leopold Report.[122] However, focused more on ungulates, the report had not analyzed bear management in detail, leaving it essentially open-ended as to manipulation or natural regulation. The decision to close the dumps brought the Service into conflict with the recommendations of John and Frank Craighead, biologists (and twin brothers) who had been studying Yellowstone's grizzlies intensively since 1959 and were recognized as the world's leading experts on this species.

The Craigheads (who were not Park Service scientists) believed that if certain precautions were not taken, closure of the dumps would threaten the grizzlies' survival in the park. They judged that since late in the previous century, when garbage dumps had first attracted grizzlies, development and use of once-primitive lands in and adjacent to the park had possibly reduced the bears' natural food supplies below what was necessary to support a viable grizzly population. But the Park Service overrode this argument. Although it had no systematic population survey of its own, it asserted that the Craigheads had underestimated the number of grizzlies in the park, and that the bears had survived in the area for millennia and could continue to do so.[123]

The dispute narrowed to whether the dumps should be closed suddenly or gradually. The Craigheads argued that a gradual, monitored closing would give the grizzlies time to adjust and thus have less impact on their population. Entwined with this concern was the factor of human safety—whether the dispersal of bears seeking food after a sudden closing would be a greater threat to campers and hikers than after a gradual closing. All parties were keenly aware of the August 1967 incidents in Glacier National Park when, on a single night and in widely separated areas, two women were mauled to death by grizzlies. These remarkably coincidental killings had brought pressure on the Service to reevaluate its bear management. After first trying gradual closing, Superintendent Anderson concluded that a quick closing was safer for both humans and bears. In the fall of 1970, he abruptly announced that the last big dump—at Trout Creek, south of Canyon Village—would be shut down.[124] Following this decision, the controversy shifted to a kind of grim, competitive watch, with both sides counting population figures year to year to see how well the grizzlies survived.

Underlying the disagreements was the question of scientific research to enable the park to make informed management decisions on the grizzlies.

Since Stephen Mather's time, the Service had used the availability of outside scientists as a rationale for not strengthening its own research capability—an attitude still pervasive in the late 1950s when the Craigheads began their studies. Indeed, their research funds (ultimately more than a million dollars) came from a variety of sources, including the National Science Foundation, National Geographic Society, Philco Corporation, and Bureau of Sport Fisheries and Wildlife. The Park Service did not support the Craigheads substantially, covering only a small fraction of the cost, much of it in the form of staff and logistical support. Work space in an unused mess hall was provided by the park concessionaire.[125] Operating with limited Service support, the Craigheads' studies became what was at that time the most in-depth natural-history research ever conducted in a national park.

Still, an acrimonious debate arose over the Craigheads' progress in publishing their research and whether the information they made available was adequate to determine the effects that dump closure would have on the grizzlies. Rejecting the Craigheads' recommendations and asserting that their research did not address the specific concerns at hand, Superintendent Anderson closed the last dump. As had happened for decades—including the termination of the elk reduction program—the Park Service made a key management decision with little scientific information of its own.[126]

The disagreements intensified the fractious professional and personal differences that had arisen between the Craigheads and certain park staff. Early in the research project, the relations had seemed cordial and supportive. But in Frank Craighead's opinion, after Anderson's and Glen Cole's arrival in the park in 1967, the situation became increasingly "characterized by mistrust, suspicion, and . . . hostility." Part of the problem stemmed from the Craigheads' use of the public media. Even before beginning their Yellowstone research, the brothers were well-known naturalists—a "glamour family within the wildlife establishment," as one writer put it. Their grizzly bear studies attracted even greater attention, giving them a public platform from which they at times criticized park management.[127]

Park management's attitude toward research (and toward the Craigheads themselves) was clearly revealed when the Craigheads requested permission to continue monitoring the dispersal of the grizzlies following final closure of the dumps. This involved tracking the animals by means of multicolored tags, which the researchers had attached to a large number of bears (as well as some elk) for identification and tracking purposes. Their request, coming at the height of acrimony between the two sides, was rejected by Superintendent Anderson, who characterized the colored tags as an unwanted intrusion into the natural scene. Supported by biologist Cole, Anderson rejected the Craigheads' request and ordered that the tags be removed from any bears captured by park rangers for management purposes, thereby thwarting research use of the tags.[128]

The superintendent asserted that the public had complained about the colored tags, pointing out to John Craighead that there had been a "great deal of comment from the park visitor attempting to photograph the wildlife in their native habitat." Anderson believed that the tagging had "reached the point where it detracts from the scenic and esthetic values," and he wanted as many tags as possible removed by the time of the Yellowstone centennial, to be celebrated in the park in the summer of 1972. Thus, the park's excuse for obstructing this final aspect of the Craigheads' research was based on the claim that the tags, in effect, decreased public enjoyment of Yellowstone. The National Academy's 1963 report had specifically recommended that the Service "avoid interference with independent research which has been authorized within the parks," citing problems that had occurred in Mammoth Cave and Shenandoah. Chaired by Starker Leopold, a science advisory committee that met in the park in September 1969 had urged that the "response of [the bears] to the elimination of garbage" be studied. Yet to Anderson and Cole, the colored ear tags on an elusive animal rarely seen by the public were an intrusion on the natural scene and had to go. The park had effectively blocked the bear dispersal research.[129]

Anger and discord surrounded this celebrated conflict over the grizzlies, and a cloud of uncertainty and distrust still remains. Reflecting on the controversy more than a decade after its onset, Nathaniel Reed, who as assistant secretary of the interior had been a close observer of the dispute, voiced his opinion that "mistakes have been made" and "neither the Craigheads nor the Park Service have a perfect record." The Service's actions were, however, more crucial than those of the Craigheads, because it had the legal responsibility and decisionmaking authority to safeguard the public trust through ensuring survival of Yellowstone's grizzlies.[130] In making its decisions, the Service rejected the advice of internationally recognized experts who had studied the bears for more than a decade. The Craigheads estimated the grizzly population to be fewer than two hundred and believed that the dump closure increased the risk that the bears would become extinct in the park. During the first two years after closure, approximately eighty-eight grizzlies were killed in or near Yellowstone, mainly to ensure human safety. Even with this number slain, the grizzlies survived; but in Frank Craighead's opinion, there had been "very little margin for error." Indeed, in 1975, shortly after what had been by far the most intensive killing of grizzlies in the park's history, the grizzly was placed on the list of threatened species, pursuant to the Endangered Species Act.[131]

In the push toward natural regulation and in a concern for safety, the Park Service had been in a sudden hurry with grizzly bear management. It seemed compelled to change a feeding policy that had existed for nearly a century, during which time it had had ample opportunity to conduct its own research on the bears but had neglected to do so.

The Service began to expand its knowledge of the grizzlies in 1973 with the initiation of a bear monitoring program. That same year the Interagency Grizzly Bear Study Team was created to undertake long-term scientific studies; it included biologists from the Park Service, the U.S. Fish and Wildlife Service, the U.S. Forest Service, and the state governments of Wyoming, Idaho, and Montana. The 1975 listing of the grizzly as a threatened species triggered a close evaluation of the bears' critical habitat and the development of a "recovery plan" for the species.[132] Grizzly habitat had already been recognized as including expansive tracts of lands surrounding the park, an area constituting the central portion of what came to be called the Greater Yellowstone Ecosystem. The Craighead studies and the grizzly bear controversy helped spawn a coordinated approach to management of this species by federal and state agencies. Although the disagreement and controversy did not end, through extensive research the new approach sought to improve understanding of the grizzly and how it might best be managed.

Forests

Concurrent with its new emphasis on the natural regulation of wildlife, the Park Service moved toward a policy of restoring natural conditions in plant communities. Previously, plant ecology had received little attention from the Service. Management of national park flora had been mainly either the domain of foresters or adjunct to the management of ungulates, with primary focus on ensuring adequate range for grazing. Despite the concerns of the wildlife biologists, control of insects, disease, and fires had continued unabated. But the new policies signaled an eventual end to *total* fire suppression and to extensive disease and insect control in park forests.[133]

Addressing the problems of traditional forest management, the Leopold Report had raised "serious question" about the wisdom of "mass application of insecticides in the control of forest insects," where "unanticipated effects on the biotic community . . . might defeat the overall management objective." Spraying, the report emphasized, should be discontinued until "research and small-scale testing have been conducted."[134]

The report also advocated a change in fire policies, viewing the "controlled use of fire" as the most "natural" means of managing vegetation. Controlled burning could help restore the prefire suppression density of forested areas, after which "periodic burning" could be "conducted safely and at low expense." Of specific concern was the situation in Sequoia and Yosemite, where areas long protected from fire had developed dense understory vegetation. In what would become a much-quoted phrase, the report stated that such overgrown areas were like a "dog-hair thicket." They were a "direct function of overprotection from natural ground fires." This accumulated fuel was "dangerous to the giant sequoias and other mature trees" because of the potential to cause abnormally hot and more damaging fires. The Leopold Committee believed this situation should be of "immense concern" to the Park Service.[135]

The Service, however, initially resisted changes in forest policies. Bolstered by continued funding from the 1947 Forest Pest Control Act, control of forest insects and disease had remained a vigorous program. In accordance with the act, a federal review board annually examined park budget requests for pest control; and in an August 1963 response to the Leopold Report, the Park Service stated that to date no park projects had been disapproved by the board. It also noted the support of the Fish and Wildlife Service and the Forest Service in planning and implementing the insecticide programs. Clearly identifying the program's goals with public enjoyment, the Service maintained that spraying insecticides in national parks was "restricted to areas of heavy public use where high value trees and the forest scene must be maintained."[136]

Especially following the 1962 publication of Rachel Carson's *Silent Spring*, pesticide use became a national concern. Nevertheless, extensive spraying continued in the national parks, covering much greater areas than were indicated in the Service's August 1963 statement on restricted use. For instance, in the mid-1960s (and in conjunction with similar efforts by the U.S. Forest Service on adjacent national forests), Grand Teton National Park began a three-year, million-dollar pesticide program to eradicate a native park insect, the bark beetle, from much of the park's backcountry. Voicing objections he had long held, the recently retired Service biologist Adolph Murie denounced spraying in the Tetons. In a 1966 *National Parks Magazine* article Murie wrote that, in the interest of "saving park scenery," spraying would "disrupt natural relationships between beetles and lodgepole pine," the host plant for the bark beetles. He placed this "destructive operation," in which Park Service crews systematically killed off native species, "in the same category" as coyote control, believing it destroyed "natural conditions and fundamental ideals" of the Service.

Murie blamed the current practices on his longtime adversaries the foresters, who were, he claimed, "frustrated" because they could not "practice their professions." He asserted that there was "little, if anything, for a forester to do" with national parks because there was no logging or other "commercial operations dealing with trees." Murie judged that many Service employees, influenced by Rachel Carson, opposed insect control programs, but that "top administrators" had been "conditioned to accept bug control as sacrosanct, normal park dogma" and were hesitant to terminate a long-standing program. Illustrating a glaring Park Service double standard, he quoted one "high ranking" official as saying that the Service would "'wring some poor woman visitor's neck for picking a flower and at the same time permit bug people to spray trees, kill large areas of vegetation and pollute the soil.'"[137]

Replying sympathetically to these concerns, Assistant Secretary of the Interior Stanley Cain asserted that the Park Service was already "changing its attitudes and programs in the direction Murie wants it to go." Yet only gradually did the changes take place. Even after the Forest Service encouraged early termination of the Grand Teton spraying, having decided it would do no good in the long run, park management stalled. The Park Service was hesitant to end the program because it benefited the local economy through the creation of jobs. But by 1968, the Service's official policies placed tight restrictions on control of native insects and forest diseases, which were recognized as "natural elements of the ecosystem." In the 1970s and 1980s, widespread use of chemical biocides was replaced by the restrictive Integrated Pest Management program. Intended to avoid use of chemicals except when absolutely necessary, this program would emphasize natural controls with minimal environmental effects, including use of naturally occurring predators and disease agents.[138]

The Park Service was equally reluctant to change its fire policies. In response to the Leopold Report, Director Conrad Wirth (surely influenced by the bureau's tradition-bound foresters) had stated that although "less intensive" fire control deserved serious consideration, "no change" in policy was "contemplated at this time." Through much of the 1960s, the goal of total suppression of forest fires in the parks remained in effect. In the opinion of Park Service fire management expert Bruce Kilgore, the public seemed "quite ready" to accept a "reasonable explanation" of new fire policies, but the Service did not. The "biggest problem was within our own agency," he observed. Certainly the Service's top forester, Lawrence Cook, resisted change, withholding for months the release of a mid-1960s study of fire ecology in Sequoia and Kings Canyon national parks that threatened the total fire suppression policy to which he adhered.[139]

Despite such reaction, fire-related research accelerated, and fire was an important catalyst for the study of plant ecology. The Service's initial research into fire, conducted by biologist William Robertson in Everglades in the 1950s, had indicated the decline of native sawgrass and pines in areas of the park where fire suppression had been vigorously conducted. This prompted the park to set fires to simulate fire's natural role. Robertson's objectives in his study suggested the necessarily close connection between understanding fire and understanding plant ecology. Setting goals that future fire researchers would also pursue, he hoped to determine the effect of fire on soils; the "effect of burning on the vegetation, including plants killed and injury to those that survive the fire"; the "recovery of the vegetation after fire"; and the "probable course of development of the vegetation in the absence of fire."[140]

Fire concerns contributed to a greater integration of plant and animal research, and thus to a broader ecological understanding of the parks. For instance, research in the 1960s on fire in the giant sequoia forests focused on a broad ecological picture with a variety of interrelated topics. Biologists studied the effects of fire on native trees and on birds and mammals, as well as on sequoia seed and cone production: they studied the chickaree squirrel's role in breaking apart sequoia cones and releasing seeds, the relationship of invertebrates to the sequoia's reproduction and life cycle, the buildup of flammable debris under sequoias and other native trees, and forest succession when fire is suppressed. Similarly, the study of fire ecology led to a deeper awareness of the ecological influences of American Indian activity than had ever before existed in the Park Service. Robertson's Everglades study had mentioned the probability of extensive impacts from prehistoric fire practices in the area. In the 1970s researchers in Sequoia determined that fires set by many generations of Indians constituted a significant part of the fire history of that part of the Sierra Nevada. Within the Service this research opened the way toward an understanding that, particularly because of fire practices, areas largely untouched by European Americans but long used by Native Americans were probably not in a truly pristine condition.[141]

Fire ecologist Bruce Kilgore believed that outside pressure helped bring a change in Park Service fire management. The Service was, he wrote, "pushed considerably by certain conservation organizations," which were especially concerned that the giant trees of Sequoia National Park might be threatened by extraordinarily hot fires resulting from accumulated, unburned forest understory. Most of all, Kilgore credited the Leopold Report with being the true catalyst for change—it was the "document of greatest significance to National Park Service [fire] policy." In response to the report and its emphasis on fire's threat to the giant trees, Sequoia National Park took the lead in changing fire management policy. By 1968 the park had launched an aggressive program to reduce the "dog-hair thickets" threatening the big trees.[142]

Fundamental aspects of the change in policy were the recognition of fire's ecological role and the acceptance of fire as a valid means of management. In contrast to the Service's long-established suppression efforts, the 1970 management policies acknowledged fire as "one of the ecological factors" affecting the preservation of native plants and animals. Under closely controlled circumstances, certain fires could be allowed to "run their course" in parks.[143] Such "prescribed burning" came to include allowing selected naturally caused fires to burn, and purposely setting fires in designated areas to simulate natural fires—especially where suppression efforts in the past had seriously altered plant ecology. All other fires, however, were to be suppressed. Accordingly, the parks began preparing "prescription" fire plans that designated which areas needed burning and under what conditions (based on factors such as forest types, moisture content of the forest, humidity, wind, topography, and weather forecasts, as well as human safety and proximity to buildings and privately owned property). Over time, prescribed burning would undergo some refinement; although criticized at times, not well understood by the public, and perennially short of staffing and funding, the program would remain.[144]

The Park Service's official policy shift toward prescribed burning came several years in advance of the Forest Service's policy change. Concerned about flammable forest debris accumulated during the decades of total suppression, Sequoia's managers moved rapidly to begin prescribed burning. In addition, the forest understory was thinned by hand to reduce heat intensity and ensure that prescription burning could indeed be contained. The prescribed burning program spread from Everglades, Sequoia, and Kings Canyon to other parks. By the mid-1970s the Service had begun implementation in a dozen parks, including Yosemite, Yellowstone, Grand

Teton, Carlsbad Caverns, Wind Cave, and Rocky Mountain. Already parks had intentionally allowed more than six hundred fires to burn, covering nearly ninety thousand acres.[145]

In a 1974 press release the Park Service defended its new fire policies, identifying the dense understory in Sequoia as a "severe threat" to the big trees because it would "provide the fuel for devastating crown fires [in the tops of trees] that would kill these ancient monarchs." The release also stated that scientists believed that "not all fires are bad" and that some fires were "absolutely necessary" to maintain the "ecosystem of a park in its proper natural balance." In 1976 the Park Service announced an agreement with the Forest Service to allow "some naturally caused fires" (those that fit the fire prescription) to cross the boundary between Yellowstone and the adjacent Teton Wilderness, in the Bridger-Teton National Forest, thus extending the program beyond park boundaries into national forests, where similar policies were beginning to take effect. This agreement foreshadowed cooperative arrangements between the Park Service and other land-managing agencies. The Service affirmed its new fire policy for parks in its 1978 *Management Policies,* which stated that most fires are "natural phenomena which must be permitted to continue to influence the ecosystem if truly natural systems are to be perpetuated."[146]

Exotic Species

In addressing the question of naturalness posed by the Leopold Report, the Service reinvigorated its efforts to eliminate—or at least reduce—populations of nonnative species from the parks. Although some exotics barely survived or were a benign presence, other highly adaptable species, such as wild goats, burros, hogs, and the prolific kudzu vine, greatly expanded their territory, altering park habitat and threatening the existence of native flora and fauna. A 1967 report listed thirty parks with active programs to eradicate or control exotic plant species, and nine parks with exotic mammal control programs. The Service's new policies briefly mentioned exotics, declaring that nonnative plants and animals would be "eliminated where it is possible to do so by approved methods" (a reflection of policies recommended in Fauna No. 1).[147]

More than with exotic plants, attempts to eradicate exotic mammals sometimes precipitated difficult political problems for the Park Service. Strenuous objection to killing nonnative mammals—intensified by political pressure and by media coverage—came mainly from two sources. Animal-rights activists sought to protect appealing species such as burros, which seriously damaged native vegetation and caused erosion of topsoil in Bandelier National Monument and Grand Canyon National Park. Hunters opposed efforts by rangers to eradicate animals like the European wild boar in Great Smoky Mountains, where the voracious animal caused extensive damage to native vegetation. Hunting organizations in Tennessee and North Carolina wanted to maintain viable populations to ensure good hunting outside park boundaries, and some also demanded participation in any killings that took place within the park.

Such interest groups, often politically well connected, put the Park Service on the defensive; and the threat of litigation stimulated research to document habitat destruction by nonnative species. In the mid-1970s a reduction program was initiated at Bandelier National Monument. As recalled by biologist Milford Fletcher, the Service believed that, of the parks affected by burros, Bandelier had gathered the most scientific data on damage to soils and vegetation and could thus make the best legal case for eliminating burros. The program was promptly contested in court by the Fund for Animals. The court ruled that the Service had a legal mandate to remove such destructive exotic animals. Following an agreement to allow the Fund to attempt live removal—an effort that proved unsuccessful—the park completed eradication of the burro population. By contrast, passionate denunciation of the proposed shooting of burros in Grand Canyon led to a successful removal program of live trapping and transplanting, again largely undertaken by the Fund for Animals. This effort was supplemented by limited shooting (supported by the court decision at Bandelier) and by fencing off areas where burros might reenter the park.

Similarly, managers at Great Smoky Mountains initiated research on the wild boar population and the boar's effects on park habitats. In this instance, however, belligerent opposition by North Carolinians to rangers shooting wild boars in the park prompted the Service to devise a split policy. Under pressure, it discontinued killing the animals in the North Carolina part of the park, relying mainly on trapping and removal of the wild boars (aided in some instances by private individuals supervised by park staff) and on fencing. Rangers continued to shoot boars on the Tennessee side of the park.[148]

Faced with angry, outspoken opposition to the control of certain exotic mammals, yet aware of the damage the animals were inflicting on natural resources, the Park Service at times seemed caught in a no-win situation. Efforts to reach an acceptable compromise sometimes seemed awkward at best, as illustrated not only in Great Smoky Mountains, but especially in the attempt to control feral goats in Hawaii Volcanoes National Park in the

1970s. Despite decades of reduction (more than seventy thousand goats had been killed since the park was established in 1916), the goats maintained a high population, about fifteen thousand in 1970.[149] By the early 1970s, local pressure prompted the Park Service to allow hunters to participate in the reduction. To some, this highly unusual agreement to allow public hunting seemed justified because the animals being killed were exotics. (Contrary to the situation in Great Smoky Mountains and Hawaii Volcanoes, the public hunting issues in Yellowstone and Grand Teton involved both native animals and a federal law specifically allowing public hunting in Grand Teton as a means of population control.) However, even with hunting, reduction in Hawaii Volcanoes was not significantly diminishing the goat population, and an unfunded fencing program did not promise a solution in the near future. Thus, some park staff viewed the agreement to allow public hunting in the park as ensuring perpetual, "sustained-yield recreation" for the hunters, as then–park ranger Donald W. Reeser later stated.[150]

Moreover, in October 1970, to quell the hunters' apprehension that an ambitious proposed fencing program would jeopardize their opportunity to hunt in the park, Director Hartzog made a public promise that he had "no intention of exterminating goats from Hawaii Volcanoes National Park." The Service adopted the position that it wanted to "control goats, not to eliminate them." Reeser recalled that after Hartzog's pronouncement a perpetual "goat ranching operation loomed on the horizon as [the park's] new goal." Submitting to local pressure, the Park Service had strayed far afield from its official policy of eliminating nonnative species "where it is possible to do so."[151]

The strategy of reducing but not eliminating the goats was a clear instance of disregard of policy in an effort to achieve a political solution. The Service got into even greater difficulty when its new goat policy drew criticism from conservation groups, angry that Hawaii's native resources were being sacrificed to the goats and to the hunters' lobby.[152] Park Service leadership then resorted to the argument that "it is conceivable" that the goats benefited the park by keeping some exotic plants from spreading. In June 1971 Hartzog wrote to Anthony Wayne Smith, head of the National Parks and Conservation Association, that "some of [the exotic plants] may be held in a state of equilibrium by the pressure of the exotic goat."[153]

Yet the director did not even have the support of his own on-site scientific staff. Hawaii Volcanoes biologist Ken Baker characterized as "poor thinking" the idea to "perpetuate goats as biological controls on exotic plants." Baker saw this as an attempt to "evade the issues and . . . not really tell it like it is." He feared that the situation actually amounted to public hunting "in perpetuity," and that "a rose by any other name is still a rose. What we have is nothing more than public hunting." So long as there are goats in the park, he declared, "we are only kidding ourselves about 'restoring and maintaining natural ecosystems.'" In line with Baker's thoughts, the park superintendent, Gene J. Balaz, wrote to the National Parks and Conservation Association in early May 1971 (six months after Hartzog had disavowed any intention of eliminating the goats) that "the aim of this program is to reduce the number of goats—any goats." In June, with Balaz's determination to remove the goats at odds with Hartzog's statements, the director abruptly removed Balaz from the Hawaii Volcanoes superintendency.[154]

Only when the park moved determinedly ahead (at first on its own) with a fencing and killing program aimed at eliminating the goat population was the Service able to correct its course and stay in line with official policy. In late 1971 newly arrived park superintendent G. Bryan Harry stubbornly pushed forward a major fencing program, drawing heavily from the park's annual maintenance funds, until the Washington office finally agreed to provide special funding to construct the fences. Indeed, given the goats' reproductive capacity, almost certainly the only long-range solution lay in fencing important habitats, coupled with killing the goats in and near the areas being fenced. A three-thousand-acre tract enclosed by July 1972 became, Donald Reeser recalled, the "first area of goat range that had been made lastingly free of goats" since the park's establishment. A dramatic recovery of vegetation after exclusion of the goats provided substantial reason for continuing the program. By 1980, goats in Hawaii Volcanoes National Park were, in Reeser's words, "virtually gone." A few remain today in remote, unfenced areas.[155]

With greater understanding of nonnative species in the parks came realization of the magnitude and persistence of the exotics issue. Certain exotics can be successfully contained in some parks but, overall, the problem will never go away and will continue to create management quandaries. In 1981 Olympic National Park managers initiated a program to rid the park of appealing but habitat-destructive mountain goats, on the grounds (hotly disputed by opponents of the program) that the goats are not native to the park, although they occur naturally very close by. Even at Isle Royale National Park, where efforts have been made to preserve wolf populations, managers have had to confront the question of whether wolves (as well as moose, their chief prey) are truly native to the park.[156]

The State of the Parks Reports

Reflecting on the National Park Service's fragmented, ambiguous natural resource management programs, a work session on Science and the NPS at the 1978 superintendents conference deliberated on the serious deficiencies in scientific research. In somewhat milder language than the National Academy of Sciences had used in 1963, the session participants reported that there were "significant gaps" in basic knowledge of natural resources, that "decisions are being made on the basis of inadequate information," and that natural resource management programs "desperately need strengthening."[157] This admission by the superintendents appeared in an internal paper and, given past experience, could be absorbed within the bureau without a ripple.

The next year, however, a National Parks and Conservation Association report—the "NPCA Adjacent Lands Survey"—appeared in the winter and spring issues of in the association's magazine. It would lead to a major investigation of Park Service science and natural resource management. The survey, based on an assessment of park conditions, emphasized the great variety and potency of influences originating outside national park boundaries and threatening the remaining integrity of the parks' natural conditions. Such "external threats" (as they became known) included air and water pollution, clear-cutting, and intensive development. The association warned that the parks were being treated like "isolated islands" and that unless the external threats were seriously confronted, traditional efforts to preserve the parks from within would be "rendered meaningless."[158]

As with the Leopold and National Academy reports, this catalyst for improving science in park management came mainly from outside the Park Service. The Service had already been made aware of perils from activity near park boundaries, especially in cases such as alterations to the South Florida water system that affected the Everglades and clear-cutting adjacent to Redwood National Park. Indeed, the impacts of logging on contiguous lands had prompted a declaration in the Redwood National Park Expansion Act of 1978 encouraging protection of national parks from threats outside their boundaries.[159] Heretofore the Service had never analyzed the external threats collectively as a special type of problem for the parks, nor had it aroused the public to their seriousness and national scope. The National Parks and Conservation Association's 1979 study significantly enhanced awareness of such factors, ultimately leading to the emergence of external threats as not only an enduring part of the Park Service's lexicon and its policy and budget deliberations, but also as a widespread public concern.

As a result of the association's report and subsequent lobbying efforts, Congressmen Phillip Burton and Keith G. Sebelius, ranking members of the House Subcommittee on National Parks and Insular Affairs, requested that the Service make its own study of the condition of the parks. Assigned to compile the study, Roland Wauer, head of the natural resource management office in Washington, devised a questionnaire on park conditions and polled all superintendents. The ensuing report, entitled *State of the Parks—1980: A Report to the Congress*, amplified the National Parks and Conservation Association's study and, in addition to external threats, included data on problems originating within the parks, such as those caused by management actions or park use by visitors.[160] This document prompted the most significant boost to scientific resource management in the parks since the National Academy and Leopold reports.

Under Wauer's direction, State of the Parks was both comprehensive and candid. Rivaling in tone the National Academy study, the report noted that internal and external threats were causing "significant and demonstrable damage," which, unless checked, would "continue to degrade and destroy irreplaceable park resources." In many instances such degradation was deemed "irreversible." Among numerous specifics, State of the Parks revealed that "aesthetic degradation," air and water pollution, encroachment of nonnative plant and animal species, impacts of visitor use (wildlife harassment, off-road vehicles, and trail erosion, among others), and park operations (including "suppression of natural fires, misuse of biocides, employee ignorance") constituted the most damaging types of impacts. Although many threats resulted from activities within the parks, more than half came from external sources, such as commercial and industrial development and air and water pollution.[161]

In truth, the Park Service had not realized the variety and magnitude of the threats—an indication of the deficiency of its research programs. Seventy-five percent of the threats, the report stated, were "*inadequately documented.*" And "very few" parks had the baseline information "needed to permit identification of incremental changes" that could be affecting the integrity of natural resources. The report cited a situation that had in fact existed since the founding of the National Park Service. It noted that the "priority assigned to the development of a sound resources information base has been very low compared to the priority assigned to meeting construction and maintenance needs. Research and resources management activities have been relegated to a position where only the most visible and severe problems are addressed." The document concluded with an admission that the Service's scientific resource management efforts were "completely inadequate to cope effectively" with the many problems affecting the parks' resources. The Park Service, it stated, "publicly calls attention to this serious deficiency."[162]

These remarkably critical observations represented the scientific (rather than the traditional) perspective from within the National Park Service, and voiced the frustrations of those who had long advocated a strong scientific research program to inform park management. The report received attention in the national press, which "alarmed" some high-level officials in the Service and in the Interior Department, as Wauer recalled. He added that, having second thoughts after they "realized the visibility" of the report, Service leaders began "playing down" State of the Parks because they thought it "made the National Park Service look bad."[163]

State of the Parks made specific proposals for improving natural resource management. These included a "comprehensive inventory" of natural resources, programs to monitor changes in the parks' ecology, individual park plans for managing the resources, and increased staffing and training in science and natural resource management. But the document contained no firm commitment by the Park Service that it would act on the proposals. Indeed, the proposal section read as if it were prepared by individuals who had no power to enforce change, only to recommend it. Believing that the Park Service was vacillating, and with no specifics on how the proposals would be implemented, Wauer feared the Service might let the State of the Parks effort "fade away" unless Congress required action. His subsequent contacts with National Parks and Conservation Association representatives and with congressional staff soon prompted a request by Congressmen Burton and Sebelius for the Service to prepare a "mitigation report" documenting the exact steps by which the bureau's own proposals would be realized.[164]

In January 1981, following a period of intense data gathering, the Park Service submitted its mitigation report to Congress, as the second State of the Parks report. Articulating a complex, ambitious plan, the document included several significant points. As an immediate step, the Service pledged to prepare a list of the most crucial threats, which would receive the highest priority for funding in upcoming fiscal years. In addition, the Park Service would complete its resource management plans for each park by December 1981. This planning effort, long under way but never finished, would strengthen justifications for future budget submissions to

Congress. The resource management plans were to document the general condition of the resources, the necessary research, and possible management actions necessary to respond to particular problems. Finally, the Park Service promised a greatly expanded training program, to give superintendents and other personnel a better grasp of natural resource needs. Perhaps most important, through special training the Service would develop a stronger, more professional cadre of natural resource managers.[165] Together, these basic approaches constituted the most comprehensive, systemwide strategy yet devised by the Service to address the parks' natural resource problems.

Subsequent to the Leopold and National Academy reports of 1963, scientists had struggled for two decades to gain an effective role in national park management. Handicapped by a lack of experience in bureaucratic affairs, the scientists were the chief proponents of the ecological point of view in the Service—but they were confronted by leadership that embraced traditional practices and lacked a commitment to ecological management principles.

The Service had continued to respond to the pragmatic pressures of park operational needs, and the science programs never received the steady, continuing support given, for instance, to law enforcement in the 1970s, with greatly increased funding, personnel, and training throughout the park system. Nor did science get large, permanent facilities and consistent high-level support, as did interpretation. The scientists' role in the Denver Service Center's far-reaching planning, design, and construction programs remained extremely weak. And, in striking contrast to that of other key Service functions, the organizational status of the scientists fluctuated for a decade and a half before achieving long-term stability at a high level.

In the absence of sustained commitment from Park Service leadership, a strong push from outside the Service in the late 1970s and early 1980s finally motivated an earnest reconsideration of science in the national parks. With the Service having always operated without a specific science mandate from Congress, commitments made to Congress in the State of the Parks reports served as a kind of substitute—a nonlegislative scientific mandate. To establish accountability in its renewed effort, the Service pledged to submit progress reports to Congress. In addition, as noted in the second State of the Parks report, the National Academy of Sciences agreed to plan an "in-depth study" of the Park Service's science program—

to undertake a repeat performance of its 1963 study, which had been, in effect, suppressed. The State of the Parks endeavors thus gave the Park Service, as an internal paper proclaimed during preparation of the first report, a "golden opportunity" to set "new directions of conservation leadership."[166] In 1981, with prodding from Congress, the National Park Service had a renewed opportunity to revise its traditional priorities and develop an ecological perspective on park management.

CHAPTER 7

A House Divided: The National Park Service and Environmental Leadership

The National Park Service is a large, complex, and geographically dispersed agency with strong traditions in both its policies and its management styles. It will not be transformed quickly or easily.—THE VAIL AGENDA, 1993

In looking back on the span of national park history, it is the infusion of an ecological and scientific perspective that constitutes the most substantive difference between late-nineteenth-century and late-twentieth-century natural resource management in the parks. As long as management emphasized little more than preserving park scenery, it did not require highly specialized data and an in-depth understanding of the parks' natural phenomena. The emergence of ecological concerns, however, necessitated scientific research in the parks as the only real means of comprehending the mysteries of the complex natural systems under the Service's care. This need fostered a slow buildup of Park Service ecological expertise, principally scientists and natural resource management professionals. Beginning mainly with the environmental era of the 1960s and 1970s, scientific and ecological factors became the chief criteria by which the Park Service's natural resource management—and much of its overall management—has since been judged. The State of the Parks reports reflected such criteria; and proponents of the reports looked forward to improvements in resource management.

In October 1991, a decade after the State of the Parks reports were issued in the early 1980s, a major conference on the national parks was held in Vail, Colorado, to commemorate the Service's seventy-fifth anniversary. Attended by several hundred experts from inside and outside the Service, the Vail conference reviewed the status of national park management and deliberated on future prospects. The meeting focused on several topics of special concern, among them the question of "environmental leadership"—by what means should the Service "embrace a leadership role" in sound ecological (and cultural) management? A draft report prepared in advance of the conference cited shortcomings in natural resource management and noted that in recent decades authorities had repeatedly called for a "strong science component" in the Park Service. But, the draft report acknowledged, the bureau's reaction had been "sporadic and inconsistent, characterized by alternating cycles of commitment and decline."[1]

Indeed, since the advent of the Leopold Report in 1963, critics had seized on reports by experts as a means of pressuring the Park Service to undertake resource management that was truly informed by science. Through these reports the Service was, in effect, being exhorted to assume "environmental leadership" in public land management. Among more than a dozen such efforts were studies in 1967 and 1972 by the Conservation Foundation; the 1977 report by Starker Leopold and Durward Allen; the two State of the Parks reports; and several studies by the National Parks and Conservation Association, including a multivolume work in 1988 and a special report in 1989. Seeking to enhance recognition of its 1989 effort (entitled *National Parks: From Vignettes to a Global View*), the association billed it as the successor to the Leopold Report.[2]

Like the Leopold study before them, these reports were promoted by scientists and environmentalists and presented a strong proscience message. Similarly, the 1991 draft Vail report recommended "sound ecological management" backed by a well-funded research program as a means of asserting environmental leadership. Following the Vail conference, two additional reports expressed an urgent need for park management based on scientific research. In 1992 the National Academy of Sciences issued *Science and the National Parks*, an extended analysis of the role and status of science in the Service, and the academy's first report on the parks since its 1963 effort; and in 1993 came official publication of the Vail Agenda, the summary report of the findings and recommendations of the Vail conference.[3]

Had the Park Service's response been resolute rather than "sporadic and inconsistent," there would have been little cause for such repeated, intense scrutiny. The reports (including those generated from within the Service itself) amounted to a litany of criticism and demands for improved scientific resource management. But the Park Service had responded with its own litany—of promises to make substantive changes. Although the Service had increased its scientific efforts, its reluctance over a long period of time to address the issue forthrightly and establish a truly "strong science component" makes its promises seem largely rhetorical.

Building an Environmental Record

One of the significant changes repeatedly recommended in the reports was for the Service to inventory the parks' natural resources and monitor their condition over time. Without such data, a scientific understanding of the parks could not be achieved and any Park Service claim to leadership in environmental affairs would be seriously undermined. Virtually every report emphasized the need for this information—and the Service accrued a considerable history of promises, each followed by resistance and procrastination.

Long before the external reports began to appear in 1963, the Park Service had declared its intention to inventory and monitor species. Made official policy in 1934, Fauna No. 1's wildlife recommendations included the charge to undertake for each park a "complete faunal investigation . . . at the earliest possible date." Although making little progress, the Service repeated its commitment to this task through the 1930s and during World War II—for instance in a February 1945 report on research. Such declarations became more common in the environmentally conscious 1960s. The 1961 internal document "Get the Facts, and Put Them to Work" recognized the need for a "continuous flow of precise knowledge" about park resources. Two years later, Director Conrad Wirth stated that the insistence of the National Academy Report on inventorying and monitoring in the parks was a "basic recommendation"—that it would "be implemented as rapidly as possible." And in October 1965, the Service reiterated its commitment to prepare "an inventory of existing biotic communities" in the parks.[4]

Fifteen years later, the Service issued its first State of the Parks report, aimed at gaining congressional support and funding for the Service's resource management and science programs. The report admitted that there was a "paucity of information" on park conditions and called for "comprehensive inventory" and "comprehensive monitoring." Several large parks, such as Great Smoky Mountains, Shenandoah, Everglades, and Yellowstone, did begin to make some headway. Responding to 1980 legislation for Channel Islands National Park that called for an analysis of species to determine "their population dynamics and probable trends as to future numbers and welfare," Channel Islands developed an ambitious inventorying and monitoring program. In addition, the Service substantially increased its monitoring capabilities for air and water quality in the parks.

This progress was offset by widespread neglect—in spite of the need for data to comply with the National Environmental Policy Act and especially with the Endangered Species Act. In a 1988 commentary on inventorying and monitoring, journalist Robert Cahn, awarded a Pulitzer Prize for an earlier analysis of national park issues, reported that "possibly the greatest failure" in Park Service history was the bureau's not having gained "solid knowledge" about park resources through "systematically identifying them and regularly determining their condition." A similar charge appeared in the 1989 National Parks and Conservation Association report on park management. And the same 1991 Vail conference draft document that exhorted the Park Service to "embrace a leadership role" in environmental affairs noted that more often than not the Service knew "little about the actual resources parks contain, their significance, degree of risk, or response to change." As had others before it, this document urged a "comprehensive program" to inventory and monitor park resources. In 1993, six decades after Fauna No. 1 and three decades after the Leopold and National Academy reports, this entreaty was repeated in the Vail Agenda.[5]

Such vacillating, sporadic support as was given the inventorying and monitoring programs mirrored to some degree the Service's overall response to the State of the Parks reports, the principal attempt of the 1980s to bring a scientific, ecological perspective to national park management. By 1982, only a year after the later of the two reports appeared, the Service's response had begun to "lose its focus," as William Supernaugh noted in a detailed study of the State of the Parks results. Reorganizations and shifting priorities in the Washington office reflected a weakening of resolve by the Park Service directorate. Supernaugh, who was in Washington during the early 1980s, recalled that the professional staff there had been "transferred, restructured, given contradictory assignments and unclear instructions, ignored, rewarded, had their jobs abolished, transferred or redescribed, served under three Associate Directors in two separate organizational lines, [and] served through a succession of Division Chiefs." He stated that the "end result" had been "an overwhelming lack of consistency of purpose and continuity of direction."

This situation was exacerbated by the lack of support from Interior Department officials under Secretary James G. Watt. Progress reports to Congress and other internal follow-up reporting procedures recommended by State of the Parks were soon abandoned. In the face of opposition from Secretary Watt, a National Academy study of science in the parks (also called for by State of the Parks) was postponed, not to be undertaken until the close of the decade.[6]

Other elements of the State of the Parks reports were in fact addressed. Prompted by the reports, the Service developed a variety of training courses in the 1980s to improve its natural resource capability, providing superintendents and other employees with a better grasp of ecological management principles and environmental law. This effort remained strong into mid-decade; then, with decreasing budgets and competition from other programs, the variety of courses and the number of trainees declined markedly between 1987 and 1993.

Most ambitious of the training efforts was a long-term course designed for the parks' natural resource managers. Begun in 1982 and modeled on training developed in the Southwest Region in the 1970s, this course initially extended over a two-year period, during which the students divided their time between training and their regular assignments. But, like other natural resource training, it was soon cut back. In 1986 the program encountered stiff resistance from the regional directors, who sought to shift the funds to other uses. In reaction, Director William Penn Mott resorted to different funding sources and scaled down the course. With the concurrence of many natural resource managers who wished to improve the course, the Service began a reevaluation in the early 1990s. Soon, however, this effort was subsumed (and the long-term course was suspended) in an extensive reevaluation of all training programs.[7]

State of the Parks helped prompt increases in funding and staffing for scientific research and natural resource management. The decentralization of these programs, the lack of a Servicewide system for detailed tracking of funds, the perennially vague distinction between research and resource management, and park management's frequent shifting of funds all serve to make calculation of increases uncertain. However reliable they are, budget data from the time of State of the Parks through the early 1990s indicate that the overall natural resource management budget (including research) quadrupled between fiscal years 1980 and 1993, from approximately $23 million to just over $95 million. During the same thirteen-year period, the research portion of that budget doubled, from about $10 million to $20 million. (These figures do not reflect the declining value of the dollar.) By 1993, natural resource activities amounted to about 9.23 percent of the Service's operating budget of $1.03 billion, while scientific research remained quite low—just under 2 percent (not in addition to, but included within, the 9.23 percentage).

Between fiscal years 1980 and 1993, Servicewide natural resource management staffing increased more than fivefold, from just above 200 positions to 1,164. Throughout this period many of the positions were part-time and entailed duties other than natural resource management. Also, a large number were filled by "technicians"—individuals who undertook many resource management activities but generally did not direct them. There was little increase in the number of research scientists: among the total natural resource management positions for 1993 were 100 researchers—about the number estimated by State of the Parks for 1980.[8]

Thus, by the 1990s, attempts to improve the Park Service's scientific resource management through training, funding, and staffing had met with only partial success. Even with the increases in funding and staffing since State of the Parks, the National Academy in its 1992 report asserted that the Service's science program was "unnecessarily fragmented and lacks a coherent sense of direction, purpose, and unity"—an echo of the academy's statement thirty years previously that the science effort was "fragmented" and lacked "continuity, coordination, and depth." In 1993 David A. Haskell, chief of resource management at Shenandoah National Park, commented that the "critically needed focus on science as the basis for park management has not occurred." He added that there was "no definite signal" that the Service had "made the commitment to become a resource stewardship agency."[9]

Yet Haskell himself, working with Superintendent John W. (Bill) Wade, had built up Shenandoah's natural resource management program to include a sizable contingent of ecologists, biological technicians, and data management personnel. The park's inventorying and monitoring of flora and fauna, begun in the mid-1980s, made steady progress; so did its air and water quality programs. Among the ecological processes that the park began to monitor were stream aquatic habitat, watershed acidification, forest response to gypsy moth infestations, and deciduous forest watershed dynamics. Integrated into park operations, Shenandoah's natural resource program became one of the most effective in the system—and the kind of expertise and inquiry it utilizes today is mirrored by that of Channel Islands, Sequoia–Kings Canyon, and Yellowstone, among other parks.

While these parks made substantial advancement, many did not. Despite gains in recent years, a 1995 Washington office report noted a need for properly trained natural resource management specialists throughout the park system. Overall, the Service had an average of just over one per park; but the report added that, in truth, "there are a few specialists in a handful of parks and no specialists in many parks."[10] Trained resource managers were distributed unevenly throughout the system, and programs as strong as Shenandoah's were the exception rather than the rule.

This unevenness was but one factor accounting for the variability in quality of the Park Service's resource management during the 1980s and early 1990s. At Carlsbad Caverns, for instance, after the 1986 discovery of the vast lower regions of Lechuguilla Cave, the Service took a strong preservation stance, rejecting proposals to open the cave to tourism. Instead, it determined Lechuguilla's pristine natural habitat to be worthy of scientific exploration and study rather than subjecting it to the kind of intensive public use permitted in other of the park's caves. Exploration revealed a variety of rare geological, paleontological, and biological features, plus more than eighty miles of passages, making it the seventh-longest known cave in the world and the deepest limestone cave in the country. Well before the discovery, the park had established a special cave resource management position, the first such position in the Service. The incumbent, Ronal C. Kerbo, gained the park's support for having Lechuguilla officially designated a wilderness cave, which would have been the first designation of this kind. The effort proved unsuccessful, however, owing to both internal and external doubt about separately designating subsurface wilderness beneath an existing wilderness designation on the surface. Lechuguilla's prominence and the management debates it engendered helped prompt the Service to create a national cave management specialist position to assist park managers across the system with similar issues.[11]

On another front, continuing well into its fourth decade, the study of wolves and moose at Isle Royale has become the longest-running research ever conducted on mammalian predator-prey relationships in a national park, and perhaps in the world. Begun in 1958 by Purdue University biologist Durward Allen, the study (conducted mostly during winter) has been continued by his former doctoral student Rolf O. Peterson, a professor at Michigan Technological University. One of the most highly regarded research efforts in the national parks, the program has endured mainly because of the initiative and determination of Allen and Peterson. Although it never established a wildlife biologist position at the park, the Service at times provided significant logistical, funding, and political support. To prevent disturbance of both wolves and moose, and of the research itself, management officially closes the park each year from November 1 to April 15. And following a decline in the island's wolf population in the 1980s, the park made an important shift in policy. Anticipating considerable public scrutiny, it abandoned its long-standing adherence to natural regulation of the wolf and moose populations, which prohibited direct interference with either species (the researchers' observation of the wolves has always been conducted principally from aircraft during the winter months). Instead, the park approved a blood-sampling and radio-tracking program for the wolves in hope of determining reasons for their population decline.[12]

In Yellowstone, the natural regulation policy for the northern elk herd has never been rescinded, and it remains the source of recurring, heated controversy. Following the 1967 moratorium on reducing the number of elk, extensive research on the northern range got under way, addressing a broad variety of ecological questions. With some exceptions, Park Service biologists have maintained that natural regulation is working. Other scientists, mostly outside the Service, have asserted that the policy is destructive of range habitat; and in the mid-1980s, writer Alston Chase made a stinging attack on the environmental consequences of the park's natural resource management, particularly with regard to elk. Similarly, the National Academy in its 1992 report on park science criticized the Service for the "deteriorating condition of the northern range." The academy pointed out that the controversy over range condition "stems in large part from the lack of long-term data." Enlarging the perspective to the national parks in general, the academy observed that "substantial and sustained" research efforts were necessary to detect changes in habitat.[13] The Park Service's long neglect of science had crippled its recent research efforts and thus the credibility of its natural resource programs.

Somewhat like Alston Chase in his critique, in 1993 ecologist Karl Hess, Jr., charged that the failure to control Rocky Mountain National Park's elk population and to implement an approved prescribed fire program had caused serious modification of that park's ecological conditions. Both Chase and Hess believed that the ecological problems they discerned were a direct result of traditional management attitudes. In Chase's view, managers were so focused on visitor safety and protection and so indifferent to science that they embraced destructive resource management practices. Hess found parallel circumstances in Rocky Mountain, and both writers argued that the dominant park management culture had co-opted the scientists' independence and initiative.[14]

The impact of elk on habitats in Yellowstone and Rocky Mountain, the monitoring of air and water quality in Shenandoah, and the decline of Isle Royale's wolf population reflect not just the plight of park resources, but also the fact that parks are part of larger ecological systems and are readily affected by external influences. Even as deep beneath the surface and seemingly isolated as Lechuguilla Cave is, its exceptionally pristine qualities face potential external threats such as contamination by oil and gas development on nearby lands and unregulated entry and use should the cave be found to extend to areas outside Carlsbad Caverns National Park. Parks like Redwood, Big Thicket National Preserve, and Indiana Dunes National Lakeshore, which have small, fragmented land bases, are particularly vulnerable to the effects of nearby land uses.

With ecologically disastrous activity immediately adjacent to many parks—for example, clear-cutting next to Olympic and Redwood, and agriculture and hydrological manipulation upstream from Everglades—the Park Service has long been in situations badly in need of broader, more cooperative solutions to resource management issues. External threats highlighted in the State of the Parks reports to Congress were prompted by the National Parks and Conservation Association's 1979 study of adjacent lands. These reports sought to create a greater sense of urgency, systemwide, regarding uses of neighboring lands; subsequently, the Service made efforts to improve cooperation with local and regional land managers, both public and private. As this broader, more inclusive approach to resource management began to take on more specifically scientific aspects (such as the interest in preserving gene pools and biological diversity), it became subsumed under the term "ecosystem management." This phrase, employed by the Service's biologists since at least the 1960s, was adopted by management as a concept for addressing local and regional resource issues jointly with other land managers. Still loosely defined, it remains more a concept than a reality—the focus of frequent rhetorical flourish as well as serious deliberation.[15]

Perhaps the most prominent ecosystem management effort is in the Greater Yellowstone Ecosystem—a vast area (also not precisely defined) surrounding Yellowstone and Grand Teton national parks, and including national forests and wildlife refuges, as well as lands under the control of state and local governments and the private sector. The Greater Yellowstone Coordinating Committee, consisting of the chief federal land managers in the area, has sought to establish common ground in the management of grizzly bears, elk, wolves, fire, and tourism, among other concerns.[16] As fires swept vast expanses of Yellowstone and surrounding lands in the summer of 1988, the committee coordinated suppression activity; subsequently, it coordinated rehabilitation efforts. It also undertook a postfire assessment and review of fire policies in and around the park (a review was mandated not only for Yellowstone but also for those national parks, forests, and other federal lands with fire management programs). At issue were the appropriateness of Yellowstone's existing policies and the degree to which they had been implemented. Like other parks, Yellowstone had to update its fire management plan. Ultimately, although the revisions refined the tactics of fire programs in Yellowstone and other parks, they mainly vindicated existing fire policies and left their principles largely intact, including use of both natural and management-ignited prescribed fires.[17]

Greater Yellowstone is the scene of another highly controversial ecosystem management issue—the reintroduction of gray wolves, eradicated long ago from the park through aggressive predator control. A recovery effort for threatened and endangered species under the authority of the Endangered Species Act, the reintroduction project was conducted by the Park Service and U.S. Fish and Wildlife Service, with active support from other public and private interests. It culminated in the initial release of wolves in the park in March 1995. The basic goal of the program is to establish wolves in the Yellowstone ecosystem sufficiently that they are no longer listed as an endangered or threatened species in the area.[18] Widespread public interest in the recovery is engendered not only by the controversy invariably surrounding wolves, but also by the wilderness symbolism of the wolf and the effort to restore a key element of primeval Yellowstone.

As with fire policy and wolf recovery, ecosystem management relies on cooperative arrangements to influence regional land planning and use—an exceptionally difficult enterprise in an era of highly polarized debates over land use. Also involving complex working relationships with outside interests, the Service's "partnership" programs—with their roots in the parks and recreation assistance provided during the 1920s, and under Conrad Wirth in the 1930s—became particularly prominent (and acquired the "partnership" designation) in the 1980s. Through promoting parks and recreation projects to be developed and managed jointly with other public entities or with the private sector, these programs represent another Park Service effort at local and regional cooperation.[19]

The Vail Agenda

The National Park Service's success in ecosystem management will be determined in part by the level of commitment to ecological preservation within the parks themselves. By any measure, going into the 1991 Vail conference the National Park Service had not achieved a distinguished record in scientific resource management, despite six decades of prodding from concerned professionals within the bureau and three decades of pressure from external groups. Sponsored by a number of corporations, charitable trusts, and national park support organizations, the conference was intended as a means for the Service to undertake "constructive criticism, self-examination, and commitment to greater responsibility." In addition to "environmental leadership," the other "broad areas of concern" addressed by the conference were organizational renewal (analyzing personnel and career concerns and related aspects of the Service), park use and enjoyment, and resource stewardship. The last was especially pertinent to the question of environmental leadership.

However, the published report of the conference, the Vail Agenda (drafted mainly by Service staff), revealed the substantial shortcomings of self-examination and criticism, even when external authorities on national parks were involved, as they were at Vail. The most important review of the Service's management and operations since State of the Parks, the report nevertheless presented a confused analysis of what the bureau's true focus had been, and of what had and had not been accomplished in three-quarters of a century. For instance, perhaps in an effort to assure the environmental community that the Service was right-thinking, the report declared flatly that to "preserve and protect park resources has from the beginning been the primary goal of the National Park Service"—a statement that overlooked seventy-five years of mainly tourism-oriented management and sixty years of refusal to adopt a truly ecological perspective. The report also claimed the highest of resource management credentials for the Park Service, stating that it had "long been acknowledged as the country's leader in resource preservation" and was "being looked to as a model of conservation and preservation management" worldwide.[20] Such remarks demonstrate a clear presumption of environmental leadership by the Service, specifically in resource management.

In contradiction to such self-commendation, the Vail report sharply criticized the Service for "sporadic and inconsistent" support of science—overall, it was "extraordinarily deficient" in scientific matters and even "in danger of becoming merely a provider of 'drive through' tourism." In such comments, however, the report implied a positive record of *past* accomplishments, stating that the Park Service was "no longer" a leader in natural resource and environmental issues, and that it must "regain" its "former stature." This could be achieved by "reestablishing . . . respect and credibility" within the professional resource community.[21] These statements about *regaining* former status glossed over several decades of often harsh criticism leveled by scientists and other natural resource professionals of the Service's very failure to have achieved such status.

As in many previous instances, official rhetoric blurred the Park Service's response to criticism. It also obscured differences between the bureau's true historic strengths and its demonstrated weaknesses. The Vail Agenda combined claims of excellence with admissions of serious negligence—and in so doing it failed to distinguish tourism-oriented park management from scientific, ecologically based resource management. The Agenda's statement that the Service should strengthen its world leadership in "park affairs" reflected a more accurate understanding of the bureau's accomplishments and status in the field of general park management—that which is focused mainly on tourism, including attracting, accommodating, educating, and managing visitors.[22] Certainly, the Service was "being looked to," even internationally, as a "model" of general park development and management; yet it was admittedly "deficient" in scientific and ecological matters. The desire to regain its "former stature" more properly harkened back to the pre-environmental era of the New Deal years and even early Mission 66, halcyon days of park development for recreational tourism. At that time, before national concerns about ecological preservation escalated, the Service had enjoyed high status—not just with the general public but also with conservation groups. Then, such groups were less confrontational, rarely questioned natural resource management policies in the parks, and focused on the appearance of park development rather than its ecological impacts.

The chief strengths of the Vail conference may have been its recommendations concerning park use and enjoyment and organization; but regarding natural resources, the Vail Agenda broke little if any new ground. Most of its recommendations reflected those of previous studies, such as the perennial call for inventorying and monitoring park resources. Others included addressing external threats, improving cooperation with universities and with managers of neighboring public or private lands, educating the public on environmental issues, increasing and professionalizing Park Service staff (in part through better training for both park managers and specialists), increasing funding for science and natural resource management, and securing a legislative mandate for scientific research in the parks. The Agenda was, it acknowledged, confronting "challenges" that were "long-standing"—in truth, problems that a reluctant Park Service had never confronted wholeheartedly.[23]

Yet the Vail Agenda revealed that such reluctance might continue, particularly in light of the Service's refusal to give full-faith compliance to the National Environmental Policy Act, considered by many to be the keystone of environmental legislation. Addressing the important topic of how the Park Service might make "wise decisions regarding park use and enjoyment," the Agenda called into question the increased "legislative requirements" for the Service—especially the public involvement requirements stemming from the National Environmental Policy Act and related laws. Although it recommended that the Service "improve the public involvement process," its discussion of the issue actually showed little genuine enthusiasm. Perhaps reflecting on prior experience, it stated that the Service would accept increases in public involvement "either willingly or by legal and political coercion." Further vacillating on the matter of compliance with such legislation, it declared uncritically that "many park managers view the resource base as their client rather than society, and would prefer to make decisions about resources with little interference from the public that owns them." Indeed, the Agenda noted that the conferees were "unsure and divided" on this issue. But, implying support of the park managers' views, the report added that many believed that "there is already too much public involvement in NPS decision making."[24]

In the section on environmental leadership, the Vail Agenda recommended that the Park Service become "the most environmentally aware agency in the U.S. government," noting that it could demonstrate this through "leading by example." Yet the Agenda itself refrained from leading by example by not insisting on full-faith compliance with the National Environmental Policy Act, one of the country's most important environmental laws. As recently as 1990, the year before the Vail conference, an internal Park Service magazine published a special issue on this act, which included commentary by outside experts on the Service's record of compliance with the act. All of the commentators found fault. Most commonly criticized were the "attitudes of park managers and decision-makers," in the words of University of Utah law professor and environmental law authority William J. Lockhart. His impression was that far too many times the Park Service approached compliance "grudgingly—with the intent merely of going 'through the hoops,'" and he recommended "managerial humility" as a means of achieving "meaningful compliance." Similarly, Jacob J. Hoogland, Washington-based head of the Service's environmental compliance programs, noted deep-seated indifference, observing that the act had not produced a change in "the *attitudes* of . . . the National Park Service."[25]

Truly, the Park Service's irresolute compliance with the National Environmental Policy Act reflected a pattern—a long history of ambivalence toward the environmental movement, marked by failure to lead at crucial times. Cooperating with the Bureau of Reclamation in planning Colorado

River Basin reservoirs (including the Echo Park dam proposal in Dinosaur National Monument), the Service had helped bring about the Echo Park conflict, then was reduced to a negligible role in the final decision not to build the dam. It had withheld genuine support for passage of the 1964 Wilderness Act, then implemented it less than enthusiastically. And despite questions raised long before by its own biologists, the Service did not provide "leadership by example" through decisively curtailing pesticide application in the parks after pesticide use had become a major national issue in the 1960s. With such weak responses, Park Service management had remained largely out of step with the environmental movement.[26]

It is significant also that well before the Vail conference posed the question of leadership by the Park Service, the role of the national parks in environmental affairs had diminished nationwide. Such issues as elk and grizzly bear management in Yellowstone, the discovery and management of Lechuguilla Cave, the decline of the wolf population on Isle Royale, and the 1988 fires in Yellowstone at times brought national park management front and center among public environmental concerns. Yet in recent decades other issues, such as population growth, pesticides, toxic landfills, depletion of natural resources, accelerated loss of species, global warming, and clean air and water, have intensified—a reflection of the ever-expanding interests of environmentalists, far beyond specifically park-related matters.[27] The parks are not forgotten, but other concerns dominate; and although many of these concerns affect the parks, they are much broader in scope. Moreover, given the political circumstances within which the Park Service has had to function and survive, and given its fundamental interest in accommodating the public, it has always been unlikely that it would become a leading national voice on environmental issues not closely related to the parks. As a bureau of the executive branch, it has been very cautious in speaking out publicly on the specific actions of other federal (or state) land management bureaus. This has been true even when lands adjacent to parks are involved.[28] By contrast, the Service has been much more assertive in promoting environmental awareness in a broad, generic way, principally through its interpretive programs.

National Park Service Culture and Recreational Tourism

In contrast to its record in natural resource management, the Park Service truly can claim leadership in the field of recreational tourism—the development and management of parks for public use, enjoyment, and education. Indeed, the Northern Pacific Railroad's backing of the 1872 Yellowstone legislation had provided an important clue to the destiny of the national parks. The ensuing development of the parks for tourism—well under way by the early twentieth century—was affirmed in 1916 by the Organic Act's mandate to ensure public enjoyment of the parks. From 1916 on, the Service's "administrative interpretation" of the act has perpetuated the emphasis on accommodating tourism. Development and construction flourished especially during the Mather era, the New Deal, Mission 66, and the Bicentennial program of the 1970s. Backed by Secretary of the Interior Watt, the Park Restoration and Improvement Program of the early 1980s funded mainly the upgrading of existing park facilities, rather than new development. Other visitor-related programs, such as interpretation and law enforcement, also grew over the decades.[29]

Furthermore, through its own persistent lobbying and that of various national and local allies, the Service secured expansion of the national park system from a handful of parks and monuments in 1916 to approximately 370 units by the mid-1990s, including historical, archeological, recreational, and a variety of other types of parks. In favorable times, and under leaders like Mather, Albright, Cammerer, Wirth, and Hartzog, the system expanded rapidly—a result of both genuine altruism and bureaucratic aggrandizement (tempered by active resistance to many proposals for parks that the Service deemed unworthy of inclusion in a national system). Beginning with the first state parks conference in 1921, the Service extended its influence to nonfederal lands, promoting the growth and development of state park systems, with notable success during the New Deal era. The Park Service's growing involvement in recreational demonstration areas, national parkways, and national recreation areas in the 1930s helped place it unquestionably at the forefront in the setting aside of recreational open space for millions of Americans—an accomplishment about which the Service has repeatedly and justifiably expressed pride.

The loss in 1962 of Director Wirth's state and local recreational programs to the Bureau of Outdoor Recreation was reversed in 1981, when Secretary Watt returned those programs to the Park Service, citing the need for more efficient government. Significantly expanded in scope since 1962 (although with sharply decreased funding beginning under Watt), the programs have bolstered the Service's authority and status in the recreation field and have become part of its overall "partnership" effort. Involving cooperation with national, state, and local entities in parks and recreation endeavors, the partnership programs are by congressional intent focused more on public use than on wildland preservation, thus reinforcing the Service's interest in recreational tourism.[30]

The long-dominant emphasis on accommodating public use in parks had a profound impact on the National Park Service, leading to the entrenchment of specific values and perceptions. With tourism and the economics of tourism being fundamental to the parks' very existence, the utilitarian, businesslike proclivities of park management (spawned in Yellowstone and other early parks) thrived as the system grew. Striving for ever more parks and better accommodations, the Service measured its success by indicators such as annual visitor counts; the increasing scope of its programs and size of the park system; and the number of new campgrounds, visitor centers, and related developments. Recreational tourism was, moreover, the chief impetus behind the diversification of the Park Service's mission. Always cherishing its identification with the large natural areas of the system, the Service nevertheless used the recreational aspects of its mandate to justify its tremendous expansion into reservoir, urban area, and parkway management, as well as assistance for state and local recreational programs.

The ever-demanding construction and development programs relating to public use of the parks ensured the ascendancy of those professions overseeing such work and greatly influenced Park Service funding and staffing priorities. From Mather's selection of engineers to fill superintendencies to the present day, the developmental professions have consistently maintained prominence within the Service's highest ranks, whether in Washington, in other central offices, or in the parks. Of fourteen directors, only two have been landscape architects—Conrad Wirth and William Penn Mott—and one an engineer—Gary Everhardt. (Even though the Service is best known worldwide for its large natural parks, no one with a professional background in natural science has ever been chosen director.) Dozens from the construction and development professions have served in other key organizational positions: as superintendents, regional directors, and associate regional directors; and as deputy, associate, and assistant directors in Washington. Logically, they have also headed such influential offices as the eastern and western design and construction centers and their successor, the Denver Service Center—which has been responsible for far more of the parks' design and construction work than any other office. As of the end of 1992, the service center alone had a work force of 773, out of a total Park Service force of about 22,700. Service center personnel included 123 landscape architects, 81 architects, 73 civil engineers, 51 general engineers, 11 electrical engineers, 19 mechanical engineers, 8 environmental engineers, 2 safety engineers, 22 engineering and architectural student trainees, 47 engineering and architectural drafting technicians, and 11 construction representatives. At the same time, 41 positions were devoted in one way or another to natural resources.[31]

In analyzing what it viewed as an "abysmal lack of response" to repeated calls for research-based management, the 1992 National Academy Report on science in the parks stated that the problem was "rooted in the culture" of the National Park Service, but made no effort to identify the cultural characteristics. The Vail Agenda, on the other hand, did attempt to define the Service's culture. Managers who could be "creative and embrace responsibility, not avoid accountability and play it safe" exemplified the culture. The Agenda further identified such positive attributes as independence, initiative, imagination, and commitment—altogether a definition so conventional that it provided no clues to the substantive values, perceptions, and attitudes of the organization and its leaders.[32]

In truth, the leadership culture of the Park Service has been defined largely by the demands of recreational tourism management and the desire for the public to enjoy the scenic parks. Since the establishment of Yellowstone and other nineteenth-century parks, managers have had to deal not only with planning, development, construction, and maintenance of park facilities, but also with ever more demanding political, legal, and economic matters such as concession operations, law enforcement, visitor protection, and the influence of national, state, and local tourism interests. Such imperatives have *driven* park management. Especially since the 1960s, deeper involvement in urban parks, greater drug and crime problems, more development on lands adjacent to parks, and the escalating political strength of concessionaires and other commercial interests have added to the pressure on management.[33]

From this evolving set of circumstances, certain shared basic assumptions began to emerge even before the Park Service was created, gained strength under Mather and his successors, and endured—some to the present. Close consideration of eight decades of National Park Service history reveals that these assumptions have long reflected the perceptions and attitudes of the Service's leadership culture: with public enjoyment of the parks being the overriding concern, park management and decisionmaking could be conducted with little or no scientific information. Appearance of the parks mattered most. Even when dealing with vast natural areas, resource management did not seem to require highly trained biological specialists—the unscientifically trained eye could judge park conditions adequately. What is more, scientific findings could restrict managerial discretion, and park managers needed independence of action. Each park was a superintendent's realm, to be subjected to minimal interference, primarily that sought by the superintendent, perhaps through the regional director. Similarly, the Service was the recognized, right-thinking authority on national park management—it could provide the kind of "environmental leadership" necessary to run the parks properly with little or no involvement from outside groups. In this regard, environmental activism was often unwelcome; and legislation such as the Wilderness Act or the National Environmental Policy Act should not interfere unduly with traditional management and operations of the Service. Moreover, natural qualities for top leadership in the Service were to be found mainly within the ranger and superintendent ranks and the developmental professions.[34]

Overall, then, the dominant Park Service culture developed a strongly utilitarian and pragmatic managerial bent. It adopted a management style that emphasized expediency and quick solutions, resisted information gathering through long-term research, and disliked interference from inside or outside the Service.

Primarily concerned with varied aspects of recreational tourism, the Park Service's leadership culture has been extremely reluctant to abandon traditional assumptions. It has long proved its persistence and adaptability in the face of repeated criticism. Much of that criticism has come from inside the Service, especially from biologists from the 1930s on, very often with support from naturalists and interpreters in the parks. Some superintendents also have been openly disapproving: the uniformed, "green blood" groups within the Park Service family have not always been of one accord. Numerous individual superintendents, in major parks such as Shenandoah, Sequoia, Yellowstone, and Channel Islands, have been recognized in recent years for their contributions to various aspects of natural resource management.[35] Nevertheless, such advances have largely depended on the chance of a particular superintendent's attitude and willingness to strive for ecologically informed management, rather than on any pervasive environmental perspective within the Park Service. Overall, the Service's rank and file has been more ecologically aware than its leaders.

Through research and careful planning, ecological preservation and recreational tourism do not have to be mutually exclusive. But in the ebb and flow of national park history, loyalty to traditional assumptions has prevented the Service from establishing unquestioned credentials as a leader in scientifically based land management.[36]

Yet, it must be noted that the emphasis on recreational tourism in the national parks has always had a statutory basis. Tourism and public use have had explicit congressional sanction since the legislation establishing Yellowstone and other early parks authorized accommodations and roads and trails to facilitate public enjoyment. This authority was strongly reaffirmed in the National Park Service Act of 1916, with its emphasis on public use. Not only did Congress not challenge the Park Service's interpretation of the act during the ensuing decades, but it also encouraged development and use—at times aggressively. The Service's remarkable success in building the national park system, developing the parks, and expanding into many new tourism-related program areas continually depended on congressional sanction and appropriation of funds. Furthermore, such congressional support surely reflected widespread public support. Public enthusiasm for the parks has been evident from the steady increase in the annual number of visits to the parks (reaching, by one estimate, 281 million per year in 1990—more than the national population) and, in recent times, the repeated designation of the Park Service as among the most popular and respected federal bureaus.[37]

Overall, then, national park management with its emphasis on tourism and use has largely reflected the values and assumptions of the Service's utilitarian-minded leadership culture. The culture has been grounded in legislative mandates. And the legislation has derived from public values and perceptions, principally the appreciation and enjoyment of the parks' scenic beauty and recreational opportunities.

In significant contrast to management for public use and enjoyment, science as a means of informing natural resource management in the national parks has never gained specific statutory authority. This fact has been acknowledged again and again in the reports on park science. For instance, the National Parks and Conservation Association's 1989 report recognized the lack of a scientific mandate, as did the National Academy's 1992 report and the Vail Agenda, which stated plainly that the Park Service "does not have any specific statutory language directing it to engage in science as part of its assigned mission."[38]

Indeed, even though the Organic Act of 1916 called for the parks to be left "unimpaired for the enjoyment of future generations," it did not mandate science as a means of meeting that goal. Seeking to preserve the parks' majestic landscapes by preventing excessive commercialism, the founders had lobbied for legislation to place protection and development of the scenic areas under federal control. Scientific research was not at all prohibited by the act; certainly by implication it could be read into the act's principal mandate. But that was not enough to convince Mather and Albright, two founders who became directors and saw little need for scientific expertise to manage parks created for scenic preservation and public enjoyment. These giants of Park Service history deeply infused their values and assumptions into the Service. Their development-oriented Lane Letter of 1918 was fundamental dogma for decades, and deemed official policy as late as the 1970s. The emergence in the 1930s of an ecological and scientific perspective and its revival in the 1960s threatened to make park management more costly, difficult, and time-consuming, thus bringing about a struggle within the Service between the more ecologically oriented and the more traditional factions. As heirs to the vision of Mather and Albright, the Service's top leadership by and large has shared the founders' apathy toward scientific resource management. Their views have prevailed; and to public expressions of ecological concern, rhetoric has been used many times to mask the deficiencies of the Service's response.

Focusing on recreational tourism, the Service neglected to push science to the forefront and make it a nonnegotiable element of park management. In an age of ecological science, the acknowledged lack of a congressionally imposed scientific mandate for national park management clearly means that ecological preservation still is not a primary concern of Congress. Without such a mandate, the Service has not seized the initiative to build sufficient science programs on its own. And the recognition that only through scientific resource management can ecological preservation in the parks be adequately addressed negates rhetorical claims that preservation has been the Service's primary goal.

In response to the National Academy's 1992 analysis of science in the parks, a high-level committee headed by scientist Paul G. Risser (who had chaired the 1992 report) and consisting of superintendents, a regional director, and other authorities inside and outside the Service, issued a report declaring that without a legislative mandate, there can be no assurance that the Park Service will make a "genuinely lasting commitment to science-based management." Noting that an "adequate science and technology program and organization" had never been established, the report added that the Service "had simply never done so, in spite of repeated authoritative urging. There is no assurance that it will do so now, on a long-term sustained basis, without statutory direction."[39] Indeed, the history of the National Park Service is the history of a bureau without a scientific mandate and unwilling to act decisively in support of science unless specifically directed to by Congress—the Service would have to be told to "make a genuinely lasting commitment." Such reluctance makes it appear that the lack of a mandate has served, in effect, as an excuse for not being resolute in scientific matters.

Despite long-standing recognition of its deficient science programs, the Park Service has remained highly popular with the public. In a 1991 study entitled *A Race Against Time*, the National Parks and Conservation Association cited polls by the Roper Organization, which indicated that the Park Service "continues to enjoy the highest public approval rating of all government agencies." Nonetheless, the association concluded by castigating the public for "ignorance and complacency" and for "acting like recreational tourists at a theme park," oblivious to the responsibility to ensure preservation of the parks.[40] Although such environmental organizations may wish that it were not so, national park management, in refusing to come to grips fully with science and ecological concerns, tends to reflect the attitudes of a public that values the parks mainly for their scenery and for the enjoyment and recreation they provide.

For many, spectacular scenery may create an impression of biological health and provide such satisfaction that little consideration is given to the parks as segments of great ecological complexes under stress. Living almost entirely in extensively manipulated and altered landscapes, the public may take for granted that unimpaired natural conditions exist, especially in the larger parks. To the untrained eye, *unoccupied* lands can mean *unimpaired* lands, even where scientists might quickly recognize that human activity has caused substantial biological change. The loss of ecological integrity may have little or no effect on the aesthetics or the general appearance of an area. Even when ecological degradation is pointed out to park visitors, the new conditions may be thought of as merely "another change in the scenery."

Even though it admits to a deficiency in scientific management, the Park Service—as host to the millions of tourists who come to the parks to enjoy nature and majestic scenery—has sought to inspire the public to a deeper understanding and appreciation of the complexities of natural history. In so doing, the Service has helped build an environmental ethic, fostering greater knowledge and concern about ecological issues nationwide. This influence has been evolving since campfire talks, nature walks, and museum displays spread throughout the park system in the 1920s and 1930s. The effort expanded over the years to include a huge and varied array of museum and visitor center exhibits, interpretive talks, guided hikes, and trailside exhibits, augmented by brochures, films, books, and other means of enlightening the public. Begun in the 1960s, Director Hartzog's environmental education programs reached out to thousands of schoolchildren, many of them underprivileged and without access to parks outside urban areas. Through its involvement with state and local parks and the more recent partnership programs, the Service has effectively advanced nature appreciation and understanding. Furthermore, the Service has extended its influence worldwide through assistance to foreign countries in the development, interpretation, and operation of parks.[41] Thus, despite limitations in ecological management, the national parks, the National Park Service, and the uniformed ranger have become symbols of a conservation and environmental ethic.

Such constructive efforts have no doubt moved the public toward a greater comprehension of environmental matters. For many in the Park Service, scenic preservation and accommodation of tourists remained the focus of their careers, even after science and ecology gained favor during the environmental movement. Yet, in an important way, their work served broad environmental purposes. For many visitors drawn to the national parks partly by their very accessibility and convenience, contemplation of the natural beauty displayed and interpreted in the parks surely has nurtured a deeper realization of the complexities of nature—aesthetic appreciation thus serving as a threshold to ecological awareness. It may be that few people develop a concern for ecology without having first acquired a heightened sense of the beauty in nature, as is fostered in the national parks.

A 1993 merger of biological research functions within the Department of the Interior and a sweeping reorganization of the National Park Service in 1995 brought substantial changes for the Service. On October 1, 1993, Secretary of the Interior Bruce Babbitt officially established the National Biological Survey (later "Service"), including scientists and support staff drawn primarily from the department's three public land-managing bureaus—the National Park Service, the U.S. Fish and Wildlife Service, and the Bureau of Land Management. (As a Department of Agriculture bureau, the Forest Service was not involved.) The Park Service's contribution to the new bureau was the equivalent of 168 full-time positions (scientists and support personnel) and approximately $20 million in base funds. Created by administrative order and thus without congressional sanction, the bureau was to foster an ecosystem management approach through coordinating biological and ecological research to address land management issues on a national, regional, and local scale. It was to be a nonadvocacy research bureau, with no responsibilities for actual land management or regulation.[42]

Although different in purpose and scope from Secretary Harold Ickes' 1940 administrative transfer of the Park Service's wildlife biologists to the Bureau of Biological Survey, the 1993 merger had a similar effect, in that it suddenly withdrew from the Park Service virtually all of its biological research capability. Science had at last achieved independence—but it was through removal, rather than by remaining in the Service and gaining independence from "operational management," as advocated beginning in the 1960s and realized to some extent until Director Hartzog suddenly placed the biologists under the regional directors in 1971. However, in the political climate of the mid-1990s the National Biological Service was weakened by funding and staffing cutbacks, which helped bring about its merger with its geological counterpart, the U.S. Geological Survey. The uncertain, changing situation increased the doubts that already existed within the Park Service about the future of its biological research.

Soon after the Service lost its research biologists, it undertook a major reorganization in response to the goals of the administration of President Bill Clinton to reduce the size of the federal bureaucracy and improve efficiency. The 1995 reorganization substantially modified the hierarchical system in place since 1937, in which parks reported to regional offices, which in turn reported to Washington. In the new arrangement the parks gained much greater autonomy: the regional offices were abolished and replaced by smaller central offices with less capability to oversee park operations; and the Washington office was sharply reduced, diminishing its oversight capabilities as well.[43]

Remaining in the Park Service after creation of the Biological Service was a sizable force of well-trained natural resource managers, their support staff, and many others of like persuasion. Still, the loss of the research biologists surely diminished the ecological and scientific perspective within national park management. Furthermore, the emancipation of the parks from the leadership and oversight of well-staffed central offices reduced the park superintendents' accountability to higher authority and to national standards of park management. Acknowledging the strong traditions of the Park Service, the Vail Agenda had noted that the Service would "not be transformed quickly or easily."[44] Indeed, although the organizational structure was quickly changed, the reorganization left the central cultural assumptions of the Service fully intact, and has even created a situation where, with less oversight and fewer constraints, traditional attitudes may be reinforced and flourish.

The organization's most deeply imbedded assumptions are far more difficult and slower to change than the organizational structure. Given the strength and persistence of ancestral attitudes within the Service, its core values are likely to outlast any one director, even one who is stubbornly determined to change them. And succeeding directors may well rescind prior modifications and reaffirm old attitudes. Even a whole generation of leaders may not succeed in changing the core values of the Park Service to establish what the Vail Agenda termed a "strong ecosystem management culture."[45] Such changes are not impossible—but they are improbable.

Beginning with the construction of Yellowstone's roads and lodges, the history of development and use of the parks for tourism extends for more than a century and reflects an entrenched perception of the purpose of national parks. Backed by the Organic Act's mandate for public use and enjoyment, early attitudes and actions of the Service created a powerful, virtually irresistible trend in national park management. But in time, the dignity and nobility of the parks, once seen largely in terms of majestic landscapes, came also to be understood in more precise scientific and ecological terms—a new and challenging perception arose within the Service, never to be fully integrated into park operations. In both philosophy and management, the National Park Service remains a house divided—pressured from within and without to become a more scientifically informed and ecologically aware manager of public lands, yet remaining profoundly loyal to its traditions.

In this era of heightened environmental concern, it is essential that scientific knowledge form the foundation for any meaningful effort to preserve ecological resources. If the National Park Service is to fully shoulder this complex, challenging responsibility at last, it must conduct scientifically informed management that insists on ecological preservation as the highest of many worthy priorities. This priority must spring not merely from the concerns of specific individuals or groups within the Service, but from an institutionalized ethic that is reflected in full-faith support of all environmental laws, in appropriate natural resource policies and practices, in budget and staffing allocations, and in the organizational structures of parks and central offices. When—and only when—the National Park Service thoroughly attunes its own land management and organizational attitudes to ecological principles can it lay serious claim to leadership in the preservation of the natural environment.

Abbreviations

A.Murie	Papers of Adolph Murie, American Heritage Center, University of Wyoming
ASLA LC	Papers of American Society of Landscape Architects, Library of Congress
Bratton	Personal files of Susan P. Bratton, National Park Service
CAHA	Cape Hatteras National Seashore files
Dennis	Personal files of John G. Dennis, National Park Service
DEVA	Death Valley National Monument files
EVER	Everglades National Park files
Falb	Personal files of Richard D. Falb, National Park Service
FLO-LC	Papers of Frederick Law Olmsted, Jr., Library of Congress
Garrison	Papers of Lemuel A. Garrison, Special Collections, Clemson University
Grinnell	Papers of George Bird Grinnell, Manuscripts and Archives, Yale University Library
GRSM	Great Smoky Mountains National Park files
Hartzog	Papers of George B. Hartzog, Jr., Special Collections, Clemson University
JHMcF	Papers of J. Horace McFarland, Pennsylvania State Archives
Kent	Papers of William Kent, Manuscripts and Archives, Yale University Library
Leopold	Papers of A. Starker Leopold, Bancroft Library, University of California at Berkeley
Leopold-FRM	Papers of A. Starker Leopold, Department of Forestry and Resource Management, University of California at Berkeley
Linn	Personal files of Robert M. Linn, Houghton, Michigan
Mather-BL	Stephen T. Mather scrapbooks, Bancroft Library, University of California at Berkeley
MVZ-UC	Museum of Vertebrate Zoology Archives, University of California at Berkeley
NPS-HC	National Park Service History Collection, Harpers Ferry, West Virginia

OLYM	Olympic National Park files
O.Murie	Papers of Olaus Murie, Conservation Manuscripts Collection, Denver Public Library
REDW	Redwood National Park files
Reeser	Personal files of Donald W. Reeser, National Park Service
RG79	Record Group 79, Records of the National Park Service, National Archives
Sprugel	Personal files of George Sprugel, Jr., Champaign, Illinois
Supernaugh	Personal files of William R. Supernaugh, National Park Service
THRO	Theodore Roosevelt National Park files
TIC	Technical Information Center, Denver Service Center, National Park Service
Waggoner	Personal files of Gary S. Waggoner, U.S. Geological Survey
Walker	Papers of Ronald H. Walker, Special Collections, Clemson University
Wauer	Personal files of Roland H. Wauer, Victoria, Texas
WS	Papers of the Wilderness Society, Conservation Manuscripts Collection, Denver Public Library
YELL	Yellowstone National Park archives
YOSE	Yosemite National Park archives

Notes

Chapter 1. Creating Tradition

1. The principal early account of the campfire discussion is found in Nathaniel Pitt Langford, *The Discovery of Yellowstone Park: Journal of the Washburn Expedition to the Yellowstone and Firehole Rivers in the Year 1870* (1905; reprint ed., Lincoln: University of Nebraska Press, 1972), ix–xvii, xix–xx, 117–118. See also Hiram Martin Chittenden, *Yellowstone National Park: Historical and Descriptive* (1895; rev. ed., Stanford, California: Stanford University Press, 1933), 69. In 1994 Death Valley National Monument was redesignated a national park and expanded to include more acreage than Yellowstone.

2. With Yosemite the federal government divested itself of responsibility for management of the reserved lands, whereas the Yellowstone Act called for the national government's involvement in park management. In addition to the commitment to federal rather than state management, the sheer size and scope of Yellowstone and the 1872 act's more fully developed national park policy statement—mandating for the first time preservation as well as use—make Yellowstone the true benchmark for the national park concept. For a discussion of the authorization of the Yosemite lands as a state-managed park, see Alfred Runte, *Yosemite: The Embattled Wilderness* (Lincoln: University of Nebraska Press, 1990), 18–21; and Richard A. Bartlett, *Nature's Yellowstone* (Albuquerque: University of New Mexico Press, 1974), 195–197.

Artist George Catlin's 1832 proposal that the federal government establish a large "*magnificent park*" on the American plains had no influence. In contrast, an 1865 suggestion by Montana's acting territorial governor, Thomas Francis Meagher, that Yellowstone become a national park, may be seen as part of the background of the park movement. The term "national park" was not used in the Yellowstone Park Act itself, but was used during debates over passage of the act. Aubrey L. Haines, *Yellowstone National Park: Its Exploration and Establishment* (Washington, D.C.: National Park Service, 1974), xxi, 45, 113, 121, 126. Catlin's proposal is mentioned in Roderick Nash, *Wilderness and the American Mind* (1967; 3rd ed., New Haven: Yale University Press, 1982), 100–101.

3. Haines, *Yellowstone National Park,* 58–59, 93–98, 109–110, 114, 120–121, 126–128, 153–155; Aubrey L. Haines, *The Yellowstone Story: A History of Our First National Park* (Yellowstone National Park: Yellowstone Library and Museum Association 1977), I, 153, 164–166, 172;

Alfred Runte, *National Parks: The American Experience* (1979; rev. ed., Lincoln: University of Nebraska Press, 1987), 44–46; and Runte, *Trains of Discovery: Western Railroads and the National Parks* (1984; rev. ed., Niwot, Colorado: Roberts Rinehart, 1990), 13–20.

4. Cooke made his remarks in a letter to Northern Pacific official W. Milner Roberts, October 30, 1871. Haines, *Yellowstone National Park,* 109–110.

5. Aubrey Haines comments in *The Yellowstone Story* (p. 105) that the June 1870 meeting between Cooke and Langford "evidently led to some understanding between them concerning the usefulness of Yellowstone exploration in the grand scheme of Northern Pacific Railroad publicity." Similarly, Richard Bartlett, in *Nature's Yellowstone* (p. 208), states that "the evidence, though fragmentary, is sufficient to credit the inspiration for the creation of Yellowstone National Park to officials of the Northern Pacific Railroad." See also Robin W. Winks, *Frederick Billings: A Life* (New York: Oxford University Press, 1991), 285–287; and Runte, *Trains of Discovery,* 13.

In fact, the evidence is slim that the altruistic campfire discussion occurred—and even if it did, it likely reflected the Northern Pacific's aspirations. The Washburn-Doane Expedition left no written record of such a discussion. Nathaniel Langford's original account of the expedition, taken from his field notes, did not even mention the conversation—nor was it mentioned in any of the known diaries or notes written by other participants. Not until thirty-five years later, when he compiled a diary and published it in 1905, did Langford produce his account of the campfire discussion, including a suspicious amount of detail in light of the length of time elapsed. By then, with several new parks created, the national park idea had attained greater popularity; thus, recognition for having helped conceive the idea may have had special appeal to Langford. At about the time Langford's new account appeared, another expedition member, Cornelius Hedges, added to his diary a brief reference to the campfire meeting—altogether a curious rush to amend the record. With such meager historical documentation, the campfire story cannot be proved or disproved. Whether or not it is rooted in historical fact, the story achieved legendary status. Haines, *The Yellowstone Story,* I, 105, 130, 137–140, 164–165; and Louis C. Cramton, *Early History of Yellowstone National Park and Its Relation to National Park Policies* (Washington, D.C.: Government Printing Office, 1932), 19. See also Runte, *National Parks,* 41–42; and Bartlett, *Nature's Yellowstone,* 198–208. Ferdinand Hayden's role in establishing Yellowstone's boundaries is mentioned in Runte, *National Parks,* 46; and Nash, *Wilderness and the American Mind,* 112.

6. Keith R. Widder, *Mackinac National Park, 1875–1895,* Reports in Mackinac History and Archaeology no. 4 (Mackinac Island State Park Commission, 1975), 6, 41–46.

7. Barry Mackintosh, *The National Parks: Shaping the System* (Washington, D.C.: Department of the Interior, 1991), 12–13, 15–17; Runte, *National Parks,* 65–66, 75–76; *Yosemite,* 54–55; and *Trains of Discovery,* 13–54; Lary M. Dilsaver and William C. Tweed, *Challenge of the Big Trees of Kings Canyon and Sequoia National Parks* (Three Rivers, California: Sequoia Natural History Association, 1990), 70–73. In *Trains of Discovery* (p. 1) Runte characterizes the close ties between the western railroad companies and the national parks as a "pragmatic alliance." National parks promised profits for the companies, and without railroad support many of the major parks "might never have been established in the first place." Richard J. Orsi, in " 'Wilderness Saint' and 'Robber Baron': The Anomalous Partnership of John Muir and the Southern Pacific Company for Preservation of Yosemite National Park," *Pacific Historian* 29 (Summer-Fall 1985), 136–152, emphasizes the conservation concerns as well as the economic interests of Southern Pacific officials in the company's support of national parks.

As examples of the size of some of the larger parks, at the time of their authorization, Sequoia comprised 161,597 acres; Yosemite, 719,622; Mt. Rainier, 207,360; Crater Lake, 159,360; Mesa Verde, 79,561; and Rocky Mountain, 229,062. With additions and deletions of park lands, the size of the parks would vary over time. Information on original acreage provided by Renee C. Minnick, National Park Service.

8. Hillory A. Tolson, *Laws Relating to the National Park Service, the National Parks and Monuments* (Washington, D.C.: National Park Service, 1933), 50, 65; Runte, *Yosemite,* 45–47, 54–55; Dilsaver and Tweed, *Challenge of the Big Trees,* 64–73. Dilsaver and Tweed note (p. 73) a

"corporate greed" factor in the creation of the Sierra parks—a reference to the Southern Pacific Railroad Company's support of both the Sequoia and the Yosemite legislation, which would enhance the company's tourism and agricultural interests. Orsi, in "'Wilderness Saint'" and 'Robber Baron'" (p. 148), credits the Southern Pacific executive and "devoted lover of wilderness" Daniel K. Zumwalt with increasing the size of Yosemite and Sequoia as originally legislated. See also Roderick Frazier Nash, *The Rights of Nature: A History of Environmental Ethics* (Madison: University of Wisconsin Press, 1989), 35. Forest and watershed protection as a factor in the 1885 establishment of New York's Adirondack Forest Preserve (later State Park) is discussed in Frank Graham, Jr., *The Adirondack Park: A Political History* (New York: Alfred A. Knopf, 1978), 70–71, 76–77, 88–91; and Nash, *Wilderness and the American Mind,* 104–105, 116–121.

9. In *National Parks: The American Experience* (pp. 14–47) Alfred Runte discusses "monumentalism," a term used to describe the urge to create national parks in areas of the most grand and unusual scenery. He also examines the "cultural nationalism" factor in the park movement, asserting that Americans looked to the monumental scenery of the parks as an affirmation that their young nation was not inferior to Europe. A history of the early national park years is also found in William C. Everhart, *The National Park Service* (New York: Praeger Publishers, 1972), 3–21.

10. Platt and Sully's Hill are discussed in John Ise, *Our National Park Policy: A Critical History* (Baltimore: The Johns Hopkins Press, 1961), 139–142. See also Palmer H. Boeger, *Oklahoma Oasis: From Platt National Park to Chickasaw National Recreation Area* (Muskogee, Oklahoma: Western Heritage Books, 1987). Ise (p. 139) asserts that Sully's Hill was "unworthy" of being a national park. Similarly, Horace M. Albright, the second director of the National Park Service, viewed Platt and Sully's Hill as "totally lacking in national park qualifications. . . . established because of the parochial enthusiasm of local politicians." Horace M. Albright, as told to Robert Cahn, *The Birth of the National Park Service: The Founding Years, 1913–1933* (Salt Lake City: Howe Brothers, 1985), 5; see also 223.

11. Hal Rothman, *Preserving Different Pasts: The American National Monuments* (Urbana: University of Illinois Press, 1989), 34–116; and Ronald F. Lee, *The Antiquities Act of 1906* (Washington, D.C.: National Park Service, 1970), 73–76, 87–96. See also Harold K. Steen, *The U.S. Forest Service: A History* (Seattle: University of Washington Press, 1976), 98–100. Robert W. Righter, in "National Monuments to National Parks: The Use of the Antiquities Act of 1906," *Western Historical Quarterly* 20 (August 1989), 281–301, analyzes the sometimes deliberate use of the national monument designation as a first step toward creating national parks.

12. Much later, and with the precedent of size long established, President Jimmy Carter would use the Antiquities Act to establish vast national monuments in Alaska.

Today, units of the national park system have approximately two dozen different designations—including national parks, national monuments, national preserves, and national recreation areas, plus a bewildering variety of designations for historic and prehistoric sites. Throughout this study the terms "national park" and "park" are used interchangeably to refer to units of the national park system, whatever their individual designation, unless otherwise specified.

13. Theodore Roosevelt, "Wilderness Reserves: The Yellowstone Park," from *Outdoor Pastimes of an American Hunter* (New York: Charles Scribner's Sons, 1905), quoted in Paul Schullery, ed., *Theodore Roosevelt: Wilderness Writings* (Salt Lake City: Peregrine Smith Books, 1986), 148–149.

14. Tolson, *Laws Relating to the National Park Service* (1933), 26–27. Unless specifically noted, as in this instance, italics in quoted material appear in the original.

15. John F. Reiger, *American Sportsmen and the Origins of Conservation* (1975; rev. ed., Norman: University of Oklahoma Press, 1986), 102–111; Michael P. Cohen, *The History of the Sierra Club, 1892–1970* (San Francisco: Sierra Club Books, 1988), 6, 24; Nash, *Wilderness and the American Mind,* 131.

16. Soon after creation of the National Park Service, its first director, Stephen T. Mather, expressed an interest in the dunes along Lake Michigan's shoreline as a possible national park.

However, Mather seems to have been primarily interested in the area's recreational potential for serving Chicago, his hometown. *Report of the Director of the National Park Service to the Secretary of the Interior for the Fiscal Year Ending June 30, 1920* (Washington, D.C.: Government Printing Office, 1920), 85–86.

17. John Muir and Yosemite National Park are discussed extensively in Michael P. Cohen, *The Pathless Way: John Muir and American Wilderness* (Madison: University of Wisconsin Press, 1984); see especially 260–273, 302–310, 323–338. See also Nash, *Wilderness and the American Mind,* 125–129, 161–181; Runte, *Yosemite,* 81–82; Stephen Fox, *John Muir and His Legacy: The American Conservation Movement* (Boston: Little, Brown, 1981), 3–7, 82–83, 139–147; and Cohen, *Sierra Club,* 6, 22–33. For a discussion of late-nineteenth-century developments in American ecological science, see Donald Worster, *Nature's Economy: A History of Ecological Ideas* (1977; 2nd ed., Cambridge: Cambridge University Press, 1994), 195–197, 205–220.

18. Langford, *The Discovery of Yellowstone Park,* 117. Recreational tourism in national parks would come to include a great diversity of activities—for instance, sightseeing, automobile touring, camping, museum going, attending campfire talks, fishing, horseback riding, wilderness hiking and camping, motorboating, snowmobiling, even downhill skiing.

19. The wording of the Yellowstone Park Act is found in Tolson, *Laws Relating to the National Park Service* (1933), 26–27.

20. Anne Farrar Hyde, *An American Vision: Far Western Landscape and National Culture, 1820–1920* (New York: New York University Press, 1990), 53–190; Richard A. Bartlett, *Yellowstone: A Wilderness Besieged* (Tucson: University of Arizona Press, 1985), 113–115; Hans Huth, *Nature and the American: Three Centuries of Changing Attitudes* (1957; rev. ed., Lincoln: University of Nebraska Press, 1972), 73–86, 106–128; Earl Pomeroy, *In Search of the Golden West: The Tourist in Western America* (New York: Alfred A. Knopf, 1957), 17–28, 112–125; and William H. Goetzman, *Exploration and Empire* (New York: Alfred A. Knopf, 1971), 406. Graham, in *The Adirondack Park* (pp. 31–44), discusses the first railroad penetration of the Adirondacks and the public and private resorts that soon spread through the area.

21. John C. Paige and Laura Soulliere Harrison, *Out of the Vapors: A Social and Architectural History of Bathhouse Row, Hot Springs National Park, Arkansas* (Washington, D.C.: National Park Service, 1988), 32–35, 72.

22. Runte, *Yosemite,* 28–44, 51–53; Robert P. Gibbens and Harold F. Heady, *The Influence of Modern Man on the Vegetation of Yosemite Valley,* California Agricultural Experiment Station Extension Service Manual, no. 36 (Berkeley: University of California, Division of Agricultural Sciences, 1964), 2–5; James Francis Milestone, "The Influence of Modern Man on the Stream System of Yosemite Valley," Master's thesis, San Francisco State University, May 1978, 77–84; Ise, *Our National Park Policy,* 71–73, 76–83. Linda Wedel Greene, in *Yosemite: The Park and Its Resources* (Denver: National Park Service, 1987), details the extensive early development in the valley. See for instance I, 114–163.

Now considered a farsighted statement of national park principles, Olmsted's 1865 report was quickly suppressed, apparently by men who did not fully agree with its philosophy or who feared it might draw funding from their own projects. The report disappeared and was not rediscovered until the early 1950s. It is reproduced in Frederick Law Olmsted, "The Yosemite Valley and the Mariposa Big Trees: A Preliminary Report (1865)," with an introduction by Laura Wood Roper, *Landscape Architecture* 43 (October 1952), 12–25. See also Laura Wood Roper, *FLO: A Biography of Frederick Law Olmsted* (Baltimore: Johns Hopkins University Press, 1973), 283–287; Runte, *Yosemite,* 28–32, 39–40; and Joseph L. Sax, "America's National Parks: Their Principles, Purposes, and Prospects," *Natural History* 85 (October 1976), 57–87. Sax analyzes the report and its implications for contemporary national park management.

23. Widder, *Mackinac National Park,* 6, 8–9, 17–26, 37, 42–46; Huth, *Nature and the American,* 146.

24. The comments were from U.S. Senator Thomas W. Tipton, of Nebraska; from a petition by the Montana State Legislature to Congress; and from the *New York Times,* February 29, 1872,

quoted in Haines, *Yellowstone National Park,* 118, 121, 126. The Yellowstone Act's wording is in Tolson, *Laws Relating to the National Park Service* (1933), 26–27.

25. Bartlett, *Nature's Yellowstone,* 115–120; Bartlett, *Yellowstone: A Wilderness Besieged,* 115, 129–131; Hyde, *An American Vision,* 251–252; Ise, *Our National Park Policy,* 33–34; Chittenden, *Yellowstone National Park,* 110–111; and Robert Shankland, *Steve Mather of the National Parks* (New York: Alfred A. Knopf, 1951), 116–117.

26. Hyde, *An American Vision,* 253–268; Bartlett, *Yellowstone: A Wilderness Besieged,* 170–186; Ise, *Our National Park Policy,* 40; Runte, *National Parks,* 94; *Trains of Discovery,* 22–23; Winks, *Frederick Billings,* 284–292; Laura Soulliere Harrison, *Architecture in the Parks: National Historic Landmark Theme Study* (Washington, D.C.: National Park Service, 1986), 61–71; Chittenden, *Yellowstone National Park,* v, 115–116, 240–253. On congressional parsimony, see for example H. Duane Hampton, *How the U.S. Cavalry Saved Our National Parks* (Bloomington: Indiana University Press, 1971), 31–39. A detailed history of road construction in Yellowstone is found in Mary Shivers Culpin, *The History of the Construction of the Road System in Yellowstone National Park, 1872–1966* (Denver: National Park Service, 1994), I. Use of the U.S. Army to manage Yellowstone or other early national parks does not seem to have been influenced by the army's experience in operating Mackinac National Park.

27. Hyde, *An American Vision,* 281–293; C. W. Buchholtz, *Man in Glacier* (West Glacier, Montana: Glacier Natural History Association, 1976), 39–56; and "The National Park as a Playground," *Journal of Sport History* 5 (Winter 1978), 22–23. Examples of similarities in leasing provisions of national park enabling legislation are found in Tolson, *Laws Relating to the National Park Service* (1933), 49 (Sequoia and General Grant), 102 (Mount Rainier), 112 (Crater Lake), 123–124 (Wind Cave), 138–139 (Glacier), 155 (Rocky Mountain), 172 (Hawaii), and 189 (Lassen Volcanic).

28. Harlan D. Unrau, *Administrative History: Crater Lake National Park, Oregon* (Denver: National Park Service, 1988), I, 220; Harrison, *Architecture in the Parks,* 186; Dilsaver and Tweed, *Challenge of the Big Trees,* 140.

29. Boeger, *Oklahoma Oasis,* 49–66, 80–107; Harrison, *Architecture in the Parks,* 40–47.

30. *Reports of the Department of the Interior, 1910* (Washington, D.C.: Government Printing Office, 1911), 57; U.S. Department of the Interior, Memorandum for the Press, December 10, 1915, typescript, JHMcF; *National Park Conference, 3rd, Berkeley, Calif., 11–13 March 1915, Proceedings* (Washington, D.C.: Government Printing Office, 1915), 14–15.

31. *National Park Conference, 1915, Proceedings,* 19–20. The kind of development designed by Daniels and future generations of national park landscape architects helps account for the distinctly different appearance of many national park villages when compared to typical American towns having little control over development.

32. Samuel P. Hays discusses Gifford Pinchot and the U.S. Forest Service's interest in gaining control of the national parks in *Conservation and the Gospel of Efficiency: The Progressive Conservation Movement, 1890–1920* (Cambridge, Mass.: Harvard University Press, 1959), 195–198.

33. Most parks had mandates similar to Yellowstone's to preserve natural conditions. As with the leasing provisions, in some legislation the wording was taken verbatim from the Yellowstone Park Act. See for example Tolson, *Laws Relating to the National Park Service* (1933), 49 (Sequoia and General Grant), 102 (Mount Rainier), 155 (Rocky Mountain), 172 (Hawaii), and 189 (Lassen Volcanic).

34. Dilsaver and Tweed, *Challenge of the Big Trees,* 44–50, 55–60. The authors note (p. 59) that in certain local areas the damage from grazing was "appalling." Runte, *Yosemite,* 49; Buchholtz, *Man in Glacier,* 31, 34–40; Ise, *Our National Park Policy,* 122, 124–125, 172.

35. John D. Varley, "A History of Fish Stocking Activities in Yellowstone National Park between 1881 and 1980," Yellowstone National Park Information Paper, no. 35, January 1, 1980, 2–3; John D. Varley and Paul Schullery, *Freshwater Wilderness: Yellowstone Fishes and Their World* (Yellowstone National Park: Yellowstone Library and Museum Association, 1983), 101–

102; Ise, *Our National Park Policy,* 128–129; Hampton, *How the U.S. Cavalry Saved Our National Parks,* 99–100, 112, 158; Runte, *Yosemite,* 65–66; Dilsaver and Tweed, *Challenge of the Big Trees,* 106; and *Annual Report of the Superintendent of National Parks to the Secretary of the Interior for the Fiscal Year Ended June 30, 1916* (Washington, D.C.: Government Printing Office, 1916), 50, 60, 72.

36. Reiger, *American Sportsmen,* 101–109.

37. Ise, *Our National Park Policy,* 47, 123–124, 178; Dilsaver and Tweed, *Challenge of the Big Trees,* 106; Runte, *Yosemite,* 90, 127. National park historian Richard Bartlett depicts predator control in the parks as having been a "spillover from the ranching frontier." The war against predators indeed reflected practices on private and public lands throughout the country, practices that would soon be underwritten by the Bureau of Biological Survey. Bartlett, *Yellowstone: A Wilderness Besieged,* 328–330. See also Thomas R. Dunlap, *Saving America's Wildlife* (Princeton: Princeton University Press, 1988), 38–40.

38. Barlett, *Yellowstone: A Wilderness Besieged,* 316–321; Ise, *Our National Park Policy,* 23–25, 62–64, 110–111, 123, 178; Hampton, *How the U.S. Cavalry Saved Our National Parks,* 39–41, 105–110, 121–129; Rieger, *American Sportsmen,* 129–133; Runte, *Yosemite,* 86, 90.

39. Curtis K. Skinner et al., "The History of the Bison in Yellowstone Park" [with supplements], 1952, typescript, YELL; George M. Wright, Joseph S. Dixon, and Ben H. Thompson, *Fauna of the National Parks of the United States: A Preliminary Survey of Faunal Relations in National Parks,* Contributions of Wild Life Survey, Fauna Series no. 1 (Washington, D.C.: Government Printing Office, 1933), 117–118. The Lamar Valley bison, an introduced herd, came from two subspecies, both different from the remnant wild herds located in other areas of the park. Although the wild herds at times interbred with the introduced bison, they were almost always left alone and never received the intensive management given those in the Lamar Valley. Margaret Mary Meagher, *The Bison of Yellowstone National Park,* National Park Service Scientific Monograph Series no. 1 (Washington, D.C.: National Park Service, 1973), 26–37.

40. Paul Schullery, *The Bears of Yellowstone* (1986; 3rd ed., Worland, Wyoming: High Plains Publishing Company, 1992), 93–96, 100; Dilsaver and Tweed, *Challenge of the Big Trees,* 146.

41. Stephen J. Pyne, *Fire in America: A Cultural History of Wildland and Rural Fire* (Princeton: Princeton University Press, 1982), 227–229; Runte, *Yosemite,* 62–65; Hampton, *How the U.S. Cavalry Saved Our National Parks,* 83, 100, 107; Orsi, "'Wilderness Saint' and 'Robber Baron,'" 155n34.

42. Dilsaver and Tweed, *Challenge of the Big Trees,* 35, 59–62; Runte, *Yosemite,* 46–47, 60–62; Wright, Dixon, and Thompson, *Fauna of the National Parks,* 33, 101, 131.

43. John W. Henneberger has detailed the history of the national park rangers in "To Protect and Preserve: A History of the National Park Ranger," 1965, typescript, copy courtesy of the author. For early ranger activity, see pp. 18–227.

Chapter 2. Codifying Tradition

1. Robert Sterling Yard, "Making a Business of Scenery," *The Nation's Business* 4 (June 1916), 10–11.

2. The National Park Service Act is found in Hillory A. Tolson, *Laws Relating to the National Park Service, the National Parks and Monuments* (Washington, D.C.: Department of the Interior, 1933), 9–11. For an extensive compilation of national park legislation and related documents, see Lary M. Dilsaver, *America's National Park System: The Critical Documents* (Lanham, Maryland: Rowman & Littlefield Publishers, 1994).

Robert B. Keiter, in his "National Park Protection: Putting the Organic Act to Work," in David J. Simon, ed., *Our Common Lands: Defending the National Parks* (Washington, D.C.: Island Press, 1988), 81, states that the Organic Act "sets forth an impressive, unambiguous resource preservation mandate." Alston Chase, in *Playing God in Yellowstone: The Destruction of*

America's First National Park (Boston: Atlantic Monthly Press, 1986), 6, comments that the National Park Service's "sole mission is preservation." Similarly, John Lemons and Dean Stout, in "A Reinterpretation of National Park Legislation, *Environmental Law* (Northwestern School of Law, Lewis and Clark College) 15 (1984), 41–65, argue that the Organic Act's primary mandate is to preserve park resources: "the purpose of natural parks is to preserve pristine ecological processes" (p. 53); and "the most basic fiduciary duties of the [National Park Service] are to reduce development and promote preservation of resources" (p. 65). By contrast, Alfred Runte, in *National Parks: The American Experience* (1979; 2nd ed. rev., Lincoln: University of Nebraska Press, 1987), emphasizes the parks' tourism and economic potential as key motivating factors for the legislation establishing the National Park Service. See his chapter entitled "See America First," 82–105.

3. J. Horace McFarland to Frederick Law Olmsted, November 11, 1907, JHMcF; *National Park Conference, 4th, Washington, D.C., January 2–6, 1917, Proceedings* (Washington, D.C.: Government Printing Office, 1917), 107. See also J. Horace McFarland to Henry S. Graves, February 21, 1911, JHMcF; and House Committee on the Public Lands, *Hearings on H.R. 434 and H.R. 8668*, 64th Cong., 1st sess., 1916, 52. In a telegram to the 1915 National Park conference, the American Civic Association depicted the parks as great scenic areas "set aside for national recreation." See *National Park Conference, 3rd, Berkeley, Calif., March 11–13, 1915, Proceedings* (Washington, D.C.: Government Printing Office, 1915), 10. The "National Park Service" designation is mentioned in Walter L. Fisher to Reed Smoot, February 6, 1912, JHMcF. McFarland's career is detailed in Ernest Morrison, *J. Horace McFarland: A Thorn for Beauty* (Harrisburg: Pennsylvania Historical and Museum Commission, 1995). For an account of the founding of the National Park Service, see 173–193.

4. McFarland recounted his initial proposal in J. Horace McFarland to Gifford Pinchot, February 13, 1911, JHMcF.

5. J. Horace McFarland to Stephen T. Mather, November 22, 1926, JHMcF; J. Horace McFarland, "The Beginnings of the National Park System," 1929, typescript, JHMcF; House Committee on the Public Lands, *Hearings on H.R. 434 and H.R. 8668*, 1916, 4. McFarland to Pinchot, February 13, 1911; J. Horace McFarland to Frederick Law Olmsted, October 13, 1910; McFarland, "The Beginnings of the National Park System," JHMcF. Robert Shankland, *Steve Mather of the National Parks* (New York: Alfred A. Knopf, 1951), 52–53.

6. McFarland to Olmsted, October 13, 1910; F. L. Olmsted, note to files, November 20, 1910, NPS-HC; Shary Page Berg, "Frederick Law Olmsted, Jr.: A Preliminary Investigation," January 1985, typescript, 8, Radcliffe College.

7. Frederick Law Olmsted to John C. Olmsted, December 19, 1910; Frederick Law Olmsted to the President and Council of the Appalachian Mountain Club, January 19, 1912; Frederick Law Olmsted to Mark Sullivan, December 19, 1910, NPS-HC.

8. Shankland, *Steve Mather,* 20, 36, 40–41.

9. Shankland, *Steve Mather,* 57–59, 62–63; Horace M. Albright, as told to Robert Cahn, *The Birth of the National Park Service: The Founding Years, 1913–1933* (Salt Lake City: Howe Brothers, 1985), 19–21.

10. Albright and Cahn, *Birth of the National Park Service,* 32–43; Donald C. Swain, *Wilderness Defender: Horace M. Albright and Conservation* (Chicago: University of Chicago Press, 1970), 38–60.

11. *National Park Conference, 1st, Yellowstone National Park, Wyo., September 11–12, 1911, Proceedings* (Washington, D.C.: Government Printing Office, 1912), 1–4, 6–9, 17–19; *National Park Conference, 2nd, Yosemite National Park, Calif., October 14–16, 1912, Proceedings* (Washington, D.C.: Government Printing Office, 1913), 5–7, 95, 137, 141; *National Park Conference, 1915, Proceedings,* 4–5, 11–12, 15–20.

12. *National Park Conference, 1911, Proceedings,* iii–iv, 1–2; *National Park Conference, 1912, Proceedings,* 5–7; *National Park Conference, 1915, Proceedings,* 4–5.

13. *National Park Conference, 1911, Proceedings,* 5. Among the many other references to American tourists going to Canada or Europe are Committee on the Public Lands, *Hearings on H.R. 434 and 8668,* 1916, 5–7; and "A National Park Service," *New York Times,* May 30, 1916, editorial section, typescript copy, 8, NPS-HC.

14. *National Park Conference, 1911, Proceedings,* 3; *National Park Conference, 1912, Proceedings,* 9. Fisher's successor, Franklin K. Lane, would reiterate these factors just before passage of the Organic Act. See Franklin K. Lane to Henry L. Myers, July 8, 1916, Kent.

15. *National Park Conference, 1912, Proceedings,* 94–96. A history of conservation and the concerns for efficiency during this period is found in Samuel P. Hays, *Conservation and the Gospel of Efficiency: The Progressive Conservation Movement, 1890–1920* (Cambridge, Mass.: Harvard University Press, 1959).

16. *National Park Conference, 1912, Proceedings,* 95. For other expressions of the need for an engineer to oversee the national parks, see *National Park Conference, 1911, Proceedings,* 111; Franklin K. Lane to Henry L. Myers, July 8, 1916, Kent; House Committee on the Public Lands, *Hearings on H.R. 434 and H.R. 8668,* 1916, 3. Horace McFarland stated in 1911 that the national parks needed the leadership of "some great landscape engineer of international reputation" who would perform in the same manner as the landscape engineers of the city parks of Boston, Minneapolis, and Kansas City. McFarland to Pinchot, February 13, 1911. And in 1914 an Interior Department official testified on the parks' need for engineers who could handle roads, sanitation, and "scenic problems." House Committee on the Public Lands, *Hearings on H.R. 104,* 63rd Cong., 2nd sess., 1914, 73. (The Robert Marshall associated with the early national parks was not the individual who championed wilderness areas on public lands and helped found the Wilderness Society in 1935.)

17. U.S. Department of the Interior, Memorandum for the Press, typescript, December 10, 1915, JHMcF. See also Albright and Cahn, *Birth of the National Park Service,* 9, 24.

18. *National Park Conference, 1915, Proceedings,* 15–17, 115.

19. Graham Romeyn Taylor, "Washington at Work, II: The Nation's Playgrounds," *American Forestry* (January 1916), clipping, n.p., Mather-BL; *National Park Conference, 1915, Proceedings,* 16–17. See also Linda Flint McClelland, *Presenting Nature: The Historic Landscape Design of the National Park Service* (Washington, D.C.: National Park Service, 1993), 96–97.

20. *National Park Conference, 1911, Proceedings,* 108–110, 113–114.

21. House Committee on the Public Lands, *Hearings on H.R. 434 and H.R. 8668,* 1916, 85. Marshall's commitment to utilitarian land uses became even more apparent when, not long after he left the national parks' general superintendency, he became a leading advocate for a huge irrigation project in California's Central Valley. For much of the remainder of his career, he devoted his energies to publicizing and promoting his "Marshall Plan" for agricultural irrigation in the valley. Biographical note on Robert Bradford Marshall, n.d., typescript, Mather-BL.

22. *National Park Conference, 1912, Proceedings,* 10; House Committee on the Public Lands, *Hearings on H.R. 434 and H.R. 8668,* 1916, 75; William Kent to R. B. Watrous, January 17, 1916, JHMcF. See also Albright and Cahn, *Birth of the National Park Service,* 37–39.

23. *Reports of the Department of the Interior, 1907* (Washington, D.C.: Government Printing Office, 1907), I, 55–56; Hays, *Conservation and the Gospel of Efficiency,* 73, 195–197; Harold K. Steen, *The U.S. Forest Service: A History* (Seattle: University of Washington Press, 1976), 114. See also Morrison, *J. Horace McFarland,* 177–178, 183–184, 188–189.

24. Gifford Pinchot to Frederick Law Olmsted, December 26, 1912, NPS-HC; H. S. Graves to J. Horace McFarland, March 30, 1916, ASLA-LC. The legislation establishing Glacier National Park is found in Tolson, *Laws Relating to the National Parks* (1933), 138–139.

25. H. Graves to William Kent, draft, March 17, 1916, Kent; Graves to McFarland, March 30, 1916.

26. McFarland to Graves, February 21, 1911; J. Horace McFarland to Gifford Pinchot, March 24, 1911, JHMcF.

27. J. Horace McFarland to Frederick Law Olmsted, April 17, 1916, FLO-LC.

28. William Kent to the Secretary of Agriculture, April 7, 1916, Kent; House Committee on the Public Lands, *Hearings on H.R. 434 and H.R. 8668,* 1916, 16. Kent's interest in the grazing provision is discussed in Albright and Cahn, *Birth of the National Park Service,* 36, 45.

29. House Committee on the Public Lands, *Hearings on H.R. 434 and H.R. 8668,* 1916, 15–17. Much later, Albright recalled that Mather was "strongly opposed, but tended to take the long view" on grazing. He added that "the important thing . . . was to get a National Park Service Act passed; the grazing provision was something we could eventually get rid of." Albright and Cahn, *Birth of the National Park Service,* 36; see also 37–39.

30. *National Park Conference, Proceedings, 1917,* 105. McFarland had earlier written that the statement of purpose had been "jealously preserved with much fighting and effort." J. Horace McFarland to H. A. Caparn, October 9, 1916, JHMcF. Frederick Law Olmsted to Frank Pierce, December 31, 1910; Frederick Law Olmsted to J. Horace McFarland, September 13, 1911, NPS-HC.

31. Tolson, *Laws Relating to the National Park Service* (1933), 10.

32. It is the author's personal observation that even today the statement of purpose is referred to regularly and routinely by National Park Service employees, and it is displayed prominently in many Park Service offices. Other sections of the act are rarely mentioned.

33. Olmsted to Pierce, December 31, 1910. The earliest draft is quoted in Olmsted's letter. Similar to this version of the national parks' purpose, Congress would declare in a law passed in 1978 that the parks were to be protected and managed in a manner that avoids "derogation of the values and purposes for which the parks were established." Redwood National Park Expansion Act, sec. 101b, Public Law 95–250, 16, United States Code. See also Michael A. Mantell, ed., *Managing National Park System Resources: A Handbook on Legal Duties, Opportunities, and Tools* (Washington, D.C.: Conservation Foundation, 1990), 14–15.

34. Olmsted to Olmsted, December 19, 1910; J. Horace McFarland to Gifford Pinchot, February 12, 1911, JHMcF; Olmsted to Pierce, December 31, 1910.

35. Olmsted to Pierce, December 31, 1910; McFarland to Graves, February 21, 1911. McFarland wrote to Pinchot that the new statement declared, "in fully framed language, the purpose of national parks in a fashion never before undertaken, and not present in any of the loosely drawn legislation under which national parks now exist." McFarland to Pinchot, February 12, 1911. See also J. Horace McFarland to Walter L. Fisher, January 2, 1912.

36. Walter L. Fisher to J. Horace McFarland, December 22, 1911, JHMcF.

37. McFarland to Fisher, January 2, 1912; Walter L. Fisher to J. Horace McFarland, January 30, 1912, JHMcF.

38. Draft of legislation appended to letter from Richard B. Watrous to Frederick Law Olmsted, October 19, 1915, NPS-HC. Although the correspondence reveals no specific evidence, it may be that inclusion of "historic" objects was intended to cover the kinds of resources protected by the 1906 Antiquities Act, which authorized creation of national monuments. By the Organic Act, the National Park Service would be mandated to administer those national monuments then under the jurisdiction of the Interior Department.

39. Olmsted to Watrous, November 1, 1915; Richard B. Watrous to Frederick L. Olmsted, November 13, 1915. NPS-HC.

40. Olmsted to Watrous, November 1, 1915.

41. Olmsted to Pierce, December 31, 1910; Olmsted to the Appalachian Mountain Club, January 19, 1912. Frederick Law Olmsted to James Sturgis Pray, February 3, 1915, NPS-HC. Expressing similar concerns about adverse development, McFarland wrote of the need for a statement to "form a bar against any possibly strained construction which might be damaging to the parks." J. Horace McFarland to Walter L. Fisher, December 19, 1911, JHMcF.

42. Olmsted proposed that the independent board be similar to those used with local park systems around the country; it should be advisory, as opposed to executive, and serve as a check against misuse of the parks by ensuring a "harmonious continuity of policy." Shortly before passage of the Organic Act, Olmsted's proposal was rejected because of the Interior Department's uneasi-

ness about the board's potential power. Less than two decades later, with the Historic Sites Act of 1935, Congress authorized a national park advisory board. Legislative draft appended to letter from Frederick Law Olmsted to J. Horace McFarland, November 21, 1911, NPS-HC; Olmsted to McFarland, September 13, 1911; Watrous to Olmsted, November 13, 1915; McFarland, "The Beginnings of the National Park System." See also Albright and Cahn, *Birth of the National Park Service,* 36. Authorization of an advisory board is discussed in Conrad L. Wirth, *Parks, Politics, and the People* (Norman: University of Oklahoma Press, 1980), 164–165.

43. Donald C. Swain, "The Passage of the National Park Service Act," *Wisconsin Magazine of History* 50 (Autumn 1966), 15–17; Albright and Cahn, *Birth of the National Park Service,* 38; Gilbert H. Grosvenor, "The Land of the Best," *National Geographic* 24 (April 1916): 327–430; Department of the Interior, *National Parks Portfolio* (Washington, D.C.: Department of the Interior, 1916).

44. House Committee on the Public Lands, *Report No. 700 to Accompany H.R. 15522,* 64th Cong., 1st sess., 1916, 2–3; House Committee on the Public Lands, *Hearings on H.R. 434 and H.R. 8668,* 1916, 5–8.

45. An account of the political strategies and promotional activities is found in Swain, "Passage of the National Park Service Act."

46. William Kent draft memorandum to John D. Muir, E. T. Parsons, Wm. F. Borby, Mrs. R. B. Colby, and members of the Society for the Preservation of National Parks, n.d. (ca. 1913), Kent. Earlier, Kent's donation of a redwood forest area north of San Francisco had led to creation of Muir Woods National Monument. Albright and Cahn, *Birth of the National Park Service,* 35. Plotting their political strategies, the National Park Service founders frequently met in Congressman Kent's Washington home on the corner of F and Eighteenth streets—the same house where plans had earlier been formulated for passage of the Hetch Hetchy legislation. Morrison, *J. Horace McFarland,* 186.

47. The quote is from Philip P. Wells, "Conservation of Natural Resources," quoted in Hays, *Conservation and the Gospel of Efficiency,* 123; see also 5, 122–127, 176–177. For comments on utilitarian and aesthetic conservationists, see Roderick Nash, *Wilderness and the American Mind* (1967; 3rd ed., New Haven: Yale University Press, 1982), 129–130, 135–139. "Wise use," as employed in the early twentieth century, is not synonymous with the policies of the "wise use" movement of the late-twentieth-century West.

48. Tolson, *Laws Relating to the National Park Service* (1933), 26, 49, 65, 102, 139.

49. Tolson, *Laws Relating to the National Park Service* (1933), 10–11; House Committee on the Public Lands, *Hearings on H.R. 434 and 8668,* 1916, 17.

50. Tolson, *Laws Relating to the National Park Service* (1933), 3, 11.

Chapter 3. Perpetuating Tradition

1. Joseph Grinnell and Tracy I. Storer, "Animal Life as an Asset of National Parks," *Science* 44 (September 15, 1916), 377.

2. Horace M. Albright to the Director, November 8, 1928, Entry 17, RG79.

3. H. Duane Hampton, *How the U.S. Cavalry Saved Our National Parks* (Bloomington: Indiana University Press, 1971), 175–179; John Ise, *Our National Park Policy: A Critical History* (Baltimore: The Johns Hopkins Press, 1961), 208. See also letter from the Secretary of War stating that the War Department was no longer able to bear the responsibilities of the Department of the Interior, and that "the time has come for the Interior Department to take over the entire handling of these parks." Secretary of War to Secretary of the Interior, May 1, 1914, quoted in House Committee on the Public Lands, *Hearings on H.R. 104,* 63rd Cong., 2nd sess., 1914, 66–69.

4. Robert Shankland, *Steve Mather of the National Parks* (New York: Alfred A. Knopf, 1951), 247–252; Horace M. Albright, as told to Robert Cahn, *The Birth of the National Park Service: The Founding Years, 1913–1933* (Salt Lake City: Howe Brothers, 1985), 64.

5. Horace M. Albright to J. Horace MacFarland, May 28, 1917, JHMcF; Albright and Cahn, *Birth of the National Park Service,* 54–57.

6. Edgar H. Schein, *Organizational Culture and Leadership* (San Francisco: Jossey-Bass Publishers, 1992), 1–2, 15, 58. Schein adds (p. 12) that an organization's "shared assumptions derive their power from the fact that they begin to operate outside of awareness," and that once the assumptions attain a history of success, they are assumed to be "right and good." Among the manifestations of an organization's culture that he notes (pp. 8–10) are espoused values and formal philosophy, the special skills of group members, the habits of thinking and the shared meanings as group members interact with one another, and "implicit rules for getting along in the organization." For a lengthy analysis of organizational culture in the U.S. Forest Service, see Herbert Kaufman, *The Forest Ranger: A Study in Administrative Behavior* (Washington, D.C.: Resources for the Future, 1986; originally published by Johns Hopkins University Press, 1960). See also Ashley L. Schiff, *Fire and Water: Scientific Heresy in the Forest Service* (Cambridge, Mass.: Harvard University Press, 1962), 1–14, 164–170.

7. Russ Olsen, *Administrative History: Organizational Structures of the National Park Service, 1917 to 1985* (Washington, D.C.: National Park Service, 1985), 10, 34–35.

8. Horace M. Albright to Arthur A Shurtleff, November 9, 1929, Entry 18, RG79; Shankland, *Steve Mather,* 9. By the early twentieth century the demand for landscape architects in the United States had increased considerably for designing resorts, country estates, campuses, city parks, state capitol grounds, subdivisions, and other developments. See Laura Wood Roper, *FLO: A Biography of Frederick Law Olmsted* (Baltimore: Johns Hopkins University Press, 1973), 397.

9. Stephen T. Mather to Horace M. Albright, January 10, 1922, YELL. See also Stephen T. Mather to Ike E. O. Pace, August 17, 1928, Entry 6, RG79; and Arthur E. Demaray to Ernest Pl. Leavitt, September 28, 1928, Entry 17, RG79. As Horace Albright later described it, the landscape architects were given "power of approval, modification or veto" over plans submitted by concessionaires. Horace M. Albright to Ray Lyman Wilbur, March 5, 1929, Entry 6, RG79.

10. Arno B. Cammerer to Frederick Law Olmsted, Jr., September 11, 1922, Entry 18, RG79; National Park Conference, 7th, Yellowstone National Park, Wyoming, October 22–28, 1923, "Minutes," typescript, 29–30, NPS-HC. See also Linda Flint McClelland, *Presenting Nature: The Historic Landscape Design of the National Park Service, 1916 to 1942* (Washington, D.C.: National Park Service, 1993), 95–96.

11. Paul P. Kiessig, "Landscape Engineering in the National Parks," December 2, 1922, typescript, YELL.

12. Thomas C. Vint to Horace M. Albright, May 22, 1929, Entry 18, RG79. Vint was a professional architect as well as a landscape architect.

13. James S. Pray, comments at February 1916 meeting of American Society of Landscape Architects, copy courtesy Denis P. Galvin; Albright and Cahn, *Birth of the National Park Service,* 104, 272; Franklin K. Lane to Stephen T. Mather, May 13, 1918, Entry 17, RG79; William C. Tweed and Laura Soulliere Harrison, "Rustic Architecture and the National Parks: The History of a Design Ethic," 1987, typescript, chapter 4, 17–18, copy courtesy of the authors; Albright to Shurtleff, November 9, 1929.

14. The quote is from National Park Service, Office Order no. 228, April 3, 1931, typescript, copy courtesy of Denis P. Galvin. See also McClelland, *Presenting Nature,* 86–112; Tweed and Harrison, "Rustic Architecture," chapter 4, 17–18; Norman T. Newton, *Design on the Land: The Development of Landscape Architecture* (Cambridge, Mass.: Harvard University Press, 1971), 535–536; and Thomas C. Vint, "National Park Service Master Plans," *Planning and Civic Comment* (April 1946), 21–24.

15. See Olsen, *Organizational Structures of the National Park Service,* 10, 40–41, 60–61, 66–67; and Vernon L. Hammons, "A Brief Organizational History of the Office of Design and Construction, National Park Service, 1917–1962," 1962, typescript, 6, NPS-HC; Newton, *Design on the Land,* 535.

16. Olsen, *Organizational Structures of the National Park Service,* 34–37; Hammons, "History of the Office of Design and Construction," 1–2; Albright to Wilbur, March 5, 1929; Shankland, *Steve Mather,* 147–162; Albright and Cahn, *Birth of the National Park Service,* 194–195; McClelland, *Presenting Nature,* 109.

17. Newton, *Design on the Land,* 534–535; Shankland, *Steve Mather,* 254–255; Tweed and Harrison, "Rustic Architecture and the National Parks," chapter 4, 17–18; Hillory A. Tolson, *Historic Listing of National Park Service Officials,* rev. Harold P. Danz (Washington, D.C.: National Park Service, 1986), 26, 94, 156.

18. The quote is from Albright and Cahn, *Birth of the National Park Service,* 136.

19. John W. Henneberger, "To Protect and Preserve: A History of the National Park Ranger," typescript, 1965, 246–254, copy courtesy of the author. Albright and Cahn, *Birth of the National Park Service,* 139.

20. *National Park Conference, 3rd, Berkeley, Calif., 11–13 March 1915, Proceedings* (Washington, D.C.: Government Printing Office, 1915), 44; Henneberger, "To Protect and Preserve," 285–286, 302–319. Ranger skills needed in national parks were quite similar to those needed in national forests. See Harold K. Steen, *The U.S. Forest Service: A History* (Seattle: University of Washington Press, 1976), 83.

21. The early history of interpretation is discussed in Barry Mackintosh, *Interpretation in the National Park Service* (Washington, D.C.: National Park Service, 1986), 7–17. See also Henneberger, "To Protect and Preserve," 182–183, 370–371; and Albright and Cahn, *Birth of the National Park Service,* 147. Beginning in Yosemite in 1914, the parks augmented the permanent ranger staffs with "seasonal" rangers who worked during the summer months, as they still do today. Henneberger, "To Protect and Preserve," 330–331; Albright and Cahn, *Birth of the National Park Service,* 142–145.

22. Henneberger, "To Protect and Preserve," 253; Albright and Cahn, *Birth of the National Park Service,* 145, 148.

23. Tweed and Harrison, "Rustic Architecture," chapter 3, 2; Shankland, *Steve Mather,* 266–267; Henneberger, "To Protect and Preserve," 319–330.

24. For example, Forest Townsley, a ranger at Platt National Park in Oklahoma, paid his own moving expenses when Yosemite offered him a ranger position in 1913. Henneberger, "To Protect and Preserve," 252.

25. Albright and Cahn, *Birth of the National Park Service,* 146; Henneberger, "To Protect and Preserve," 252–254. The rangers came under the civil service system in 1925, and the superintendents in 1931. See Shankland, *Steve Mather,* 249.

26. Henneberger, "To Protect and Preserve," 371. In some instances, an assistant superintendent was second in command. See Cameron, *The National Park Service,* 64.

27. Shankland, *Steve Mather,* 247–253; Albright and Cahn, *Birth of the National Park Service,* 63–65.

28. Shankland, *Steve Mather,* 245–252; Lary M. Dilsaver and William C. Tweed, *The Challenge of the Big Trees* (Three Rivers, California: Sequoia Natural History Association, 1990), 112.

29. Soon after resigning from the superintendency, Walter Fry began building a natural history program in Sequoia, one of the Service's most successful early efforts in this field. Dilsaver and Tweed, *Challenge of the Big Trees,* 101–103, 112.

30. Henneberger, "To Protect and Preserve," 288, 301–302. See 103–106, 289–298, on Mather's hiring of rangers who would go on to become superintendents at important national parks.

31. National Park Conference, 8th, Mesa Verde National Park, Colorado, October 1–5, 1925, "Minutes," typescript, 1, NPS-HC; Albright and Cahn, *Birth of the National Park Service,* 146, 200, 218. See also Shankland, *Steve Mather,* 263, 279–280, for discussions of superintendents conferences.

Included in a 1923 conference brochure "Sing in Yellowstone" were lyrics for a campfire song, no doubt a means by which Mather hoped to develop camaraderie among superintendents.

Entitled "Mather's Gang" and sung to the tune of "Clementine," the song contained a rollcall of the top names in national park management:

Hip hooray, hip hooray,
For the Conference in Yellowstone,
There is Crosby from Grand Canyon
And Nusbaum digging bones,
There is Thompson, Toll and Tomlinson,
And Boles from Hawaii (Ha-wa-e)
There is Cammerer and Mather
Both from Washington, D.C.
Next comes Brazell and Farquhar
From Maine Dorr doth hold forth,
White and Lewis from California,
And Karstens way up north,
Hull and Goodwin and Dr. Waring,
Frank Pinkley and Eakin, too,
Then Reusch and Horace Albright
Now our roll call is through.
But don't forget we have the ladies
And others with smiles and barks*
All banded here together
To boost for National Parks[.]
So once again now let us cheer
For the service one and all
And a big one for Steve Mather
He made the Parks a world-wide call.

*The lyricist apologized for having to complete the rhyme with "Parks."

National Park Service, "Sing in Yellowstone," 1923, NPS-HC.

32. Lane to Mather, May 13, 1918; Albright and Cahn, *Birth of the National Park Service*, 69. See 69–73 for an almost complete text of the Lane Letter.

33. Lane to Mather, May 13, 1918.

34. In 1921 Mather reiterated much of the Lane Letter's policy statements in Stephen T. Mather, "The Ideals and Policy of the National Park Service Particularly in Relation to Yosemite National Park," in Ansel Hall, ed., *Handbook of Yosemite National Park* (New York: G. P. Putnam's Sons, 1921), 77–86. Hubert Work, "Statement of National Park Policy," memorandum for the Director, March 11, 1925, typescript, NPS-HC. Perhaps because Lane's message was a threshold policy statement, it remained by far the better remembered and more influential of the two secretarial policy letters.

35. Susan L. Flader, *Thinking Like a Mountain: Aldo Leopold and the Evolution of an Ecological Attitude toward Deer, Wolves, and Forests* (Columbia: University of Missouri Press, 1974), 100–102; Steen, *The U.S. Forest Service*, 116–120; and Hal K. Rothman, "'A Regular Ding-Dong Fight': Agency Culture and Evolution in the NPS-USFS Dispute, 1916–1937," *Western Historical Quarterly* 20 (May 1989), 143, 153–155.

36. Stephen T. Mather, "Report on Do Functions of the National Park Service Overlap with Those of Other Bureaus?" 1925, typescript, NPS-HC.

37. Mather, "The Ideals and Policy of the National Park Service," 80; Stephen T. Mather, "Progress in the National Parks," *Sierra Club Bulletin* 11, no. 1 (1920), 6. See also "Director Mather Declares Parks Must Be True Recreation Centers," *New York Times*, December 12, 1919, Mather-BL.

38. The association's objectives are stated in "The Objects of the National Parks Association,"

National Parks Bulletin 1 (June 6, 1919), 8. Under Yard the association became an aggressive defender of the parks and, although it developed strong differences with Mather, provided the Service with valuable political support. See Bruce M. Kilgore, "Forty Years Defending Parks," *National Parks Magazine* 33 (May 1959), 13–16; and Shankland, *Steve Mather*, 167.

39. John Ise, *Our National Park Policy*, 198; Tweed and Harrison, "Rustic Architecture," chapter 2, 3. The quote is found in *Annual Report of the Superintendent of National Parks to the Secretary of the Interior, 1916* (Washington, D.C.: Government Printing Office, 1916), 1.

40. Albright to Wilbur, March 5, 1929.

41. Mather, "Progress in the National Parks," 5. For a discussion of the railroads' interest in the national parks, see Alfred Runte, *Trains of Discovery* (1984; 2nd ed., Niwot, Colorado: Roberts Rhinehart, 1990).

42. "The National Park-to-Park Highway," loose-leaf folder, n.d. (ca. 1925), NPS-HC. See also Shankland, *Steve Mather*, 147–151; and McClellan, *Presenting Nature*, 77–79.

43. *Report of the Director of the National Park Service to the Secretary of the Interior, 1919* (Washington, D.C.: Government Printing Office, 1919), 13; *Report of the Director of the National Park Service to the Secretary of the Interior, 1925* (Washington, D.C.: Government Printing Office, 1925), 1.

44. Mather, "The Ideals and Policies of the National Park Service," 81.

45. Shankland, *Steve Mather*, 57–59; *Report of the Director of the National Park Service, 1919*, 61–66.

46. *Report of the Director of the National Park Service to the Secretary of the Interior, 1924* (Washington, D.C.: Government Printing Office, 1924), 11–14; Work, "Statement of National Park Policy."

47. Stephen T. Mather to John R. White, February 24, 1927, NPS-HC. See Dilsaver and Tweed, *Challenge of the Big Trees*, 182–185, for a discussion of the proposed Sierra Highway, ultimately defeated in the mid-1930s.

48. Shankland, *Steve Mather*, 157–159; William E. Brown, *A History of the Denali–Mount McKinley Region, Alaska* (Santa Fe: National Park Service, 1991), 101–107, 163.

49. Shankland, *Steve Mather*, 153–160; Lloyd K. Musselman, "Rocky Mountain National Park Administrative History, 1915–1965" (Washington, D.C.: National Park Service, 1971), 85; Horace M. Albright, "National Park Planning," *American Civic Annual* 2 (1930), 52.

50. Musselman, "Rocky Mountain National Park," 90.

51. Mather, "Progress in the National Parks," 6. See also Michael P. Cohen, *The History of the Sierra Club* (San Francisco: Sierra Club Books, 1988), 9.

52. Typescript policy comment by Superintendent John R. White, attached to letter from Roger W. Toll to the Director, December 2, 1922, NPS-HC; Dilsaver and Tweed, *Challenge of the Big Trees*, 132–135.

53. Tweed and Harrison, "Rustic Architecture," chapter 3, 2–12, and chapter 4, 6–9. See also Ned J. Burns, *Field Manual for Museums* (Washington, D.C.: Government Printing Office, n.d.), 6–10.

54. Alfred Runte, *Yosemite, the Embattled Wilderness* (Lincoln: University of Nebraska Press, 1990), 154–157; Stephen T. Mather to E. O. McCormick, December 16, 1920, Entry 6, RG79. Neither golf course was built. However, when the Service acquired lands in the Wawona area of Yosemite, a golf course was already there; it remains in use today.

55. Mather, "Progress in the National Parks," 8. For further comments favorable to winter sports, see Mather, "Ideals and Policy of the National Park Service," 85; and Secretary Lane's policy letter to Mather, May 13, 1918.

56. Horace M. Albright to James V. Lloyd, February 13, 1929, Entry 17, RG79. See also Runte, *Yosemite*, 152–153. For winter sports development in Rocky Mountain National Park (inspired in part by the development going on in Yosemite), see Lloyd K. Musselman, "Rocky Mountain National Park Administrative History, 1915–1965" (Washington, D.C.: National Park Service, 1971), 171–188. Albright's quote is on p. 172.

57. Shankland, *Steve Mather,* 207–208.

58. Draft of resolution and White's comments are attached to letter from Toll to the Director, December 2, 1922.

59. Toll to the Director, December 2, 1922, and attachments.

60. The Yellowstone dam proposals are discussed in Shankland, *Steve Mather,* 212–220.

61. Lane is quoted in Albright and Cahn, *Birth of the National Park Service,* 101.

62. *Report of the Director of the National Park Service, 1919,* 48; *Report of the Director of the National Park Service, 1920* (Washington, D.C.: Government Printing Office, 1920), 21–30.

63. The Yellowstone dam proposals are discussed in Shankland, *Steve Mather,* 212–220; Ise, *National Park Policy,* 307–316; and Albright and Cahn, *Birth of the National Parks,* 100–102, 105–107, 113–114.

64. *Report of the Director of the National Park Service, 1919,* 59.

65. Robert Sterling Yard to Arno B. Cammerer, March 5, 1923, Entry 18, RG79; C. W. Buchholtz, *Man in Glacier* (West Glacier, Montana: Glacier Natural History Association, 1976), 71. See Albright and Cahn, *Birth of the National Park Service,* 118, for discussion of opposition to other dams.

66. Dilsaver and Tweed, *Challenge of the Big Trees,* 95–96. Inholdings consisted of state-owned lands as well; however, the service's biggest concern was the privately owned lands. See Ise, *National Park Policy,* 482–483.

67. Ise, *National Park Policy,* 179–181. Although it did not involve an actual inholding, Mather's personal supervision of the 1925 dynamiting of a sawmill that had been used during construction of a hotel in Glacier National Park provides a striking example of the concern for protecting park scenery from industrial intrusions. The sawmill was not being dismantled soon enough to suit Mather. See Shankland, *Steve Mather,* 209.

68. *Annual Report of the Superintendent of National Parks (1916),* 11–13. See also Ise, *National Park Policy,* 318.

69. Lane to Mather, May 13, 1918; *Report of the Director of the National Park Service to the Secretary of the Interior, 1929* (Washington, D.C.: Government Printing Office, 1929), 10–11. See also Albright to Wilbur, March 5, 1929.

70. Ise, *National Park Policy,* 42. Although 542 square miles were removed, at the same time 113 square miles of mountainous terrain were added to the park, making the total loss from these land swaps 429 square miles. Ise, *National Park Policy,* 68–70; Runte, *Yosemite,* 67–68.

71. Ise, *National Park Policy,* 215–216; see also 139–140, 286. Albright recalled being "delighted" when Sully's Hill was removed from the system. Albright and Cahn, *Birth of the National Park Service,* 276.

72. Albright to Wilbur, March 5, 1929.

73. Lane to Mather, May 13, 1918. Regarding a possible addition of the Tetons to Yellowstone, a *Saturday Evening Post* editor remarked to Albright that "the best part of Yellowstone is not yet in the park." Albright and Cahn, *Birth of the National Park Service,* 117.

74. Albright and Cahn, *Birth of the National Park Service,* 40.

75. Lane to Mather, May 13, 1918; Albright and Cahn, *Birth of the National Park Service,* 63. Zion National Park, established in 1919, brought full national park status to the existing Zion National Monument. Monument status dated from 1909.

76. The quote is from Charles Sheldon, a well-known naturalist and writer, who conceived the idea of a national park in the vicinity of Mount McKinley when he first saw the area and its wildlife. The park legislation that Sheldon promoted received strong support from organizations interested in the conservation of game animals, such as the Boone and Crockett Club and the Camp Fire Club. Brown, "*History of the Denali–Mount McKinley Region,* 75–92.

77. Albright describes the Service's support for such parks as Grand Canyon and Bryce Canyon in Albright and Cahn, *Birth of the National Park Service,* 83–84, 189; Ise, *National Park Policy,* 268.

78. For a complete listing of the parks that came into the system during the Mather era, see

Barry Mackintosh, *The National Parks: Shaping the System* (Washington, D.C.: Department of the Interior, 1991), 17, 22–23.

79. Shankland, *Steve Mather,* 221–224; "Gift-Parks—The Coming National Park Danger," *National Parks Association Bulletin* 35, (October 9, 1923), 4–5; Robert Sterling Yard, "Standards of Our National Parks," *National Parks Bulletin* 8 (April 1927), 1–4; and Kilgore, "Forty Years Defending Parks," 13–15.

80. Albright to Wilbur, March 5, 1918.

81. Stephen T. Mather, "Their 'Incomparable Scenic Grandeur,'" *National Parks Bulletin* 9 (November 1927), 5. Concerned about a "new and dangerous policy" of lax national park standards, and resisting a proposed Ouachita National Park in Arkansas, Frederick Law Olmsted, Jr., argued that some park proposals addressed "regional needs rather than . . . proper national purposes." Albright and Cahn, *Birth of the National Park Service,* 229.

82. *Report of the Director of the National Park Service, 1921* (Washington, D.C.: Government Printing Office, 1921), 32–33; Shankland, *Steve Mather,* 184–190.

83. *Report of the Director of the National Park Service to the Secretary of the Interior, 1917* (Washington, D.C.: Government Printing Office, 1917), 11.

84. Lane to Mather, May 13, 1918; Work, "Statement of National Park Policy."

85. Mather, "Do Functions of the National Park Service Overlap Those of Other Bureaus?" On Mather's interest in using experts from other bureaus, see National Park Service, "The National Park Service, Its Functions, Its Policies, Its Future," February 1925, typescript, NPS-HC; and Albright and Cahn, *Birth of the National Park Service,* 23.

An evolving specialty, scientific land and resource management was an applied, commodity-oriented, and production-oriented enterprise, emphasizing the propagation and harvest of resources such as trees and fish—the useful products of forests, lakes, and streams. Led by Cornell and Yale, a few universities had begun to develop programs in the applied science of land and resource management. Increasingly, the federal and state bureaus involved in game, fish, and forest management looked to such programs for advice and for new personnel. See A. Hunter Dupree, *Science in the Federal Government: A History of Policies and Activities* (Baltimore: Johns Hopkins University Press, 1986), 236–255; Susan L. Flader, "Scientific Resource Management: An Historical Perspective," in Kenneth Sabol, ed., *Transactions, Forty-First North American Wildlife and Natural Resources Conference* (Washington, D.C.: Wildlife Management Institute, 1976), 19–25.

86. *Report of the Director of the National Park Service to the Secretary of the Interior, 1923* (Washington, D.C.: Government Printing Office, 1923), 23; Horace M. Albright, "Our National Parks as Wild Life Sanctuaries," *American Forests and Forest Life* 35 (August 1929), 536.

87. National Park Service, "Policy on Predators and Notes on Predators," 1939, typescript, various pagination, Central Classified File 715, RG79; *Report of the Director of the National Park Service, 1924,* 10.

88. Horace M. Albright, "Game Conservation in the National Parks," paper presented at the Eleventh National Game Conference of the American Game Protective Association, New York, April 8–9, 1924, Entry 6, RG79; Albright, "Our National Parks as Wild Life Sanctuaries," 505.

89. Brown, *History of the Denali–Mount McKinley Region,* 135–148.

90. Hillory A. Tolson, *Laws Relating to the National Park Service, the National Parks and Monuments* (Washington, D.C.: Department of the Interior, 1933), 10. Park Service biologist Victor Cahalane observed in 1939 that the Service's predator control under Mather had "followed the general trend and pattern of thought of the times." Victor H. Cahalane, "The Evolution of Predator Control Policy in the National Parks," *Journal of Wildlife Management* 3 (July 1939), 235.

91. *Report of the Director of the National Park Service, 1918* (Washington, D.C.: Government Printing Office, 1918), 22; *Report of the Director of the National Park Service, 1923,* 23.

92. *Report of the Director of the National Park Service, 1929,* 25; Albright, "Our National Parks as Wild Life Sanctuaries," 536.

93. Cahalane, "The Evolution of Predator Control Policy in the National Parks," 230–231.

94. National Park Service, "Policy on Predators and Notes on Predators"; Cahalane, "Evolution of Predator Control Policy," 232–234.

95. Jay Bruce, Sr., *Cougar Killer* (New York: Comet Press Books, 1953), 135–136. Even though he did not enjoy killing animals, Mather once planned to join Bruce on a cougar hunt near Sequoia National Park. Although there is no evidence that the hunting trip took place, Mather discussed plans for it in Stephen T. Mather to John R. White, October 27, 1920, Central Classified Files, RG79. See also Shankland, *Steve Mather,* 270. Lewis is quoted in National Park Service, "Policy on Predators and Notes on Predators."

96. *Report of the Director of the National Park Service to the Secretary of the Interior, 1926* (Washington, D.C.: Government Printing Office, 1926), 14. By the early twentieth century, the Biological Survey had begun shifting from its earlier emphasis on scientific studies (which were increasingly viewed by Congress as not being useful endeavors) to law enforcement and regulatory work, such as protecting migrating waterfowl populations and controlling predators and rodents. Thomas R. Dunlap, *Saving America's Wildlife* (Princeton: Princeton University Press, 1988), 35–39; Donald Worster, *Nature's Economy: A History of Ecological Ideas* (1977; 2nd ed., Cambridge: Cambridge University Press, 1994), 262–279; Dupree, *Science in the Federal Government,* 253.

97. Cahalane, "Evolution of Predator Control Policy," 234; National Park Service, "Policy on Predators and Notes on Predators."

98. National Park Service, "Policy on Predators and Notes on Predators"; R. Gerald Wright, *Wildlife Research and Management in the National Parks* (Urbana: University of Illinois Press, 1992), 64. The Biological Survey's comprehensive predator and rodent control programs were also being subjected to widespread questioning and disapproval. Dunlap, *Saving America's Wildlife,* 48–61.

99. National Park Service, "Policy on Predators and Notes on Predators"; Wright, *Wildlife Research and Management,* 64. Toll is quoted in Musselman, "Rocky Mountain National Park," 129–130.

100. National Park Service, "Policy on Predators and Notes on Predators."

101. Cahalane, "Evolution of Predator Control Policy," 235; Dunlap, *Saving America's Wildlife,* 48–61.

102. National Park Service, "Policy on Predators and Notes on Predators."

103. Albright to Wilbur, March 5, 1929; National Park Service, "Policy on Predators and Notes on Predators"; Wright, *Wildlife Research and Management,* 64.

104. Horace M. Albright, "The National Park Service's Policy on Predatory Mammals, *Journal of Mammalogy* 12 (May 1931), 185–186.

105. Lane to Mather, May 13, 1918; *Report of the Director of the National Park Service, 1919,* 35.

106. M. F. Daum to Theodore C. Joslin, January 9, 1929, YELL.

107. Curtis K. Skinner et al., "History of the Bison in Yellowstone Park" [with supplements], 1952, typescript, various pagination, 9, YELL. Albright's quote is from *Report of the Director of the National Park Service, 1925,* 36. Woodring is quoted in Daum to Joslin, January 9, 1929.

108. Skinner, "History of the Bison in Yellowstone Park," 10. The roundups were initially held for management purposes, but later were staged for public enjoyment as well. See Wright, *Wildlife Research and Management,* 152.

109. On the Service's management of large grazing and browsing animals in Yellowstone, see Don Despain et al., *Wildlife in Transition: Man and Nature on Yellowstone's Northern Range* (Boulder, Colorado: Roberts Rhinehart, 1986), 14–57, 72–110.

110. Horace M. Albright to George Bird Grinnell, May 13, 1921, Grinnell; Daum to Joslin, January 9, 1929.

111. Albright to Grinnell, May 13, 1921.

112. See for example *Report of the Director of the National Park Service, 1919,* 34–35; and *Report of the Director of the National Park Service, 1921,* 37.

113. Daum to Joslin, January 9, 1929. See also Wright, *Wildlife Research and Management,* 117.

114. Wright, *Wildlife Research and Management,* 71; Daum to Joslin, January 9, 1929. See Despain et al., *Wildlife in Transition,* 28–32.

115. Thomas R. Dunlap, "That Kaibab Myth," *Journal of Forest History* 32 (April 1988), 61–63; Dunlap, *Saving America's Wildlife,* 65–68; Shankland, *Steve Mather,* 272–273.

116. *Report of the Director of the National Park Service, 1921,* 38; *Report of the Director of the National Park Service, 1923,* 24.

117. Musselman, "Rocky Mountain National Park," 127; *Report of the Director of the National Park Service, 1921,* 38. See also Arthur E. Demaray, "Regulations to Govern the Disposal of Wild Animals from Yellowstone National Park," October 26, 1929, Entry 19, RG79.

118. Shankland, *Steve Mather,* 269; *Report of the Director of the National Park Service, 1921,* 39; Runte, *Yosemite,* 130–134; Wright, *Wildlife Research and Management,* 152–153.

119. Skinner, "History of the Bison in Yellowstone Park"; Albright, "Our National Parks as Wildlife Sanctuaries," 507.

120. *Report of the Director of the National Park Service, 1921,* 38.

121. Albright is quoted in National Park Service, "Policy on Predators and Notes on Predators." Musselman, "Rocky Mountain National Park," 127–131.

122. Paul Schullery, *The Bears of Yellowstone* (1986; 3rd ed., Worland, Wyoming: High Plains Publishing Company, 1992), 89–108; Runte, *Yosemite,* 136–140; Dilsaver and Tweed, *Challenge of the Big Trees,* 145–146.

123. Albright, "Game Conservation in the National Parks."

124. Dilsaver and Tweed, *Challenge of the Big Trees,* 145–146; Wright, *Wildlife Research and Management,* 111–112; Schullery, *Bears of Yellowstone,* 104.

125. *Report of the Director of the National Park Service to the Secretary of the Interior, 1922* (Washington, D.C.: Government Printing Office, 1922), 39; *Annual Report of the Director of the National Park Service* (1925), 6–7.

126. *Annual Report of the Director of the National Park Service* (1926), 14; John D. Varley, "A History of Fish Stocking Activities in Yellowstone National Park between 1881 and 1980," Yellowstone National Park Information Paper no. 35, January 1, 1981, typescript, 1–3, YELL. See also John D. Varley and Paul Schullery, *Freshwater Wilderness: Yellowstone Fishes and Their World* (Yellowstone National Park, Wyoming: Yellowstone Library and Museum Association, 1983), 102–103.

127. Flader, "Scientific Resource Management," 20–21; Dupree, *Science in the Federal Government,* 236–238; David H. Madsen, "Report on Fish Cultural Activities," typescript, April 5, 1935, Central Classified File 714, RG79. The United States Commission on Fish and Fisheries was established in 1871 and was renamed the Bureau of Fisheries in 1903.

128. *Report of the Director of the National Park Service, 1929,* 26; Horace M. Albright to the Director, October 11, 1928, Entry 17, RG79.

129. The resolution of the Ecological Society, and Albright's response to Dr. A. O. Weese, January 23, 1922, are found in Charles C. Adams, "Ecological Conditions in National Forests and in National Parks," *Scientific Monthly* 20 (June 1925), 570.

130. The park never fully implemented the Tule elk introduction as originally intended, but fenced in the elk for more than a decade before removing them. Runte, *Yosemite,* 130–134.

131. Lane to Mather, May 13, 1918; Work, "Statement of National Park Policy."

132. Albright to Wilbur, March 5, 1929, 17. On the continued introduction of fish, see for example *Annual Report of the Director of the National Park Service* (1929), 109; and Varley, "A History of Fish Stocking Activities in Yellowstone National Park," 8, 9, 16, 26.

133. *Annual Report of the Director of the National Park Service* (1929), 83, 90; Wright, *Wildlife Research and Management,* 91–92, 97–98, 107.

134. The director's quote is from Mather, "Ideals and Policy of the National Park Service," 79. See also "Yosemite National Park Fire Control Plan, Season of 1928," Entry 17, RG79;

H. Duane Hampton, *How the U.S. Cavalry Saved Our National Parks,* 83, 100, 107; David M. Graber, "Coevolution of National Park Service Fire Policy and the Role of National Parks," in *Proceedings—Symposium and Workshop on Wilderness Fire, Missoula, Montana, November 15–18, 1983,* U.S. Forest Service General Technical Report, INT-182 (Ogden, Utah: Intermountain Forest and Range Experiment Station, 1985), 345; and Stephen J. Pyne, *Fire in America: A Cultural History of Wildland and Rural Fire* (Princeton: Princeton University Press, 1982), 296–297.

135. Pyne, *Fire in America,* 111–112, 297–298. As an example of skepticism about light burning, biologist Charles C. Adams, who surveyed ecological conditions in several national parks, wrote in 1925 that "from every standpoint, 'light burning' should not be practiced in our national parks." Adams added that "probably no one yet knows enough, and has the financial backing necessary, to practice light burning successfully." Adams, "Ecological Conditions," 571–573. On Forest Service adherence to tradition in fire management, see Schiff, *Fire and Water,* 15–50.

136. *Report of the Director of the National Park Service* (1926), 17.

137. John D. Coffman, "John D. Coffman and His Contribution to Forestry in the National Park Service," typescript, n.d., 35, NPS-HC; *Annual Report of the Director of the National Park Service* (1929), 20; Pyne, *Fire in America,* 298.

138. Coffman, "John D. Coffman and His Contribution to Forestry," 34–35. See also Olsen, *Organizational Structures of the National Park Service,* 36–43; and Albright and Cahn, *Birth of the National Park Service,* 194.

139. J. D. Coffman to Frank A. Kittredge, September 3, 1928, Entry 17, RG79; National Park Service, "A Forestry Policy for the National Parks," appended to Ansel F. Hall to the Director, October 29, 1928, Entry 17, RG79; *Report of the Director of the National Park Service* (1929), 20.

140. Ansel F. Hall, "Minutes of the Regional Forest Protection Board," San Francisco, February 16, 1928, typescript, Entry 17, RG79; *Annual Report of the Director of the National Park Service* (1929), 22; Dilsaver and Tweed, *Challenge of the Big Trees,* 180.

141. *Annual Report of the Director of the National Park Service* (1925), 7; *Annual Report of the Director of the National Park Service* (1926), 16, 41.

142. *Annual Report of the Director of the National Park Service* (1923), 26.

143. *Annual Report of the Director of the National Park Service* (1925), 7; *Annual Report of the Director of the National Park Service* (1926), 16, 41.

144. National Park Service, "A Forestry Policy for the National Parks," 1931, typescript, 7, Entry 18, RG79.

145. Ansel F. Hall, "Minutes of Meeting of the Regional Forest Protection Board, San Francisco, California," February 16, 1928, Entry 17, RG79.

146. Lane to Mather, May 13, 1918. See also Tolson, *Laws Relating to the National Park Service* (1933), 10–11.

147. Mather, "Ideals and Policy of the National Park Service," 83; Ise, *National Park Policy,* 302–307; Shankland, *Steve Mather,* 202; Albright and Cahn, *Birth of the National Park Service,* 59–60, 73–74.

148. Adams, "Ecological Conditions," 574, 585, 569; Dilsaver and Tweed, *Challenge of the Big Trees,* 146–147.

149. The Ecological Society's resolution and the Park Service's response are found in Adams, "Ecological Conditions," 569–570. The association's statement is in "A Resolution on the National Parks Policy of the United States," *Science* 58 (January 29, 1926), 115. See also Wright, *Wildlife Research and Management,* 37.

150. Runte, "Joseph Grinnell and Yosemite," 172–181.

151. Adams, "Ecological Conditions," 563, 584.

152. Adams, "Ecological Conditions," 568.

153. Harold C. Bryant and Wallace W. Attwood, *Research and Education in the National Parks* (Washington, D.C.: Government Printing Office, 1932), 48–50; Frank C. Brockman, "Park Naturalists and the Evolution of National Park Service Interpretation through World War II,"

Journal of Forest History 22 (January 1978), 31–32, 35, 37; Mackintosh, *Interpretation in the National Park Service,* 11–14. See also National Park Conference, 8th, 1925, "Minutes," 38–41; and Polly Welts Kaufman, "Challenging Tradition: Pioneer Women Naturalists in the National Park Service," *Forest and Conservation History* 34 (January 1990), 4–16.

154. Ben H. Thompson, "George Melendez Wright: A Biographical Sketch," *George Wright Forum* 7, no. 2 (1990), 3.

155. Albright to the Director, November 8, 1928.

156. Mather, "Ideals and Policy of the National Park Service," 80, 84.

157. These differing concepts are discussed in Samuel P. Hays, *Conservation and the Gospel of Efficiency: The Progressive Conservation Movement, 1890–1920* (Cambridge, Mass.: Harvard University Press, 1959), 189–198; and Roderick Nash, *Wilderness and the American Mind* (1967; 3rd ed., New Haven: Yale University Press, 1982), 129, 135–139, 180–181.

158. Lane to Mather, May 13, 1918.

159. The inscription is quoted in Shankland, *Steve Mather,* 291. Additional castings of this plaque were placed in many other units of the national park system in 1991 to commemorate the Park Service's seventy-fifth anniversary.

160. Adams, "Ecological Conditions," 567–568, 589–590.

Chapter 4. The Rise and Decline of Ecological Attitudes

1. Transfer of the National Park Service's wildlife biologists to the Biological Survey began in early December 1939 and was made official on January 1, 1940. *Annual Report of the Secretary of the Interior for the Fiscal Year Ending June 30, 1940* (Washington, D.C.: Government Printing Office, 1940), 165; National Park Service, "National Parks: A Review of the Year," *American Planning and Civic Annual* (1940), 34. The Bureau of Biological Survey had just been transferred from the Department of Agriculture to the Department of the Interior. In 1940 the survey would be merged with the Bureau of Fisheries to become the Fish and Wildlife Service, now known as the U.S. Fish and Wildlife Service.

2. Horace M. Albright, "The Everlasting Wilderness," *Saturday Evening Post* 201 (September 29, 1928), 28. Giving road mileage figures lower than earlier calculations, Albright perhaps did not count the more primitive roads.

3. Ben H. Thompson to Arno B. Cammerer, February 23, 1934, George M. Wright files, MVZ-UC. This statement was later included verbatim in George Wright and Ben Thompson, *Fauna of the National Parks of the United States,* Fauna Series no. 2 (Washington, D.C.: Government Printing Office, 1935), 123–124.

4. Donald C. Swain, *Wilderness Defender* (Chicago: University of Chicago Press, 1970), 192; and Swain, "The National Park Service and the New Deal, 1933–1940," *Pacific Historical Review* 41 (August 1972), 313, 316.

5. On his death, Mather left Albright and Cammerer $25,000 each, partly because he hoped the money would ensure their independence of thought as Park Service leaders. Swain, *Wilderness Defender,* 193. In "The National Park Service and the New Deal" (p. 316), Swain depicts Cammerer as a "relatively weak director," whom Secretary Ickes did not care for. In contrast to this perception, Cammerer adroitly used his talented staff to promote Park Service programs under the New Deal. George Collins, a longtime, highly placed Park Service employee, recalled that Cammerer "used Mr. Demaray and Mr. Wirth, Ben Thompson, Hillory Tolson and others to his highest and best advantage, and to theirs as well. The service had a growing reputation of efficiency and ability. I think you have to credit [Cammerer] a lot for that." George L. Collins, "The Art and Politics of Park Planning and Preservation," interviews by Ann Lage, 1978 and 1979, Regional Oral History Office, University of California, typescript, 86, NPS-HC. Ickes' disregard for Cammerer is discussed in Thomas H. Watkins, *Righteous Pilgrim: The Life and Times of Harold Ickes, 1874–1952* (New York: Henry Holt and Company, 1990), 552–555.

6. Arthur E. Demaray to Horace M. Albright, September 21, 1928, Entry 17, RG79; Joseph Dixon to H. C. Bryant, March 7, 1929, Harold C. Bryant files, MVZ-UC; Albright to the Director, October 11, 1928, Entry 17, RG79. In his October 11 memorandum, Albright mentions Wright's belief that the survey should be conducted under Park Service direction. A stronger statement—that Wright was "very anxious" that it be a Park Service project—is found in Joseph Dixon to Horace M. Albright, March 7, 1929, Horace M. Albright files, MVZ-UC.

7. Horace M. Albright to Ray Lyman Wilbur, March 5, 1929, Entry 6, RG79; Ansel F. Hall to the Director, October 17, 1928, Entry 17, RG79; and Ansel F. Hall to Horace M. Albright, November 23, 1928, Entry 17, RG79. Albright had earlier stated to Mather that two important benefits from the survey would be "widening the scope of our educational work . . . and [securing] material for the development of our museums and general educational activities." Albright to the Director, October 11, 1928. Negotiations on the survey were stalled briefly in the winter of 1929 owing to the proposal's being "unduly emphasized as a special achievement" of the Education Division. The division apparently sought too much credit. Dixon to Albright, March 7, 1929.

8. Ben H. Thompson, "George M. Wright, 1904–1936," *George Wright Forum* (Summer 1981), 1–2; Horace M. Albright to the Director, October 11, 1928; Lowell Sumner, "Biological Research and Management in the National Park Service: A History," *George Wright Forum* (Autumn 1983), 6–7; George M. Wright to Joseph Dixon, April 26, 1926, George M. Wright files, MVZ-UC.

9. Joseph Grinnell and Tracy Irwin Storer, "The Interrelations of Living Things," in *Animal Life in the Yosemite* (Berkeley: University of California Press, 1924), 38–39. On Grinnell's influence on Wright, Park Service naturalist Carl P. Russell commented in 1939 that "because of the preparation that [Grinnell] gave George Wright and through the warm friendship that existed between Dr. Grinnell and Mr. Wright, we have a Wildlife Division and a defined wildlife policy." Carl P. Russell to E. Raymond Hall, November 17, 1939, Carl P. Russell files, MVZ-UC. Grinnell's career and his influence on the ideas of Wright and other Park Service biologists are discussed in Alfred Runte, "Joseph Grinnell and Yosemite: Rediscovering the Legacy of a California Conservationist," *California History* 69 (Summer 1990), 173–181.

10. Thomas R. Dunlap, *Saving America's Wildlife* (Princeton: Princeton University Press, 1988), 70–74; Susan L. Flader, *Thinking Like a Mountain: Aldo Leopold and the Evolution of an Ecological Attitude toward Deer, Wolves, and Forests* (Lincoln: University of Nebraska Press, 1978), 28–33; and Donald Worster, *Nature's Economy* (1985; 2nd ed., Cambridge: Cambridge University Press, 1994), 214–219.

11. George M. Wright, Joseph S. Dixon, and Ben H. Thompson, *Fauna of the National Parks of the United States: A Preliminary Survey of Faunal Relations in National Parks,* Contributions of Wildlife Survey, Fauna Series no. 1 (Washington, D.C.: Government Printing Office, 1933), 4, 21.

12. Wright, Dixon, and Thompson, *Fauna of the National Parks* (1933), 4–5, 19–22.

13. Wright, Dixon, and Thompson, *Fauna of the National Parks* (1933), 23–28, 33–36, 71.

14. Wright, Dixon, and Thompson, *Fauna of the National Parks* (1933), 37–38, 44, 94, 132. For additional mention of the need to expand boundaries, see 114, 121, 126, and 131.

15. Wright, Dixon, and Thompson, *Fauna of the National Parks* (1933), 10.

16. Wright, Dixon, and Thompson, *Fauna of the National Parks* (1933), 147–148.

17. Wright's quote is found in National Park Service, "Policy on Predators and Notes on Predators," 1939, various pagination, Central Classified File, RG79.

18. Horace M. Albright, "The National Park Service's Policy on Predatory Mammals, *Journal of Mammalogy* 12 (May 1931), 185–186; Horace M. Albright, "Game Conditions in Western National Parks," November 23, 1932, typescript, YELL; Horace M. Albright, "Research in the National Parks," *Scientific Monthly* (June 1933), 489.

19. Wright, Dixon, and Thompson, *Fauna of the National Parks* (1933), 5.

20. Sumner, "Biological Research and Management," 6, 10; Arno B. Cammerer, Office Order no. 226, March 21, 1934, Entry 35, RG79. At this time a branch was administratively higher

than a division and usually included several divisions. Harold Bryant had come into the Service as a result of his efforts to promote education in the national parks and his interest in training park naturalists.

21. Horace M. Albright to Wild Life Survey, n.d. (ca. early 1932), Entry 35, RG79; Horace M. Albright, Office Order no. 234 to Superintendents and Custodians, February 29, 1932, Central Classified File, RG79; Cammerer, Office Order no. 226.

22. Victor H. Cahalane, Memorandum on General Procedure of the Wildlife Division, Branch of Research and Education, National Park Service, July 28, 1936, 6–7, Research Division Files, YELL.

23. Horace M. Albright, as told to Robert Cahn, *The Birth of the National Park Service: The Founding Years, 1913–1933* (Salt Lake City: Howe Brothers, 1985), 289; John Ise, *Our National Park Policy: A Critical History* (Baltimore: The Johns Hopkins Press, 1961), 359–363. A detailed history of the Service's involvement with the CCC is found in John C. Paige, *The Civilian Conservation Corps and the National Park Service, 1933–1942* (Washington, D.C.: National Park Service, 1985).

24. Complaints that CCC personnel were molesting wildlife and vandalizing park resources are found in, for example, Paul McG. Miller, Memorandum to be Posted on Bulletin Board, June 1, 1935, Entry 34, RG79; and A. E. Demaray to Park Superintendents and Custodians, May 4, 1936, Central Classified File, RG79. In late 1934, biologist Charles J. Spiker complained to Wright that the "havoc wrought" by the crews in Acadia National Park surpassed that in any other park in the eastern United States. The destruction of forests to allow for development at the top of Cadillac Mountain was only part of the "mutilation" of Acadia that concerned Spiker. Charles J. Spiker to Chief of the Wildlife Division, November 13, 1934, Entry 34, RG79.

25. Albright's comment that the superintendents might seek advice from the Wildlife Division was only a request. He wrote to the superintendents, "Should technical advice be desirable I hope you will call upon the Wild Life Division." Horace M. Albright, Memorandum for Field Officers, June 7, 1933, Harold C. Bryant files, MVZ-UC.

26. Sumner, "Biological Research and Management," 9.

27. Sumner, "Biological Research and Management," 9.

28. The estimate is found in Harlan D. Unrau and G. Frank Willis, *Administrative History: Expansion of the National Park Service in the 1930s* (Denver: National Park Service, 1983), 75.

29. George M. Wright to the Director, February 28, 1934, Central Classified File, RG79.

30. George M. Wright, Memorandum for the Director, December 13, 1935, Central Classified File, RG79.

31. E. Lowell Sumner, "Special Report on the Sixth Enrollment Period Program Posed for Death Valley National Monument," September 10, 1935, Entry 34, RG79. Titus Canyon almost certainly did not become a research reserve. It was not mentioned in a list of such reserves compiled in 1942; see Charles Kendeigh, "Research Areas in the National Parks," *Ecology* 23 (January 1942), 236–238. And the natural resource management office at Death Valley has no record that the canyon ever received this designation. Today the improved and maintained dirt road up Titus Canyon is probably the most popular and heavily traveled four-wheel-drive road in the park. But a current bighorn management plan calls for closing the Titus Canyon Road during the hot season so that bighorn will have undisturbed access to the spring. Personal communication with Tim Coonan, natural resource management specialist, September 30, 1991, and January 6, 1993.

32. Victor Cahalane to A. E. Demaray, September 14, 1935, Entry 34, RG79. Examples of nonconcurrence are Victor Cahalane, Memorandum for Mr. Demaray, September 14, 1935, Entry 34, RG79, relating to CCC projects in Glacier; and Cahalane, Memorandum for Mr. Demaray, September 23, 1935, Entry 34, RG79, relating to projects in Grand Canyon.

33. M. R. Tillotson to the Director, October 18, 1935, Entry 34, RG79; and Victor H. Cahalane to A. E. Demaray, September 23, 1935, Entry 34, RG79. Since plans for the trail might have been drawn up for some time (or the project could have been an afterthought to building the

trail, a kind of incremental development), it is possible that the biologists had no opportunity for an earlier review.

34. Lowell Sumner to George Wright, September 12, 1935, Entry 34, RG79.

35. R. L. McKown to Thomas C. Vint, October 8, 1935, Entry 34, RG79.

36. E. Lowell Sumner to R. L. McKown, October 10, 1935, Entry 34, RG79.

37. E. Lowell Sumner, "Special Report on a Wildlife Study of the High Sierra in Sequoia and Yosemite National Parks and Adjacent Territory," October 9, 1936, YOSE. The Sierra Club quote is found in Michael P. Cohen, *The History of the Sierra Club, 1892–1970* (San Francisco: Sierra Club Books, 1988), 86.

38. W. J. Liddle, "Final Construction Report on the Grading of Section A-1 of the Tioga Road, Yosemite Park Project 4-A1, Grading, Yosemite National Park, Mariposa and Tuolumne Counties, California," May 6, 1937, typescript, YOSE. The idea of the Tioga Road as a convenient means of crossing the mountains had also received support from a special executive committee of the Sierra Club, which studied the proposal in 1934. The committee reported that "the function of the Tioga Road must be not only to enable travelers to reach the Tuolumne Meadows and the eastern portion of the park readily and with comfort, but also to care for those who desire to use this highway as a trans-Sierra road." See "Relocation of Tioga Road: Report of the Executive Committee of the Sierra Club on the Proposed Relocation of the Tioga Road, Yosemite National Park," *Sierra Club Bulletin* 19, no. 3 (1934), 88.

39. E. Lowell Sumner to Joseph Grinnell, February 3, 1938, E. Lowell Sumner file, MVZ-UC.

40. E. Lowell Sumner, Jr., "Losing the Wilderness Which We Set Out to Preserve," 1938, typescript, NPS-HC.

41. Arno B. Cammerer, "Standards and Policies in National Parks," *American Planning and Civic Annual* (1936), 13–20.

42. Thomas C. Vint, "Wilderness Areas: Development of National Parks for Conservation," *American Planning and Civic Annual* (1938), 70.

43. Vint, "Wilderness Areas," 70, 71.

44. Related to this issue are Vint's earlier comments about the Yosemite concessionaire's proposal for a "ropeway" (or tram) to be built to take visitors from the valley floor to Glacier Point. An extended debate in the early 1930s focused mainly on how much the ropeway would intrude on park scenery, rather than on its potential impact on natural resources per se. Vint summed up his comments on the ropeway by noting the acceptability of roads as an alternative: "Roads have precedents in national parks while ropeways do not." Roads would "not be a new type of development. We know something of the effect of roads and can predict or visualize the result more easily." To Vint, the ropeway was a mechanical intrusion, different from that generally accepted in national parks. Given Vint's and the park superintendent's opposition to the ropeway proposal, a road was built, but not a ropeway. See Thomas C. Vint to the Director, November 21, 1930, Entry 17, RG79. Superintendent G. C. Thomson's objections to the ropeway are found in Thomson to the Director, November 17, 1930, Entry 17, RG79.

45. Lary M. Dilsaver and William C. Tweed, in *Challenge of the Big Trees: A Resource History of Sequoia and Kings Canyon National Parks* (Three Rivers, California: Sequoia Natural History Association, 1990), 157–196, discuss Superintendent White's efforts to protect Sequoia from certain kinds of development, including backcountry roads.

46. A. E. Demaray, memorandum to the Secretary of the Interior, n.d. (ca. spring 1935), Entry 34, RG79.

47. See for example the extended discussion of road proposals for Mt. McKinley National Park during the 1930s, in William E. Brown, *A History of the Denali–Mount McKinley Region, Alaska* (Santa Fe: National Park Service, 1991), 171–184, 194–196. Brown writes (p. 172) that "responding to the drumbeat of development and tourism boomers . . . Park Service policy-makers and planners envisioned a conventional Stateside park with a lodge at Wonder Lake, more campgrounds, and an upgraded road to accommodate independent auto-borne visitors."

48. National Park Service, "Proceedings," First Park Naturalists' Training Conference, Berkeley, California, November 1–30, 1929, typescript, 152, NPS-HC; Sumner, "Biological Research and Management," 11.

49. Victor H. Cahalane, "Activities of the National Park Service in Wildlife Conservation," n.d., ca. 1935, typescript, Central Classified File, RG79; *Annual Report of the Secretary of the Interior for the Fiscal Year Ending June 30, 1936* (Washington, D.C.: Government Printing Office, 1936), 123.

50. The most thorough reports continued the Fauna series with Nos. 3 and 4. Fauna No. 5 was begun in 1939 and published in 1944. Sumner, "Biological Research and Management," 11; Joseph S. Dixon, *Birds and Mammals of Mount McKinley National Park,* Fauna Series no. 3 (Washington, D.C.: National Park Service, 1938); Adolph Murie, *Ecology of the Coyote in the Yellowstone,* Fauna Series no. 4 (Washington, D.C.: National Park Service, 1940); Adolph Murie, *The Wolves of Mount McKinley,* Fauna Series no. 5 (Washington, D.C.: Government Printing Office, 1944).

51. Harold C. Bryant, "A Nature Preserve for Yosemite," *Yosemite Nature Notes* 6 (June 30, 1927), 46–48. John Merriam's interest in research reserves is found in Merriam to Members of the Committee on Educational Problems in National Parks, February 12, 1930, with attachments, Entry 17, RG79.

52. National Park Service, "Proceedings," First Park Naturalists' Training Conference, 169, 171–174. Albright's policy on research reserves is stated in Arno B. Cammerer to All Superintendents and Custodians, May 27, 1931, with attachment, Research Reserves file, YOSE. The Fauna No. 1 quote is in Wright, Dixon, and Thompson, *Fauna of the National Parks* (1933), 147.

53. Designations such as primitive, primeval, wilderness, virgin, and roadless were at times used in association with the reserves. See for instance Director to Wild Life Survey, March 4, 1932, Entry 35, RG79; and Arno B. Cammerer, "Maintenance of the Primeval in National Parks," ca. 1934, typescript, NPS-HC.

54. George M. Wright to the Director, March 14, 1932, Entry 35, RG79.

55. Director to Wild Life Survey, March 4, 1932; George M. Wright, "Research Areas," 1933, typescript, Entry 34, RG79; Kendeigh, "Research Areas in the National Parks," 236–238.

56. Wright to the Director, March 14, 1932; Wright, "Research Areas"; Thompson to Cammerer, February 23, 1934; and U.S. National Park Service, Wild Life Division, "Report for February, 1934," Classified File, RG79. Comments on buffer zones for the national parks are also found in Wright and Thompson, *Fauna of the National Parks* (1935), 109.

57. Victor H. Cahalane to George M. Wright, September 7, 1935, Entry 34, RG79.

58. H. W. Jennison, Memorandum for Superintendent J. R. Eakin, July 21, 1936, Balds file, GRSM.

59. J. R. Eakin to the Director, July 27, 1936, Balds file, GRSM; Frank E. Mattson, Memo for Mr. Eakin, July 27, 1936, Balds file, GRSM.

60. H. W. Jennison, Memorandum for Superintendent J. R. Eakin, July 21, 1936, Balds file, GRSM; Eakin to the Director, July 27, 1936.

61. A. E. Demaray to J. R. Eakin, September 4, 1936, Balds file, GRSM.

62. The Park Service itself would acknowledge in 1963 that the reserves were "dormant" and that many of the areas had "remained 'on the shelf,' awaiting a more favorable period for their utilization." This statement came at the very time Park Service leaders were withholding genuine support for the proposed Wilderness Act because they did not want restrictions placed on their administrative discretion to control national park backcountry. Sumner, "Biological Research and Management," 10–11. In his history of wildlife management, R. Gerald Wright states that there is "no evidence" that the reserves were ever used as intended. Wright, *Wildlife Research and Management in the National Parks* (Urbana: University of Illinois Press, 1992), 19–20. The 1963 statement is found in Conrad L. Wirth, Memorandum to All Field Offices, April 15, 1963, NPS-HC.

63. Wright to the Director, March 14, 1932. Keith R. Langdon, a natural resource management specialist in Great Smoky Mountains National Park, commented on the considerable value

Andrews Bald and other research reserves could have had for today's efforts to understand and manage the park's natural resources. If the park had maintained the reserves as originally intended, he stated, we would be "in the cat bird's seat." Personal communication with Keith R. Langdon, July 18, 1991.

64. Wright, Dixon, and Thompson, *Fauna of the National Parks* (1933), 4, 147–148.

65. Wildlife Division to the Director of the National Park Service, "Report upon Winter Range of the Northern Yellowstone Elk Herd and a Suggested Program for Its Restoration," February 28, 1934, reprinted in Wright and Thompson, *Fauna of the National Parks* (1935), 85; Douglas B. Houston, *The Northern Yellowstone Elk: Ecology and Management* (New York: Macmillan Publishing Co., 1982), 24–25. Don Despain et al., *Wildlife in Transition: Man and Nature on Yellowstone's Northern Range* (Boulder, Colorado: Roberts Rinehart, 1986), 22–24. See also Arno B. Cammerer to Joseph Grinnell, December 10, 1934, with attachment, Arno B. Cammerer files, MVZ-UC; and Victor H. Cahalane, "Wildlife Surpluses in the National Parks," in *Transactions of the Sixth North American Wildlife Conference* (Washington, D.C.: American Wildlife Institute, 1941), 357–358. Douglas Houston's detailed analysis of the management of the park's northern elk herd, *The Northern Yellowstone Elk,* 12–15, contradicts the idea that a population crash occurred in 1917–20.

66. Dunlap, *Saving America's Wildlife,* 69; Wright and Thompson, *Fauna of the National Parks* (1935), 85–86.

67. George M. Wright to H. E. Anthony, March 15, 1935, George M. Wright files, MVZ-UC. Victor Cahalane later indicated that outside support for the reduction program existed, but that there was "constant protest by a few local organizations." He was not specific, however, about which organizations or individuals supported or opposed reduction. Victor H. Cahalane, "Elk Management and Herd Reduction—Yellowstone National Park," *Transactions of the Eighth North American Wildlife Conference* (Washington, D.C.: American Wildlife Institute, 1943), 95–97.

68. Olaus J. Murie to Ben H. Thompson, December 27, 1934, Entry 7, RG79; Adolph Murie to Victor H. Cahalane, July 26, 1936, YELL.

69. Wright, Dixon, and Thompson, *Fauna of the National Parks* (1933), 118. Albright mentions securing private funds for Rush's research in Horace M. Albright to the Director, October 18, 1937, Central Classified File, RG79.

70. Wildlife Division to the Director, "Report upon Winter Range of the Northern Yellowstone Elk Herd," 85–86; Arno B. Cammerer, Memorandum for Assistant Secretary Walters, November 21, 1933, Central Classified File, RG79. The Park Service also saw overgrazing as a "landscape problem," and Fauna No. 2 advocated close cooperation between the wildlife biologists and landscape architects to address this concern. Wright, Dixon, and Thompson, *Fauna of the National Parks* (1933), 109–120. It does not appear that the landscape architects became much involved.

71. Joseph Grinnell to Arno B. Cammerer, December 26, 1934, Arno B. Cammerer files, MVZ-UC. Grinnell thus voiced elk management policies much like those the Service would put into effect in the late 1960s, more than three decades after the reduction program had begun.

72. Cammerer to Grinnell, December 10, 1934. A list of annual elk "removals" from 1923 to 1979, including those taken by hunters near the park, is in Houston, *Northern Yellowstone Elk,* 16–17.

73. Wright to Anthony, March 15, 1935; Murie to Cahalane, July 26, 1936; Rudolph L. Grimm, "Northern Yellowstone Winter Range Studies," 1938, typescript, 28–29, YELL. Although convinced that the range was still overgrazed, Grimm perceived that some "range recovery" had occurred, particularly in the two years just before he wrote his report. However, he credited "favorable climatic conditions" (the end of the drought), rather than the elk reduction program, as the "agency most responsible for the improvement of the range plant cover" (p. 27).

74. National Park Service, *Wildlife Conditions in National Parks, 1939,* Conservation Bulletin no. 3, Washington, D.C., 1939, 8. Other parks that eventually initiated limited control programs included Yosemite and Sequoia. Wright, *Wildlife Research and Management,* 77–78.

Reduction of the elk population is discussed in Karl Hess, Jr., *Rocky Times in Rocky Mountain National Park* (Niwot, Colorado: University Press of Colorado, 1993), 15–19.

75. Joseph Grinnell to Arno B. Cammerer, January 23, 1939, Arno B. Cammerer files, MVZ-UC.

76. Also, both Fauna No. 1 and Fauna No. 2 recommended reestablishing bison in Glacier National Park, in cooperation with local Indian tribes. Wright, Dixon, and Thompson, *Fauna of the National Parks* (1933), 117, 147; and Wright and Thompson, *Fauna of the National Parks* (1935), 59–60.

77. Harlow B. Mills to Ben Thompson, June 21, 1935, Entry 34, RG79; Skinner, "History of the Bison in Yellowstone Park." For figures on carrying capacity, see Curtis K. Skinner et al., "History of the Bison in Yellowstone Park" [with supplements], 1952, typescript, various pagination, YELL; M. R. Daum to Theodore C. Joslin, January 9, 1929, YELL; and Margaret Mary Meagher, *The Bison of Yellowstone National Park,* National Park Service Scientific Monograph Series no. 1 (Washington, D.C.: National Park Service, 1973), 32.

78. Wright and Thompson, *Fauna of the National Parks* (1935), 59.

79. Specifically regarding elk, Wright noted the situation in Mount Rainier, where nonnative elk from Yellowstone had been transplanted. As a result, in his opinion, it would be "impossible ever to realize the restoration of the native Roosevelt elk to the park." George M. Wright to Arno B. Cammerer, January 18, 1935, Central Classified File, RG79.

80. Edmund B. Rogers to the Director, December 10, 1937, YELL; *Annual Report of the Secretary of the Interior for the Fiscal Year Ending June 30, 1939* (Washington, D.C.: Government Printing Office, 1939), 280–281; *Annual Report of the Secretary of the Interior* (1940), 180–181; Arno B. Cammerer to the Secretary of the Interior, February 6, 1936, YELL. Tolson's brother, Hillory, was a member of the Park Service directorate.

81. Palmer H. Boeger, *Oklahoma Oasis: From Platt National Park to Chickasaw National Recreation Area* (Muskogee, Oklahoma: Western Heritage Books, 1987), 107, 111–112, 135–137; *Annual Report of the Secretary of the Interior for the Fiscal Year Ending June 30, 1935* (Washington, D.C.: Government Printing Office, 1935), 198; Ise, *National Park Policy,* 584.

82. Horace M. Albright, "Our National Parks as Wild Life Sanctuaries," *American Forests and Forest Life* 35 (August 1929), 507.

83. George M. Wright to the Director, December 19, 1931, Entry 35, RG79.

84. Joseph Grinnell to Arno B. Cammerer, November 9, 1933, Arno B. Cammerer files, MVZ-UC.

85. Skinner, "History of the Bison in Yellowstone Park"; Rudolph L. Grimm, "Report on Antelope Creek Buffalo Pasture" (1937), typescript, YELL.

86. In 1945 Victor Cahalane recalled that the Park Service "practiced very limited control of wolves and coyotes in our Alaska areas from about 1932 to 1939 or 1940." Victor H. Cahalane to Mr. Drury, March 14, 1945, Entry 7, RG79. See also Brown, *A History of the Denali–Mount McKinley Region,* 198.

87. National Park Service, "Policy on Predators and Notes on Predators" (1939), typescript, various pagination, Central Classified File 715, RG79.

88. Albright, "The National Park Service's Policy on Predatory Mammals," 185.

89. National Park Service, "Policy on Predators and Notes on Predators"; Wright, Dixon, and Thompson, *Fauna of the National Parks* (1933), 147.

90. The quote is from National Park Service, "Policy on Predators and Notes on Predators." Wright and Thompson, *Fauna of the National Parks* (1935), 71.

91. Curtis K. Skinner to Dr. Mills, March 12, 1935, YELL.

92. Frank W. Childs, "Report on the Present Status of Wildlife Management in Yellowstone National Park with Suggested Recommendations for Future Treatment," April 19, 1935, YELL. There was also interest among Yellowstone's staff in restoring some of the park's extirpated species. Naturalist Assistant Harlow B. Mills wrote to Ben Thompson in 1935: "As a policy I can see no great obstacle in the way of our, at least, attempting the introduction of cougar and wolves into the

Park. They were a vital part of the picture at one time, a picture which can never be the same in the Park in their absence. This should be done, I realize, with considerable forethought and care, but I believe that it should be done, nevertheless." Harlow B. Mills to Ben Thompson, June 21, 1935, Entry 34, RG79. This approach was in accord with the recommendations of Fauna No. 1 that "any native species which has been exterminated from the park area shall be brought back if this can be done." See Wright, Dixon, and Thompson, *Fauna of the National Parks* (1933), 148.

93. Murie, *Ecology of the Coyote,* 16; Sumner, "Biological Research and Management," 14.

94. C. A. Henderson to David Canfield, November 21, 1935; and David Canfield to C. A. Henderson, November 30, 1935, Entry 34, RG79. Victor H. Cahalane, "Evolution of Predator Control Policy in the National Parks," *Journal of Wildlife Management* 3 (July 1939), 236.

95. David Madsen, Memorandum for the Director, May 20, 1939, Entry 36, RG79. See also Susan R. Shrepfer, *The Fight to Save the Redwoods: A History of Environmental Reform* (Madison: University of Wisconsin Press, 1983), 61–63.

96. Joseph Grinnell to Arno B. Cammerer, April 10, 1939, Central Classified File, RG79.

97. Horace M. Albright to the Director, National Park Service, October 18, 1937, Central Classified Files, RG79.

98. Murie, *Ecology of the Coyote,* 146–148.

99. Thomas Dunlap, in *Saving America's Wildlife,* 75, indicates that some Park Service officials "wanted to fire" Murie. Alston Chase, in *Playing God in Yellowstone: The Destruction of America's First National Park* (Boston: Atlantic Monthly Press, 1986), 126–128, describes the "fierce Park Service resistance" that Murie faced during the coyote controversy. Lowell Sumner, in "Biological Research and Management," 15, recalled that, following the coyote study, "Murie's findings, and his personal concepts of ecological management of park resources, continued to be unpopular in various administrative circles." Given that Murie was quickly assigned to a similar study of wolves in Mt. McKinley National Park, however, it is obvious that he had support in high places, very likely from Director Cammerer himself.

100. Horace M. Albright to A. B. Cammerer, January 11, 1939, Central Classified Files, RG79; *Annual Report of the Secretary of the Interior* (1939), 282.

101. Murie, *Wolves of Mount McKinley,* xiii–xv; Albright to Cammerer, January 11, 1939. Murie's wolf study is discussed in Brown, *A History of the Denali–Mount McKinley Region,*" 198.

102. Madsen to the Director, May 20, 1939.

103. John D. Varley, "Record of Egg Shipments from Yellowstone Fishes, 1914–1955," Yellowstone National Park Information Paper no. 36, May 1979, YELL.

104. David H. Madsen, "Report on Fish Cultural Activities," April 5, 1935, Central Classified File, RG79; Sumner, "Biological Research and Management," 9.

105. Wright, Dixon, and Thompson, *Fauna of the National Parks* (1933), 63.

106. David H. Madsen to Arno B. Cammerer, October 6, 1933, Central Classified File, RG79; David H. Madsen, "A National Park Service Fish Policy," n.d., ca. early 1930s, typescript, Entry 36, RG79; and Madsen, "Outline of a General Policy of Handling the Fish Problem in the National Parks," May 10, 1932, typescript, Central Classified File, RG79. The records do not indicate whether Madsen was first detailed to the Park Service in 1928 or in the early 1930s.

107. Wright, Dixon, and Thompson, *Fauna of the National Parks* (1933), 148, 63.

108. Arno B. Cammerer, Office Order no. 323, April 13, 1936, Entry 35, RG79.

109. Cammerer, Office Order no. 323, April 13, 1936; *Annual Report of the Secretary of the Interior* (1936), 124.

110. John D. Varley, "A History of Fish Stocking Activities in Yellowstone National Park between 1881–1980," Yellowstone National Park Information Paper no. 35, January 1, 1981, typescript, 9, 13, 17, 19, 21, 26, 52–53, YELL. The stocking of Mammoth Beaver Ponds took place in 1936, very likely after the park had received the new fish policy issued by Cammerer in mid-April of that year. In the case of McBride Lake, also in the Yellowstone drainage, exotic rainbow trout were introduced in 1936, where previously only native cutthroat trout had existed. Varley, "History of Fish Stocking," 17.

111. Varley, "Record of Egg Shipments"; *Annual Report of the Secretary of the Interior for the Year Ending June 30, 1937* (Washington, D.C.: Government Printing Office, 1937), 44. As another example of fish production and shipment during the 1930s, the collection of approximately sixty million trout eggs in one year from several unspecified national parks, with about half of them being shipped to various states, is mentioned by Cammerer in *Annual Report of the Secretary of the Interior* (1936), 124.

112. Carl P. Russell, "Opportunities of the Wildlife Technician in National Parks," paper presented at the North American Wildlife Federation conference, St. Louis, Missouri, March 1, 1937, typescript, NPS-HC. Victor H. Cahalane, "Thoughts on National Park Service–Bureau of Fisheries Agreement," draft, August 4, 1939, Entry 36, RG79. Cahalane accepted that the Service would continue its dependency on other agencies for fish culture work. And Director Cammerer had reported in 1937, the year after the new fish policy was issued, that cooperation was closer "than ever before" between the Service and the Bureau of Fisheries and state game departments. It became even closer in 1940, with the transfer of the biologists to the Bureau of Biological Survey and the survey's subsequent merger with the Bureau of Fisheries. Cahalane, "Thoughts on National Park Service-Bureau of Fisheries Agreement"; *Annual Report of the Secretary of the Interior* (1937), 44.

113. Paige, *The Civilian Conservation Corps,* appendix A, 162. The National Park Service Act authorized the Service to "sell or dispose of timber in those cases where . . . the cutting of such timber is required in order to control the attacks of insects or diseases or otherwise conserve the scenery." Hillory A. Tolson, *Laws Relating to the National Park Service, the National Parks and Monuments* (Washington, D.C.: Department of the Interior, 1933), 10.

114. *Annual Report of the Secretary of the Interior* (1933), 157. Some national park areas were particularly affected by prefire development. On the north rim of the Grand Canyon, fire protection preparations by the CCC included improvement of existing roads and construction of primitive fire-access roads and trails, lookout towers, warehouses, a fire cache, maintenance shops, residences, telephone lines, and water ponds. Stephen J. Pyne, *Fire in America: A Cultural History of Wildland and Rural Fire* (Princeton: Princeton University Press, 1982), 300.

115. John D. Coffman, "John D. Coffman and His Contribution to Forestry in the National Park Service," n.d., typescript, 36–39, NPS-HC. Because of the CCC's heavy emphasis on forestry, Coffman was also given the huge responsibility of overseeing CCC operations within the national parks. In 1936 the director consolidated oversight of these operations with the Service's state parks assistance program (also funded by the CCC). This expanded office, combining all CCC-related national and state park work, was supervised by Assistant Director Conrad Wirth. Coffman was left free to concentrate on directing forestry management in the parks, which continued to rely on CCC manpower and money. See Coffman, "John D. Coffman and His Contribution to Forestry," 44; Conrad L. Wirth, *Park, Politics, and the People* (Norman: University of Oklahoma Press, 1980), 118; and Paige, *The Civilian Conservation Corps,* 39–40, 48, 115.

116. "A Forestry Policy for the National Parks," approved by Horace M. Albright, May 6, 1931, typescript, Entry 18, RG79.

117. Wright, Dixon, and Thompson, *Fauna of the National Parks* (1933), 33.

118. U.S. Office of National Parks, Buildings and Reservations, "Instructions for Superintendents of Eastern National Park ECW Camps and CW Projects Concerning Roadside Clean-up, Fire Hazard Reduction, Brush Disposal," chapter 9, 3, supplement no. 7 to *Forest Truck Trail Handbook* (Washington, D.C.: U.S. Forest Service, 1935); George M. Wright to the Director, February 28, 1934, Central Classified File, RG79.

119. Adolph Murie, Memorandum for Ben H. Thompson, August 2, 1935, Entry 34, RG79.

120. L. F. Cook, Memorandum for the Chief Forester, August 28, 1935, Entry 34, RG79.

121. L. F. Cook, Memorandum for the Chief Forester, August 28, 1935, Entry 34, RG79. Riley McClelland, correspondence with the author, September 2, 1993.

122. Murie to Thompson, August 2, 1935; Cook to Chief Forester, August 28, 1935. In Cammerer's 1939 annual report, the director discusses the fire prevention and fire protection work

undertaken with CCC funds and enrollees. *Annual Report of the Secretary of the Interior* (1939), 272–275.

123. *Annual Report of the Secretary of the Interior for the Year Ending June 30, 1933* (Washington, D.C.: Government Printing Office, 1933), 180–181.

124. George M. Wright to Arno B. Cammerer, August 1, 1935, Entry 35, RG79; Victor H. Cahalane to A. E. Demaray, September 23, 1935, Entry 34, RG79. For comments on CCC involvement in insect and disease control, see Paige, *The Civilian Conservation Corps*, 101–103.

125. Adolph Murie to George M. Wright, March 26, 1935, Entry 34, RG79. Similar statements regarding insect control are found in Harlow B. Mills to Ben Thompson, June 21, 1935.

126. Cook to Chief Forester, August 28, 1935; and *Annual Report of the Secretary of the Interior* (1939), 272–274. For similar comments made earlier by Cammerer, see *Annual Report of the Secretary of the Interior* (1937), 42–43.

127. Sumner, "Biological Research and Management," 13.

128. Russ Olsen, *Administrative History: Organizational Structures of the National Park Service, 1917 to 1985* (Washington, D.C.: National Park Service, 1985), 63. Under Coffman the Park Service also provided considerable training in forest protection, including techniques in fire, insect, and disease control. In many parks, rangers, park naturalists, and maintenance staffs all received this training. John W. Henneberger, "To Protect and Preserve: A History of the National Park Ranger," 1965, typescript, copy courtesy of the author, 307.

129. Tom Ela, interview with the author, January 26, 1989; Arthur Wilcox, interview with the author, March 17, 1992.

130. As an example of the growing strength of the forestry programs, a list of 137 professionally trained foresters in the National Park Service by 1952 shows most of them in key positions. Robert N. McIntyre, "A Brief History of Forestry in the National Park Service," March 1952, typescript, appendix A, NPS-HC.

131. The quote is from National Park Service, "Growth of the National Park Service under Director Cammerer," 1936, typescript, 1, Entry 18, RG79. See also Unrau and Willis, *Expansion of the National Park Service*, for a detailed account of Park Service growth and expansion in the 1930s.

132. Wirth, *Parks, Politics, and the People*, 73–74. See also Paige, *The Civilian Conservation Corps*, 38–39; and Unrau and Willis, *Expansion of the National Park Service*, 77. Arno B. Cammerer, "History and Growth—the National Park Service" (1939), typescript, 4, NPS-HC.

133. Wirth, *Parks, Politics, and the People*, 75–76, 88; Olsen, *Organizational Structures of the National Park Service*, 52–53.

134. The Park Service's CCC programs are discussed in Wirth, *Parks, Politics, and the People*, 94–127; and Ise, *Our National Park Policy*, 363–364.

135. National Park Service, "Growth of the National Park Service," 5. For discussion of the survey, see Wirth, *Parks, Politics, and the People*, 172–173; and Ise, *Our National Park Policy*, 364.

136. A 1936 internal report stated that the Service had "sponsored" the legislation. National Park Service, "Growth of the National Park Service under Director Cammerer," 5. See also *Annual Report of the Secretary of the Interior* (1935), 183. Conrad Wirth mentions in his autobiography that the act (reprinted in the book) was passed "at the request of the National Park Service through the Department of the Interior." Wirth, *Parks, Politics, and the People*, 166–168; Unrau and Willis, *Expansion of the National Park Service*, 109–120.

137. National Park Service, *A Study of the Park and Recreation Problem of the United States* (Washington, D.C.: Government Printing Office, 1941), see for example 122–132; Wirth, *Parks, Politics, and the People*, 150, 192–193. In 1937 Cape Hatteras National Seashore, on the North Carolina coast, became the first of these areas to come into the national park system. Others followed, mainly in the 1960s and 1970s. See Barry Mackintosh, *The National Parks: Shaping the System* (Washington, D.C.: National Park Service, 1991), 81–84.

138. Unrau and Willis, *Expansion of the National Park Service*, 144–145. The quote is from *Annual Report of the Secretary of the Interior* (1937), 55.

139. The recreational demonstration areas are discussed in Unrau and Willis, *Expansion of the National Park Service,* 129–143; Paige, *The Civilian Conservation Corps,* 117–118; and Wirth, *Parks, Politics, and the People,* 176–190. Wirth's promotion of the program is also discussed in Herbert Evison and Newton Bishop Drury, "The National Park Service and Civilian Conservation Corps," interviews by Amelia Roberts Fry, Berkeley, California, October 24, 1962, and April 19 and 26, 1964, typescript, 64, NPS-HC. Cammerer's quote is in *Annual Report of the Secretary of the Interior* (1936), 104.

140. Ise, *Our National Park Policy,* 465–466; Wirth, *Parks, Politics, and the People,* 184–186. In the 1940s Park Service Director Newton Drury at first opposed the North Dakota unit, believing the eroded lands were definitely below national park standards, then accepted it once its status as a memorial to Theodore Roosevelt was agreed on. David Harmon, *At the Open Margin* (Medora, North Dakota: Theodore Roosevelt Nature and History Association, 1986), 13–21.

141. National Park Service, "Growth of the National Park Service" (1936), 3.

142. Albright and Cahn, *Birth of the National Park Service,* 245, 285–286. Albright recalled (p. 286) his belief that "acquisition of the military parks situated in many eastern states would bring a much larger constituency and much broader base, and thus the Park Service would be perceived as a truly national entity." For a list of the sites managed by the National Park Service prior to the reorganization by President Roosevelt, see Mackintosh, *Shaping the System,* 16–17, 22–23.

143. Background to the reorganization and a list of sites brought into the national park system in August 1933 are in Mackintosh, *Shaping the System,* 24–43. See also Ise, *Our National Park Policy,* 352–353.

144. Wirth, *Parks, Politics, and the People,* 163–166; Mackintosh, *Shaping the System,* 49.

145. National Park Service, "Growth of the National Park Service," 2; Unrau and Willis, *Expansion of the National Park Service,* 60–64; *Annual Report of the Secretary of the Interior* (1936), 135.

146. Albright and Cahn, *Birth of the National Park Service,* 314; Mackintosh, *Shaping the System,* 26; Olsen, *Organizational Structures of the National Park Service,* 61. The director expressed a desire to return to the "National Park Service" designation in his 1933 annual report. See *Annual Report of the Secretary of the Interior* (1933), 192.

147. Unrau and Willis, *Expansion of the National Park Service,* 153–155; *Annual Report of the Secretary of the Interior* (1937), 38.

148. Louis C. Cramton, Memorandum for the Secretary, June 28, 1932, Entry 18, RG79.

149. Cramton, Memorandum for the Secretary, June 28, 1932. The reconnaissance team included superintendents Roger W. Toll (Yellowstone), M. R. Tillotson (Grand Canyon), and P. P. Patraw (Bryce Canyon and Zion). The Organic Act's wording is in Tolson, *Laws Relating to the National Park Service* (1933), 10.

150. George Collins, in "The Art and Politics of Park Planning and Preservation, 1920–1979," interview by Ann Lage, recalled that Demaray, Wright, and Thompson supported Wirth in his quest for control of recreation management at Lake Mead.

151. George M. Wright to Joseph Grinnell, August 29, 1934, George M. Wright files, MVZ-UC; George M. Wright, "Wildlife in National Parks," *American Planning and Civic Annual* (1936), 62. Aware of Wright's abilities, Grinnell wrote to him that given the significance of the recreational study he could think of "no one better fitted than . . . yourself to guide and direct along this important line." Joseph Grinnell to George M. Wright, August 18, 1934, George M. Wright files, MVZ-UC.

152. George M. Wright to Col. John R. White, June 23, 1935, Entry 34, RG79.

153. Collins, "The Art and Politics of Park Planning and Preservation," 52.

154. The organizational charts are found in Olsen, *Organizational Structures of the National Park Service,* 42–61. Conrad Wirth recalled that the superintendents were at first "adamant" in their opposition to establishing regional offices, concerned that they would encroach on the superintendents' authority and affect their lines of communication with the director. The superintendents also feared that the new offices would be headed by men who had risen through the ranks

of the CCC, rather than the Park Service. Wirth, *Parks, Politics, and the People,* 119. Also see Cammerer, "History and Growth of the National Park Service," 5. In early 1937 the Park Service established its travel division to fill, as Cammerer put it, "a long-indicated need for a national clearing house of information on recreational and travel opportunities . . . and to stimulate interest therein both at home and abroad." The division soon opened an office on Broadway in New York City. *Annual Report of the Secretary of the Interior* (1937), 35–36. See also Swain, "National Park Service and the New Deal," 318.

155. National Park Service, "Growth of the National Park Service under Director Cammerer," 1–3. Numerous parks were authorized during the New Deal era, including Everglades and Big Bend national parks, Blue Ridge and Natchez Trace national parkways, and Joshua Tree, Organ Pipe Cactus, and Capitol Reef national monuments. Mackintosh, *Shaping the System,* 58–59.

156. National Park Service, "Growth of the National Park Service under Director Cammerer," 4. Further discussion of appropriations during the New Deal is found in Unrau and Willis, *Expansion of the National Park Service,* 75–76.

157. National Park Service, "Growth of the National Park Service under Director Cammerer," 4. Almost certainly, many of these individuals were not fully trained professionals but nevertheless were working in some aspect of those fields.

158. William G. Carnes, "Landscape Architecture in the National Park Service," *Landscape Architecture* (July 1951), copy attached to Hillory A. Tolson, Memorandum to Washington Office and All Field Offices, February 15, 1952, NPS-HC; Sumner, "Biological Research and Management," 9.

159. *Annual Report of the Secretary of the Interior* (1936), 99; *Annual Report of the Secretary of the Interior* (1937), 34.

160. Newton Drury discusses Ickes' interest in Moses' becoming director in Newton Bishop Drury, "Parks and Redwoods, 1919–1971," interviews by Amelia Roberts Fry and Susan Schrepfer, 1959–1972, typescript, 352–353, NPS-HC. Ickes' quote is found in T. H. Watkins, *Righteous Pilgrim,* 578. See also Swain, "National Park Service and the New Deal," 329–330. Cammerer died of a heart attack in April 1941, less than a year after stepping down to the regional director's position in Richmond. Horace M. Albright, "Reminiscences," interview by William T. Ingerson, Oral History Research Office, Columbia University, 1962, typescript, 543, NPS-HC.

161. Schrepfer, *The Fight to Save the Redwoods,* 56–64. On the other hand, some opposed the Kings Canyon legislation because a national park would *restrict* use and development. Details of the complicated campaign to establish Kings Canyon National Park are found in Dilsaver and Tweed, *Challenge of the Big Trees,* 197–214; and Ise, *Our National Park Policy,* 396–404. See also George M. Wright to John R. White, June 23, 1935, Entry 34, RG79, for Wright's comments on the Forest Service's "treating the Kings Canyon areas as a national park . . . enforc[ing] practically the same rules for its preservation." Wright saw the Forest Service's efforts as an encroachment on traditional Park Service management practices, and thus as one of the "gravest dangers" facing the Park Service.

162. "Wanted: A National Primeval Park Policy," *National Parks Bulletin* 13 (December 1937), 13, 26; William P. Wharton, "Park Service Leader Abandons National Park Standards," *National Parks Bulletin* 14 (June 1938), 5.

163. William P. Wharton, "The National Primeval Parks," *National Parks Bulletin* 13 (February 1937), 3–4.

164. Watkins, *Righteous Pilgrim,* 554–555; Horace M. Albright, comments in *American Planning and Civic Annual* (1938), 31–32.

165. Harold L. Ickes to William P. Wharton, May 2, 1939; and William P. Wharton to Harold Ickes, April 10, 1939, Kent. George M. Wright, "The Philosophy of Standards for National Parks, *American Planning and Civic Annual* (1936), 25.

166. Joseph Grinnell to George M. Wright, April 16, 1935, George M. Wright files, MVZ-UC.

167. Sumner, "Biological Research and Management," 15; Ben H. Thompson to Joseph Grinnell, November 9, 1936, Ben H. Thompson file, MVZ-UC. Thompson did not identify the unit interested in absorbing the wildlife biologists. In 1980 the George Wright Society, dedicated to excellence in resource management in protected areas and on other public lands, was founded in Wright's honor.

168. A. E. Demaray to the Acting Secretary, Department of the Interior, August 30, 1938, Central Classified File, RG79; Sumner, "Biological Research and Management," 15.

169. Watkins, *Righteous Pilgrim,* 587–590; National Park Service, "A Review of the Year," 34.

170. Ben H. Thompson to E. Raymond Hall, June 13, 1939, handwritten, Ben H. Thompson files, MVZ-UC.

171. Arno B. Cammerer and Ira H. Gabrielson, Memorandum for the Secretary of the Interior, November 24, 1939, Central Classified File, RG79; Sumner, "Biological Research and Management," 15.

Chapter 5. The War and Postwar Years

1. National Academy of Sciences, National Research Council, "A Report by the Advisory Committee to the National Park Service on Research," August 1, 1963, typescript, 31.

2. Even Stephen Mather had had experience with the parks prior to becoming director, having assumed oversight of the national parks for Secretary of the Interior Franklin K. Lane in January 1915. Drury's work with the Save the Redwoods League is discussed in Susan R. Schrepfer, *The Fight to Save the Redwoods: A History of Environmental Reform, 1917–1978* (Madison: University of Wisconsin Press, 1983), 23–76.

3. For comments on Drury see John Ise, *Our National Park Policy: A Critical History* (Baltimore: The Johns Hopkins Press, 1961) 3, 443–444; and Ronald Foresta, *America's National Parks and Their Keepers* (Washington, D.C.: Resources for the Future, 1984), 48–49.

4. *Annual Report of the Secretary of the Interior, for the Fiscal Year Ended June 30, 1945* (Washington, D.C.: Government Printing Office, 1945), 207; *Annual Report of the Secretary of the Interior for the Fiscal Year Ended June 30, 1943* (Washington, D.C.: Government Printing Office, 1943), 218–219; Conrad L. Wirth, *Parks, Politics, and the People* (Norman: University of Oklahoma Press, 1980), 225; Lary Dilsaver and William C. Tweed, *Challenge of the Big Trees* (Three Rivers, California: Sequoia Natural History Association, 1990), 188.

5. *Annual Report of the Secretary of the Interior* (1943), 217–218; Wirth, *Parks, Politics, and the People,* 225–226. Drury's concerns about logistical and communications problems likely to result from the move to Chicago are mentioned in National Parks Association, press release 47, February 5, 1942, Kent.

6. Newton B. Drury, "What the War Is Doing to National Parks and Where They Will Be at Its Close," *Living Wilderness* 9, May 1944, 11; Newton B. Drury to Secretary of the Interior, July 28, 1950, Entry 19, RG79.

7. Newton B. Drury to Assistant Secretary Doty, August 2, 1950, Entry 19, RG79; Carl P. Russell, "The Trusteeship of the National Park Service," *Transactions, Illinois State Academy of Science* 36 (September 1943), 19. See also Dilsaver and Tweed, *Challenge of the Big Trees,* 188–189.

8. Newton B. Drury to the Secretary of the Interior, December 18, 1942, NPS-HC; *Annual Report of the Secretary of the Interior* (1943), 197; Drury to Assistant Secretary Doty, August 2, 1950.

9. Charles W. Porter, ed., "National Park Service War Work, December 7, 1941, to June 30, 1944 [and] Supplement, June 30, 1944, to October 1, 1945," typescript, 5–6, NPS-HC. The Sitka spruce issue in Olympic is discussed in Russell, "Trusteeship of the National Park Service," 19–21; Drury, "What the War Is Doing to National Parks," 12; and Ise, *Our National Park Policy,* 450.

10. Porter, "National Park Service War Work," 7–9; and Newton B. Drury to the Secretary of the Interior, December 18, 1942. Carsten Lien, in *Olympic Battleground: The Power Politics of*

Timber Preservation (San Francisco: Sierra Club Books, 1991), 215–231, recounts how the mind-set of the Park Service's own foresters, plus the ties between the park's managers and the local chambers of commerce and fraternal organizations, helped undermine Drury's resolve to prevent Olympic's forests from being cut.

11. Porter, "National Park Service War Work," 9–11; *Annual Report of the Secretary of the Interior* (1943), 208; Ise, *Our National Park Policy*, 392–395; Lien, *Olympic Battleground*, 226–231.

12. Newton B. Drury, "National Park Service Reports on Results of Its Study of Olympic National Park Boundaries," March 18, 1947, typescript, OLYM. See also National Park Service, "Study of Olympic National Park Boundaries," March 18, 1947, OLYM.

13. Irving M. Clark, "Protect Olympic Park!" *Living Wilderness*, June 1947, 2, 6. Drury is quoted in "Olympic Park Boundaries Defended," *Living Wilderness* 12 (Winter 1947–48), 6. See also *Annual Report of the Secretary of the Interior for the Fiscal Year Ending June 30, 1947* (Washington, D.C.: Government Printing Office, 1947), 337; Lien, *Olympic Battleground*, 232–253; and Irving Brant, *Adventures in Conservation with Franklin D. Roosevelt* (Flagstaff, Arizona: Northland Publishing Company, 1988), 277–286.

14. The conflict over Olympic's salvage timber program is detailed in Lien, *Olympic Battleground*, 268–298. By the mid-1950s, sales of windblown timber in Olympic had enabled the park to acquire about thirty-six hundred acres of inholdings—just over one-third of such acreage when the program began. Porter, "National Park Service War Work," 12–13; Hillory A. Tolson to Assistant Secretary Lewis, August 2, 1955, Records of Conrad Wirth, RG79. In the parks for brief assignments (usually during the summer vacation period) and largely free of pressure to conform to mainline Park Service thinking, seasonal naturalists sometimes took a critical view of management practices in the parks. See Lien, 285–286.

15. Newton B. Drury to the Under Secretary, January 14, 1943, Entry 19, RG79; Newton B. Drury, "National Park Service Grazing Policy," *National Parks Magazine* 78 (July–September 1944), reprint, n.p.; Carl P. Russell, Memorandum for the Director, January 21, 1943, Entry 19, RG79.

16. Porter, "National Park Service War Work," 18–19.

17. Porter, "National Park Service War Work," 18–19; Russell, "Trusteeship of the National Park Service," 23–25. *Annual Report of the Secretary of the Interior for the Fiscal Year Ending June 30, 1944* (Washington, D.C.: Government Printing Office, 1944), 209–210. John R. White, "Flowers for Cattle: The Demand of Stockmen," *National Parks Magazine* 17 (July–September 1943), 4–10. Newton B. Drury, "The California Drought and the Resultant Pressure for Grazing in the Sierra National Parks," March 29, 1948, Entry 19, RG79. The Park Service considered, but did not pursue, one other possibility for supplying meat to support the war: shipping to military bases the carcasses taken during big-game reduction programs. The amount of meat available from reduction of elk, deer, and bison would have been a negligible factor in meeting wartime needs.

18. The variety of ranger duties during this period are discussed in Lemuel A. Garrison, *The Making of a Ranger: Forty Years with the National Parks* (Salt Lake City: Howe Brothers, 1983), 261; and John W. Henneberger, "To Protect and Preserve: A History of the National Park Ranger," typescript, 402, 429–431, courtesy of the author.

19. See for example support of the elk reduction program by Chief Biologist Victor H. Cahalane, in Cahalane, "Wildlife Surpluses in the National Parks," in *Transactions of the Sixth North American Wildlife Conference* (Washington, D.C.: American Wildlife Institute, 1941), 360–361; and Cahalane, "Elk Management and Herd Regulation—Yellowstone National Park," in *Transactions of the Eighth North American Wildlife Conference* (Washington, D.C.: American Wildlife Institute), 1943, 99–100.

20. Newton B. Drury to C. N. Feast, March 21, 1945, YELL; Hillory A. Tolson, Memorandum for the Regional Directors, April 3, 1944, with attachment, "Is Hunting the Remedy?" YELL. G. A. Moskey to Julius M. Peterson, January 17, 1944, Central Classified File 715, RG79.

21. Edmund B. Rogers, Memorandum for the Regional Director, July 8, 1943, YELL. Especially with his U.S. Forest Service work in the Southwest, Leopold had acquired in-depth knowledge of the relationship of ungulates with their habitat. Susan L. Flader, *Thinking Like a Mountain: Aldo Leopold and the Evolution of an Ecological Attitude toward Deer, Wolves and Forests* (Lincoln: University of Nebraska Press, 1978), 14–18, 36–121. Yellowstone elk management during this period is discussed in Douglas B. Houston, *The Northern Yellowstone Elk: Ecology and Management* (New York: Macmillan Publishing Co., 1982), 15–21, 100–101; and Michael B. Coughenour and Francis J. Singer, "The Concept of Overgrazing and Its Application to Yellowstone's Northern Range," Natural Resource Ecology Lab, Colorado State University, 1989, 4–7.

22. Newton B. Drury to Alden Miller, November 13, 1943, Newton B. Drury files, MVZ-UC. The proposed new population level for the Lamar Valley herd was higher than the approximately one hundred head suggested in 1935 by biologist Harlow Mills. Harlow B. Mills to Ben H. Thompson, June 21, 1935, Entry 34, RG79.

23. Drury to Miller, November 13, 1943. See also Carl P. Russell, "Comment on Rogers' Memorandum Regarding Bison Reduction" (with attachment including comments by Chief Biologist Victor H. Cahalane), August 24, 1943, Newton B. Drury files, MVZ-UC; Curtis K. Skinner et al. "History of the Bison in Yellowstone Park" [with supplements], 1952, typescript, various pagination, YELL. The Yellowstone bison populations during this period are discussed in Margaret Mary Meagher, *The Bison of Yellowstone National Park*, National Park Service Scientific Monograph Series no. 1 (Washington, D.C.: National Park Service, 1973), 26–33.

24. Horace M. Albright to Newton B. Drury, October 29, 1943, Entry 19, RG79.

25. Chief Naturalist, Memorandum for Mr. Drury, November 3, 1943, Newton B. Drury files, MVZ-UC; Horace M. Albright to Newton B. Drury, December 1, 1943, Entry 19, RG79. Local tourism interests, such as the Dude Ranchers Association, were also apprehensive over the bison reduction. See Skinner, "History," supplement, 1942 to 1947, n.p.

26. Newton B. Drury to Horace M. Albright, December 8, 1943, Entry 19, RG79. Drury's interpretation of the 1916 Organic Act surely galled Albright, who, having helped draft the act, never questioned that he understood its intent.

27. Drury to Albright, December 8, 1943.

28. Horace M. Albright to Newton B. Drury, December 13, 1943, Entry 19, RG79; Skinner, "History," supplement, 1942 to 1947; *Annual Report of the Secretary of the Interior* (1944), 221; Horace M. Albright to Newton B. Drury, February 25, 1944, Entry 19, RG79. An alternative means of resolving the overgrazing issue was to remove all bison from the Lamar and Yellowstone river drainages—a suggestion made by Cahalane, but never implemented and perhaps never conveyed to the displeased Albright. See Skinner, "History," supplement, 1942 to 1947.

29. Horace M. Albright to Newton B. Drury, September 7, 1944, Entry 19, RG79. Albright's interest in retaining the Buffalo Ranch is indicated in his letter of September 7, 1944, as well as in Albright to Drury, December 1, 1943. The Service's intention to remove the ranch buildings is noted in Skinner, "History," supplement, 1942 to 1947.

30. *Annual Report of the Secretary of the Interior* (1944), 221; Newton B. Drury, Memorandum for the Secretary of the Interior, September 15, 1944, YELL; Robert W. Righter, *Crucible for Conservation: The Creation of Grand Teton National Park* (Niwot: Colorado Associated University Press, 1982), 131–132.

31. Adolph Murie, *The Wolves of Mount McKinley*, Fauna of the National Parks of the United States, Fauna Series no. 5 (Washington, D.C.: Government Printing Office, 1944), xv–xvii; William E. Brown, *A History of the Denali–Mount McKinley Region, Alaska* (Santa Fe: National Park Service, 1991), 196–201.

32. Brown, *History of the Denali–Mount McKinley Region*, 198–199; Victor H. Cahalane to Newton B. Drury, March 14, 1945, copy from the files of William E. Brown.

33. Adolph Murie, "A Review of the Mountain Sheep Situation in Mount McKinley National Park, Alaska, 1945," attached to Newton B. Drury to Alden H. Miller, January 10, 1946, Newton B. Drury Files, MVZ-UC. Drury's comments on the wolf-sheep controversy are in Drury, interview,

"Parks and Redwoods, 1919–1971," 362; and *Annual Report of the Secretary of the Interior for the Fiscal Year Ending June 30, 1946* (Washington, D.C.: Government Printing Office, 1946), 327.

34. Horace M. Albright, "The National Park Service's Policy on Predatory Mammals," *Journal of Mammalogy* 12 (May 1931), 185–186; and George M. Wright, Joseph S. Dixon, and Ben H. Thompson, *Fauna of the National Parks of the United States: A Preliminary Survey of Faunal Relations in National Parks,* Contributions of Wild Life Survey, Fauna Series no. 1 (Washington, D.C.: Government Printing Office, 1933), 147.

35. Brown, *History of the Denali–Mount McKinley Region,* 199–200; Aldo Leopold to J. Hardin Peterson, June 13, 1946, O. Murie.

36. Brown, *History of the Denali–Mount McKinley Region,* 199–200. For a commentary on the Service's predator policy during this period, see Victor H. Cahalane, "Predators and People," *National Parks Magazine* 22 (October–December 1948), 5–12.

37. Newton B. Drury to Alden H. Miller, December 18, 1945, with attachment, Newton B. Drury files, MVZ-UC. Problems with bear feeding during this period are discussed in Paul Schullery, *The Bears of Yellowstone* (1986; 3rd ed., Worland, Wyoming: High Plains Publishing Co., 1992), 104–108; Dilsaver and Tweed, *Challenge of the Big Trees,* 180; and Alfred Runte, *Yosemite: The Embattled Wilderness* (Lincoln: University of Nebraska Press, 1990), 174.

38. Drury to Miller, December 18, 1945; Victor H. Cahalane to Aldo Leopold, May 16, 1942, Central Classified File, RG79.

39. Drury to Miller, December 18, 1945; *Annual Report of the Secretary of the Interior* (1943), 209; Schullery, *Bears of Yellowstone,* 106–108.

40. Albright to Drury, December 13, 1943; Albright to Drury, February 25, 1944; Albright to Drury, September 7, 1944. Victor Cahalane recalled believing that the national park concessionaires also sought to continue the bear shows. Knowing the shows would keep people in the parks until dark, the concessionaires hoped that the visitors would rent overnight accommodations. Victor H. Cahalane, interview with the author, February 25, 1992.

41. Newton B. Drury to Horace M. Albright, October 30, 1945, Entry 19, RG79; *Annual Report of the Secretary of the Interior* (1944), 222.

42. *Annual Report of the Secretary of the Interior* (1945), 214; Dilsaver and Tweed, *Challenge of the Big Trees,* 262; Schullery, *Bears of Yellowstone,* 106; *Annual Report of the Secretary of the Interior for the Fiscal Year Ending June 30, 1952* (Washington, D.C.: Government Printing Office, 1952), 365.

43. To compensate for manpower shortages, the Service used teenage boys and older men in its firefighting crews. Wartime shortages of manpower and equipment are discussed in *Annual Report of the Secretary of the Interior* (1945), 212; and *Annual Report of the Secretary of the Interior* (1946), 324.

44. *Annual Report of the Secretary of the Interior* (1947), 336. See also *Annual Report of the Secretary of the Interior for the Fiscal Year Ending June 30, 1949* (Washington, D.C.: Government Printing Office, 1949), 320; *Annual Report of the Secretary of the Interior* (1946), 324; *Annual Report of the Secretary of the Interior for the Fiscal Year Ending June 30, 1950* (Washington, D.C.: Government Printing Office), 327; Harold K. Steen, *The U.S. Forest Service: A History* (Seattle: University of Washington Press, 1976), 282. An early 1960s account of the use of the Forest Pest Control Act's funds for pesticides is given in Director, National Park Service to Secretary of the Interior, August 9, 1963, attachment, 8, NPS-HC.

45. See Emil F. Ernst and Charles R. Scarborough, "Narrative Annual Forestry Report of Yosemite National Park for the Calendar Year 1953," typescript, YOSE; and Yellowstone National Park, Superintendent's Annual Report, 1956, typescript, n.p., YELL. An early record of Sumner's concerns about DDT is in E. Lowell Sumner to Victor H. Cahalane, May 14, 1948, Central Classified File, RG79. See also Runte, *Yosemite,* 176–178.

46. National Park Service, *Information Handbook: Questions and Answers Relating to the National Park Service and the National Park System,* In-Service Training Series (Washington, D.C.: National Park Service, 1957), 54; Stephen J. Pyne, *Fire in America: A Cultural History of*

Wildland and Rural Fire (Princeton: Princeton University Press, 1982), 302; Bruce M. Kilgore, "Restoring Fire to National Park Wilderness," *American Forests* 81 (March 1975), 17.

47. Orthello L. Wallis, "Management of Sport Fishing in National Parks," *Transactions of the American Fisheries Society* 89 (April 1960), 234–238; Orthello L. Wallis, "Management of Aquatic Resources and Sport Fishing in National Parks by Special Regulations," July 22, 1971, typescript, 2, 6–14, Dennis; Orthello L. Wallis, "Development and Success of Catch-and-Release Angling Programs," paper presented at the Ninety-first Annual Meeting of the American Fisheries Society, September 14, 1961, typescript, 6–8, Advisory Board on Wildlife and Game Management files, MVZ-UC; John D. Varley, "A History of Fish Stocking Activities in Yellowstone National Park between 1881 and 1980," Yellowstone National Park Information Paper no. 35, January 1, 1981, typescript, III, YELL; John D. Varley and Paul Schullery, *Freshwater Wilderness: Yellowstone Fishes and Their World* (Yellowstone National Park: Yellowstone Library and Museum Association, 1983), 104–105: Paul Schullery, "A Reasonable Illusion," *Rod and Reel* 5 (November–December 1979), 44–54; National Park Service, "Position Paper: Findings and Recommendations on Fisheries Management Policies in the National Park Service," typescript, January 13, 1987, Supernaugh; Richard H. Dawson, "Assessment of Fisheries Management Options in National Parks," typescript, n.d. (ca. 1987), Supernaugh; National Park Service, *A Heritage of Fishing: The National Park Service Recreational Fisheries Program* (Washington, D.C.: National Park Service, n.d., ca. 1990), 2–3.

48. *Annual Report of the Secretary of the Interior* (1944), 229. The "exile" to Chicago had been, as Drury saw it, "a severe and expensive handicap, now happily ended." *Annual Report of the Secretary of the Interior for the Fiscal Year Ending June 30, 1948* (Washington, D.C.: Government Printing Office, 1948), 363. The Fish and Wildlife Service had also been moved to Chicago; thus, Cahalane had been able to continue working closely with Park Service personnel. Cahalane, interview with the author, February 25, 1992.

49. Carl P. Russell, Memorandum for the Director, March 23, 1944, NPS-HC. Following up on Russell's recommendations, Drury noted in his 1944 annual report that research to back up park planning was "essential to intelligent administration" of parks. He added that one benefit would be to "determine the extent of permanent impairment that may result from development of tourist facilities and heavy use of park areas." *Annual Report of the Secretary of the Interior* (1944), 218–219.

50. Dorr G. Yeager, "Comments on the Impairment of Park Values in Zion National Park," March 23, 1944, NPS-HC.

51. National Park Service, "Research in the National Park System, and Its Relation to Private Research and the Work of Research Foundations," February 10, 1945, typescript, 2, 4–5, 8–12, NPS-HC.

52. National Park Service, "Research in the National Park System, A Narrative Statement on Policy and Research Administration Prepared for the President's Scientific Research Board," April 4, 1947, appended to Newton B. Drury, Memorandum for Thomas B. Nolan, April 7, 1947, NPS-HC.

53. Lowell Sumner, "Biological Research and Management in the National Park Service: A History," *George Wright Forum* 3 (Autumn 1983), 16; "Research in the National Park System, A Narrative Statement," 5; National Park Service, "Wildlife Resources of the National Park System: A Report on Wildlife Conditions—1948," 18, appended to Hillory A. Tolson, Memorandum for All Field Offices, February 18, 1949, Central Classified Files, RG79.

54. National Park Service, "Wildlife Resources of the National Park System: A Report on Wildlife Conditions—1949," Records of Conrad L. Wirth, RG79.

55. Cahalane, interview with the author, February 25, 1992; William G. Carnes, "Landscape Architecture in the National Park Service," *Landscape Architecture* (July 1951), reprinted and attached to Hillory A. Tolson, Memorandum to Washington Office and All Field Offices, February 15, 1952, typescript, NPS-HC.

56. Wright, Dixon, and Thompson, *Fauna of the National Parks of the United States* (1933),

147; Drury to Miller, November 13, 1943. See also *Annual Report of the Secretary of the Interior* (1944), 221.

57. Lowell Sumner, "Wildlife Management," paper presented at the National Park Service Conference, Yosemite National Park, October 18, 1950, Entry 19, RG79; Conrad L. Wirth to Harold E. Crowe, January 7, 1958, NPS-HC.

58. Wright, Dixon, and Thompson, *Fauna of the National Parks* (1933), 147.

59. Carl P. Russell, Memorandum for the Director, March 23, 1944; Olaus J. Murie to Newton B. Drury, January 11, 1951, O. Murie.

60. The Mission 66 goals are discussed in Ronald F. Lee to Regional Directors, September 8, 1958, NPS-HC; Cahalane, interview with the author, February 25, 1992; Sumner, "Biological Research and Management," 17. Cahalane's successor was Gordon Fredine.

61. Cahalane, interview with the author, February 25, 1992; Conrad L. Wirth to Horace M. Albright, November 5, 1956, Records of Conrad L. Wirth, RG79; Summary Minutes, 38th Meeting, Advisory Board on National Parks, Historic Sites, Buildings and Monuments, Washington, D.C. and Gettysburg National Military Park, Penn., April 23–26, 1958, O. Murie.

62. E. T. Scoyen, memorandum to Washington Office and All Field Offices, April 21, 1958, NPS-HC.

63. Cahalane, interview with the author, February 25, 1992; National Park Service, "Get the Facts, and Put Them to Work," October 1961, typescript, 4, NPS-HC.

64. Olaus J. Murie to Lowell Sumner, August 2, 1958, O. Murie. Another widely respected ecologist, University of Michigan professor Stanley A. Cain, told the Sixth Biennial Wilderness Conference that the Park Service had no "basic ecological research" program, adding that research "fails to approach at all closely the fundamental need of the Service itself." Stanley A. Cain, "Ecological Islands as Natural Laboratories," paper presented at the Sixth Biennial Wilderness Conference, San Francisco, March 20–21, 1959, typescript, 10, NPS-HC.

65. Daniel B. Beard, Memorandum to Chairman, Management Improvement Committee, October 28, 1960, NPS-HC. The committee focused on historical and archeological research as well as natural history.

66. Beard to Chairman, Management Improvement Committee, October 28, 1960; "Report of the National Park Service Mission 66 Frontiers Conference," Grand Canyon National Park, Arizona, April 24–28, 1961, typescript, 131, NPS-HC; W. G. Carnes, "A Look Back to Look Ahead," April 24, 1961, typescript, THRO.

67. National Park Service, "Get the Facts and Put Them to Work," 2.

68. National Park Service, "Get the Facts and Put Them to Work," 1, 26.

69. National Park Service, "Get the Facts and Put Them to Work," 5; Stagner, interview with the author, April 15, 1989. See also R. Gerald Wright, *Wildlife Research and Management in the National Parks* (Urbana: University of Illinois Press, 1992), 24–26.

70. Sumner, "Biological Research and Management," 18; Wright, Dixon, and Thompson, *Fauna of the National Parks* (1933), 147.

71. Raymond Gregg, "A Perspective Report on the National Park Service Program of Interpretation," *American Planning and Civic Annual* (1947), 31.

Carl Russell, the Service's chief naturalist, was himself a Ph.D. biologist. He had been a close associate of George Wright in Yosemite during the 1920s and continued to support the wildlife biologists. Among the many naturalists, some dedicated individuals, such as Arthur Stupka at Great Smoky Mountains National Park, Edwin McKee at Grand Canyon, and Frank Brockman at Mount Rainier, performed significant research of value to interpretation as well as to natural resource management. See Lowell Sumner, "Biological Research and Management," 16–17. Victor Cahalane remembered that the naturalists who had a background in biology were "often very helpful and cooperated" with the wildlife biologists—some even had "all four feet" in wildlife management. Cahalane, interview with the author, February 25, 1992. See also Hillory A. Tolson, Memorandum for the Regional Directors, February 26, 1944, Central Classified File 715, RG79; Natt N. Dodge, Memorandum for the Regional Director, Region Three, October 27, 1944, Entry 19,

RG79; and William R. Supernaugh, "The Evolution of the Natural Resource Specialist: A National Park Service Phenomenon," paper prepared for the Department of Parks, Recreation and Environmental Education, Slippery Rock State University, August 25, 1987, typescript, copy courtesy of the author.

72. Robert N. McIntyre, "A Brief History of Forestry in the National Park Service," March 1952, appendix A, typescript, NPS-HC. The Service had encouraged numerous colleges and universities to provide academic training in forestry as preparation for those wishing to enter the ranger ranks, even issuing a suggested curriculum in 1944 in anticipation of postwar staffing needs. National Park Service, "Recommended College Preparation for Students Planning to Enter the National Park Service through the Park Ranger Civil Service Examination," U.S. Department of the Interior, NPS files, MVZ-UC; *Annual Report of the Secretary of the Interior* (1945), 220.

73. Russ Olsen, *Administrative History: Organizational Structures of the National Park Service, 1917–1985* (Washington, D.C.: National Park Service, 1985), 76–83; Garrison, *Making of a Ranger,* 252–255; Henneberger, "To Protect and Preserve," 405–409. Wirth's creation of the new ranger branch stemmed in part from his desire to give the parks better representation in Washington. To this end he brought in as his new associate director the veteran ranger and park superintendent, Eivind T. Scoyen, who was even born in a national park (Yellowstone). Wirth's actions significantly enhanced the voice of the rangers and superintendents in high-level decision-making. Wirth, *Parks, Politics, and the People,* 290–291.

In a late-1940s reversal of prior arrangements, divisions had been designated the primary organizational units, to oversee any number of branches. Olsen, *Organizational Structures of the National Park Service,* 67–69.

74. Victor H. Cahalane to E. Raymond Hall, October 24, 1957, attached to Victor H. Cahalane to David Brower, October 24, 1957, O. Murie. Wirth was persuaded to make this merger by a former Park Service forester who had transferred to the office of the secretary of the interior. (The documents do not reveal this individual's name.) Wirth had to seek approval for this reorganization from the secretary's office, since in the mid-1950s the department had assumed authority to review and approve changes in the Service's organizational structure. See Olsen, *Organizational Structures of the National Park Service,* 76.

75. Cahalane to Hall, October 24, 1957.

76. E. Raymond Hall to Conrad L. Wirth, February 12, 1958, O. Murie. Also, Hall had deplored Park Service forestry practices in a letter to Director Wirth. E. Raymond Hall to Conrad Wirth, January 16, 1958, NPS-HC.

77. Conrad L. Wirth, Memorandum to Washington Office and All Field Offices, February 10, 1958, NPS-HC; and Olsen, *Organizational Structures of the National Park Service,* 80–81; John M. Davis to the Director, January 3, 1958, Records of Conrad L. Wirth, RG79.

78. Wirth, Memorandum to Washington Office and All Field Offices, February 10, 1958; *Annual Report of the Secretary of the Interior for the Fiscal Year Ending June 30, 1959* (Washington, D.C.: Government Printing Office, 1959), 341.

79. Wirth to Washington Office and All Field Offices, February 10, 1958; Scoyen to Washington Office and All Field Offices, April 21, 1958. See also *Annual Report of the Secretary of the Interior for the Fiscal Year Ending June 30, 1958* (Washington, D.C.: Government Printing Office), 312, 315.

80. Foresta, *America's National Parks,* 50; Yellowstone National Park, "Superintendent's Annual Report, 1946," typescript, YELL; *Annual Report of the Secretary of the Interior* (1947), 327.

81. Garrison, *Making of a Ranger,* 257; *Annual Report of the Secretary of the Interior* (1945), 224. The Service's advance planning for postwar development is discussed in *Annual Report of the Secretary of the Interior* (1943), 215–216; *Annual Report of the Secretary of the Interior* (1944), 218; and *Annual Report of the Secretary of the Interior* (1945), 225–226.

82. *Annual Report of the Secretary of the Interior* (1947), 327–328; *Annual Report of the Secretary of the Interior* (1949), 302; Ise, *Our National Park Policy,* 455. Drury wrote to the

United States Chamber of Commerce in April 1948 that the parks immediately needed "all types of construction . . . including employee housing; concessioners; facilities; water, sewer, electric, and communication systems; campgrounds; museums; comfort stations; roads and trails, etc." Newton B. Drury to D. J. Guy, April 27, 1948, Entry 19, RG79. See also Newton B. Drury, "The Dilemma of Our Parks," *American Forests* 55 (June 1949), 6–11.

83. Ronald Foresta described Drury as being "constantly restrained by his stringent sense of bureaucratic propriety and his dislike of the rough and tumble world of Washington politics" and added that in conservation politics Drury "always seemed more aware of the weakness of his position than its strength." Foresta, *America's National Parks,* 48–49. See also Ise, *Our National Park Policy,* 443.

84. Newton B. Drury to Richard M. Leonard, May 12, 1948, Entry 19, RG79. The exact date of Drury's remark about having no money is not given; it is recalled by David R. Brower in his autobiography, *For Earth's Sake* (Salt Lake City: Peregrine Smith Books, 1990), 220. Michael P. Cohen, in *History of the Sierra Club,* 126, cites the club's November 1948 statement that the Park Service could not be relied on to protect backcountry because it was so dedicated to intensive public use of the parks.

85. *Annual Report of the Secretary of the Interior for the Fiscal Year Ending June 30, 1941* (Washington, D.C.: Government Printing Office, 1941), 280; *Annual Report of the Secretary of the Interior* (1944), 219–220; *Annual Report of the Secretary of the Interior* (1945), 215; Conrad L. Wirth, "The Aims of the National Park Service in Relation to Water Resources," *American Planning and Civic Annual* (1952), 11–14; Ise, *Our National Park Policy,* 467–469.

86. H. W. Bashore, Memorandum for the Secretary, January 5, 1945, NPS-HC.

87. Newton B. Drury, Memorandum for the Secretary, January 25, 1945, NPS-HC.

88. Michael W. Straus, Memorandum for the Secretary, February 6, 1945, NPS-HC.

89. Newton B. Drury to Charles G. Sauers, December 13, 1950, Entry 19, RG79; *Annual Report of the Secretary of the Interior* (1946), 343; *Annual Report of the Secretary of the Interior* (1948), 353; and Newton B. Drury, Memorandum for the Director's Office and All Field Offices, April 30, 1948, Entry 19, RG79. Drury's attitudes toward recreation areas are also discussed in Newton B. Drury, Memorandum to the Secretary, January 13, 1947, Entry 19, RG79.

90. *Annual Report of the Secretary of the Interior* (1949), 327. In 1957 the Service would cede recreation management at Millerton Lake to the State of California.

Drury was also willing for some less spectacular natural areas to be removed from the national park system. During his tenure Wheeler and Mount of the Holy Cross national monuments in Colorado were returned by the Park Service to their previous administrators. In addition to lacking sufficient scenic qualities, Wheeler National Monument had remained inaccessible and attracted few visitors, causing the Park Service to lose interest. The appeal of the Mount of the Holy Cross was diminished when rock slides caused the right arm of the cross to slump, changing the appearance of the cross. Rejecting the idea of shoring up the arm as an inappropriate way of treating a symbol of God's work, the Park Service ultimately agreed to return the area to the U.S. Forest Service. These revealing occurrences are discussed in Ferenc M. Szasz, "Wheeler and Holy Cross: Colorado's 'Lost' National Monuments," *Journal of Forest History* 21 (July 1977), 139, 144.

91. Drury later claimed that the agreement was intended to foster cooperation and did not actually recommend redesignation from monument to national recreation status. Owen Stratton and Phillip Sirotkin, *The Echo Park Controversy* (University: University of Alabama Press, 1959), 36–38. See also Susan Rhoades Neel, "Newton Drury and the Echo Park Dam Controversy," *Forest and Conservation History* 38 (April 1994), 57–58.

92. Stratton and Sirotkin, *The Echo Park Controversy,* 38–40; Neel, "Newton Drury," 60–62; Mark W. T. Harvey, *A Symbol of Wilderness: Echo Park and the American Conservation Movement* (Albuquerque: University of New Mexico Press, 1994), 11–34, 61–65. Drury discussed the status of the threats to parks from dam construction in *Annual Report of the Secretary of the Interior* (1948), 338–340.

93. Newton B. Drury to Charles W. Davis, May 18, 1950, Entry 19, RG79; Newton B. Drury,

Memorandum for the Regional Director, Regional Four, June 3, 1948, Entry 19, RG79; Newton B. Drury to Morris Cooke, July 3, 1950, Entry 19, RG79.

94. Harvey, *A Symbol of Wilderness*, 81–89; *Annual Report of the Secretary of the Interior* (1950), 303–305. Drury indicated (p. 305) that the Park Service's efforts to defeat the Echo Park dam during the spring of 1950 were carried out with Secretary Chapman's full knowledge.

95. Herbert Evison and Newton Bishop Drury, "The National Park Service and Civilian Conservation Corps," interview by Amelia Roberts Fry, Berkeley, California, 1963, typescript, 119, NPS-HC.

The circumstances of Drury's resignation are still controversial. Yet the fact that Chapman forced him to resign is substantiated in documentary evidence. On December 13, 1950, the secretary formally notified Drury that he had to leave the directorship of the National Park Service to take over "advisory duties as Special Assistant to the Secretary as of January 15 next," and that he would have to accept a lower salary. Two days earlier Drury had prepared a handwritten note for his own files, in which he quoted from a conversation with Chapman, who told him, "I expect you to take the other position or resign." Oscar L. Chapman to Newton B. Drury, December 13, 1950; Newton B. Drury, note to files, December 11, 1950, WS. Rejecting Chapman's demand and the offer of a lower-paying position, Drury resigned effective April 1.

Chapman's reasons for forcing Drury out of office are less clear. His stated justification was that he could make "fuller utilization" of Drury's talents in a "department-wide capacity," in which he would help smooth over the many conflicts among the various Interior Department bureaus—an assignment for which Drury seems to have been ill suited. The secretary also claimed that since Associate Director Arthur Demaray had served the Park Service loyally and competently, he should be rewarded with the directorship for a short while before his already-announced retirement. Chapman was willing to force out one longtime employee to benefit another.

Chapman's explanation in 1973 to former Park Service Director George B. Hartzog, Jr., was perhaps more candid, and also was in accord with Drury's statement on the matter. In his autobiography Hartzog recalled a luncheon conversation in which Chapman "expressed his strong view that Drury had been disloyal to him during the fight over the [Echo Park] dam in that Drury was not vigorous in his support of the secretary's decision." Chapman to Drury, December 13, 1950; Charles G. Sauers to Oscar Chapman, February 3, 1951, WS; George B. Hartzog, Jr., *Battling for The National Parks* (Mt. Kisco, New York: Moyer Bell, 1988), 83.

96. Even before Demaray's appointment, Wirth had been told by Secretary Chapman that he would succeed Demaray as Park Service director. Wirth, *Parks, Politics, and the People*, 285.

97. Stratton and Sirotkin, *The Echo Park Controversy*, 51–97; Roderick Nash, *Wilderness and the American Mind* (1967; 3rd ed., New Haven: Yale University Press, 1982), 209–219; Foresta, *America's National Parks*, 51. Other perspectives on the Echo Park controversy may be found in, for example, Elmo R. Richardson, *Dams, Parks, and Politics: Resource Development and Preservation in the Truman-Eisenhower Era* (Lexington: University of Kentucky Press, 1973), 63–67; Irving Brant, *Adventures in Conservation with Franklin D. Roosevelt* (Flagstaff, Arizona: Northland Publishing Co.), 308–310; and Harvey, *A Symbol of Wilderness*. Foresta (pp. 50–51) viewed Drury's handling of the Echo Park affair as "probably his greatest failing as director," because Drury's failure to oppose the dam from the beginning of his tenure as director not only allowed the dam proponents to consolidate their position but also seriously weakened the Service's ultimate role in defending its own park lands. In Foresta's opinion (p. 51), the Service "lost mastery of its own house. The fate of a unit of the National Park System was decided by the interplay of public interest groups and their congressional allies on one side and the Bureau of Reclamation and its allies on the other. The [Park Service] was a bit player in the drama."

98. Wirth, "Aims of the National Park Service," 12–14.

99. On Glen Canyon, see Russell Martin, *A Story That Stands like a Dam: Glen Canyon and the Struggle for the Soul of the West* (New York: Henry Holt and Company, 1989); Harvey, *A Symbol of Wilderness*, 280–282, 298–301; and Cohen, *History of the Sierra Club*, 177–179.

100. Wirth, "Aims of the National Park Service," 15. Russell Martin, in *A Story That Stands*

like a Dam, 45–47, discusses the 1930s proposal to create Escalante National Monument along the Colorado and Green rivers through southern Utah. See also pp. 228–229.

101. The differences in management policy between national parks and national recreation areas are discussed in *Annual Report of the Secretary of the Interior* (1948), 353; *Annual Report of the Secretary of the Interior* (1950), 304; and Newton B. Drury to the Director's Office and All Field Offices, April 30, 1948, Entry 19, RG79.

102. Stratton and Sirotkin, *The Echo Park Controversy,* 95; and Foresta, *America's National Parks,* 52. Conrad Wirth had stated earlier that the Park Service intended to make Dinosaur more accessible, noting that if the Service could "get a few people to the [canyon] Rim, so they can see what we have, it will help win that battle." Conrad L. Wirth to William Voigt, November 21, 1951, Records of Conrad Wirth, RG79.

103. Wirth, *Parks, Politics, and the People,* 234; Lemuel A. Garrison, "Practical Experience Gained from Standards, Policies and Planning Procedures in National Parks," paper presented at the First World Conference on National Parks, 1962, typescript, NPS-HC.

104. *Annual Report of the Secretary of the Interior for the Fiscal Year Ending June 30, 1956* (Washington, D.C.: Government Printing Office, 1956), 308.

105. Wirth's quote is found in *Annual Report of the Secretary of the Interior for the Fiscal Year Ending June 30, 1952* (Washington, D.C.: Government Printing Office, 1952), 353.

106. Bernard DeVoto, "Let's Close the National Parks," *Harper's Magazine* 207 (October 1953), 49–52. Similarly, DeVoto's article "Shall We Let Them Ruin Our National Parks?" in the *Saturday Evening Post,* July 22, 1950, attacked dam proposals in the West that threatened national parks. This piece helped arouse public opposition to the Echo Park dam proposal. Harvey, *A Symbol of Wilderness,* 95–103.

107. Foresta, *America's National Parks,* 50. Until the Korean War ended, Wirth probably would have had little chance of getting such a program under way, although he does not say as much in his autobiography. See Wirth, *Parks, Politics, and the People,* 234–238.

108. Wirth, *Parks, Politics, and the People,* 238–239. See also *Annual Report of the Secretary of the Interior* (1956), 299–300.

109. The planning strategies and the meeting with President Eisenhower are discussed in Wirth, *Parks, Politics, and the People,* 239–257. The President's final comment after Wirth's presentation was, as Wirth recalled it, "This is a good project; let's get on with it" (p. 256).

110. Conrad L. Wirth, Memorandum no. 2, Mission 66, to Washington Office and All Field Offices, March 17, 1955, NPS-HC; Garrison, *Making of a Ranger,* 256–258.

111. National Park Service, "Mission 66: To Provide Adequate Protection and Development of the National Park System for Human Use," January 1956, typescript, YELL; National Park Service, *Our Heritage: A Plan for Its Protection and Use* (Washington, D.C.: National Park Service, 1955); Lemuel A. Garrison, "Guiding Precepts, Mission 66," August 29, 1955, NPS-HC. See also Wirth, *Parks, Politics, and the People,* 258–260.

112. Garrison, *Making of A Ranger,* 257; Wirth, *Parks, Politics, and the People,* 262, 266–270.

113. Carnes, "Landscape Architecture in the National Park Service"; Wirth, *Parks, Politics, and the People,* 62–63, 240, 249–250. See also Thomas C. Vint, "National Park Service Master Plans, *Planning and Civic Comment* 12 (April 1946), 21–24 ff.

114. Carnes, "Landscape Architecture in the National Park Service"; Olsen, *Organizational Structures of the National Park Service,* 76–77; Wirth, *Parks, Politics, and the People,* 292; and Vernon L. Hammons, "A Brief Organizational History of the Office of Design and Construction, National Park Service, 1917–1962," typescript, 6, NPS-HC.

115. Carnes, "Landscape Architecture in the National Park Service."

116. Among Park Service directors, few relished bureaucratic power as much as Wirth. Reflecting on Wirth's personal power and the leadership clique that developed under him, Park Service veteran Russ Olsen recalled how important it was to be a member of Wirth's carpool, where opportunities existed twice daily to influence the director's thinking on special issues. From north of the District of Columbia, the Park Service luminaries traveled south to enter Rock Creek

Parkway, then turned left on Virginia Avenue, north on 18th Street, and left into the Interior building parking garage, to Wirth's parking space, A-5. Olsen, interview with the author, February 26, 1990.

117. Devereux Butcher, "Resorts or Wilderness?" *Atlantic Monthly* 107 (February 1961), 46–47. For an example of Butcher's earlier objections to national park architecture, see Devereux Butcher, "For a Return to Harmony in Park Architecture," *National Parks Magazine* 26 (October–December 1952), 150–157.

118. Butcher, "Resorts or Wilderness?" 47–48; "A Sky-post for the Smokies," *National Parks Magazine* 33 (February 1959), inside cover. The Park Service received many letters objecting to the sky-post, but its designers believed they had created "not a monster, but a tower of pleasing and lasting significance"—that even conservationists could not live in "'cocoons,' avoiding machine-made and manufactured products" such as the reinforced concrete structure on top of Clingman's Dome. John B. Cabot to R. A. Wilhelm, March 10, 1959, GRSM.

119. Weldon F. Heald, "Urbanization of the National Parks," *National Parks Magazine* 35 (January 1961), 8.

120. *Annual Report of the Secretary of the Interior* (1958), 304–305; William C. Tweed and Laura Soulliere Harrison, "Rustic Architecture and the National Parks: The History of a Design Ethic," 1987, typescript, chapter 8, 4–5, copy courtesy of the authors.

121. Tweed and Harrison, "Rustic Architecture and the National Parks," chapter 8, 1–6; discussion with Marshall Gingery, July 28, 1989. Laura Harrison, coauthor of "Rustic Architecture and the National Parks," commented that the architects who abandoned the rustic architecture styles during Mission 66 did not want the Service to be tied to rustic "Hansel and Gretel" cottages. Interview with Laura Harrison, February 26, 1991.

122. National Park Service, *The National Park Wilderness* (Washington, D.C.: National Park Service, 1957), 9, 10, 22, 24–25. An earlier, condensed version of the brochure, with different wording in places, is found in Howard R. Stagner, "Preservation of Natural and Wilderness Values in the National Parks," *National Parks Magazine* 31 (July–September 1957), 105–106, 135–139.

123. Cohen, *Sierra Club,* 149. Cohen refers to the "traditional Club strategy—encouraging recreational use of a threatened area." He notes also (p. 181) the similar use of Sierra Club "outings" into the Hetch Hetchy Valley in the early part of the century, in an effort to save it from inundation. Efforts by conservationists to attract attention and visitors to Dinosaur are discussed in Harvey, *A Symbol of Wilderness,* 156–173, 236–243.

124. Ise, *Our National Park Policy,* 395; Foresta, *America's National Parks,* 53–54. Foresta mentions the Stevens Canyon Road in Mount Rainier National Park as another example of road building to help protect a park. See also Stratton and Sirotkin, *Echo Park Controversy,* 95.

125. David R. Brower, "'Mission 65' Is Proposed by Reviewer of Park Service's New Brochure on Wilderness," *National Parks Magazine* 32 (January–March 1958), 4–6.

126. Olaus J. Murie to Conrad L. Wirth, February 6, 1958, O. Murie; Olaus J. Murie to Conrad L. Wirth, December 10, 1957, NPS-HC.

127. Conrad L. Wirth to John B. Oakes, February 12, 1958, NPS-HC; Conrad L. Wirth to Olaus J. Murie, February 14, 1958, NPS-HC.

128. Wirth, *Parks, Politics, and the People,* 359. See also Runte, *Yosemite,* 194–197, for a discussion of the Tioga controversy. Runte indicates that in addition to its own motivations for wanting a wider, more modern road, the Park Service had been under steady pressure from commercial and community interests in the Owens Valley to the east of Yosemite—interests that would benefit from increased tourist traffic through the area.

129. The change in club strategy toward more frequent confrontation with public land managers is discussed in Cohen, *Sierra Club,* 204, 207, 234–238, 249–252. Foresta, in *America's National Parks,* 59–62, 70–71, examines the changing attitudes among conservationists.

130. Ansel Adams, "Yosemite—1958: Compromise in Action," *National Parks Magazine* 32 (October–December 1958), 167, 170–172, 190. Ansel Adams, "Tenaya Tragedy," *Sierra Club Bulletin* 43 (November 1958), 4. For views similar to those of Adams, but less angry, see Anthony

Wayne Smith, "The Tioga Road," *National Parks Magazine* 33 (January 1959), 10–13; and "Yosemite's Tioga Highway," *National Parks Magazine* 32 (July–September 1958), 123–124.

131. Wirth, *Parks, Politics, and the People,* 358; Ansel Adams to Harold Bradley, Richard Leonard, and David Brower, July 27, 1957, in Mary Street Alinder and Andrea Gray Stillman, *Ansel Adams: Letters and Images, 1916–1984* (Boston: Little, Brown and Company, 1988), 247.

132. Wirth, *Parks, Politics, and the People,* 262; George Alderson, "Instant Roads in the National Parks," *Sierra Club Bulletin* 54 (January 1969), 14; Horace M. Albright to Max K. Gilstrap, April 2, 1956, Records of Conrad L. Wirth, RG79.

133. Brown, *History of the Denali–Mount McKinley Region,* 218–222.

134. Nash, *Wilderness and the American Mind,* 220–221. An attempt to pass wilderness legislation in the late 1930s had failed.

135. Brower, "'Mission 65'," 4; Garrison, *Making of a Ranger,* 260.

136. Brower, "'Mission 65'," 4–6, 45.

137. Conrad L. Wirth to Bruce M. Kilgore, February 18, 1958, Records of Conrad Wirth, RG79. Kilgore, a scientist, later became an employee of the Park Service.

138. Conrad L. Wirth to the Washington Office and All Field Offices, February 27, 1959, NPS-HC. Identical comments on zones and corridors are found in L. F. Cook, "Zoning of Areas of Use," 1961, typescript, NPS-HC.

139. Conrad L. Wirth, "Wilderness in the National Parks," *Planning and Civic Comment* 24 (June 1958), 7.

140. The quote is from Garrison, "Practical Experience Gained from Standards."

141. Cook, "Zoning of Areas of Use"; Garrison, *Making of a Ranger,* 260.

142. Wirth, *Parks, Politics, and the People,* 360–361. Other than these remarks, Wirth's extensive account of his years as Park Service director refers only in the most cursory way to the lengthy and involved campaign to enact wilderness legislation. See 283, 328, and 386. Wirth does, however, acknowledge the act's importance (p. 329) and discusses the meaning of wilderness, including quotes from the 1957 wilderness brochure (p. 385).

143. The quote is found in Ise, *Our National Park Policy,* 650. Additional indications of the Service's opposition to the wilderness bill appear in a number of sources. Cohen, *Sierra Club,* 133, mentions that the Park Service (along with the Forest Service, Bureau of Land Management, and Bureau of Sport Fisheries and Wildlife) wanted to retain full control over backcountry areas, without interference from wilderness legislation. For comments on Horace Albright's reluctance to support the bill, see Cohen, 230–231. Albright stated at the spring 1958 meeting of the National Parks Advisory Board that "this whole wilderness bill plan is such a futile project—so much money, time and effort put into it and there is not a chance of the bill going through Congress." Excerpts from minutes of Advisory Board meetings, April 23–26, 1958, typescript, NPS-HC. See also Nash, *Wilderness and the American Mind,* 226; Foresta, *America's National Parks,* 69; and Craig W. Allin, *The Politics of Wilderness Preservation,* Contributions in Political Science, ed. Bernard K. Johnpoll, no. 64 (Westport, Connecticut: Greenwood Press, 1982), 110–111.

144. Howard R. Stagner, interview with the author, April 15, 1989. Former Park Service manager John W. Henneberger also recalled the Service's reluctance to support the wilderness bill. John W. Henneberger, interview with the author, June 17, 1989.

145. Wirth to Crowe, January 7, 1958.

146. Wilderness Society, *The Wilderness Act Handbook* (Washington, D.C.: Wilderness Society, 1984), 5–6; Hillory A. Tolson, *Laws Relating to the National Park Service, the National Parks and Monuments* (Washington, D.C.: Department of the Interior, 1933), 10.

147. Wirth, *Parks, Politics, and the People,* 281–283; Ise, *Our National Park Policy,* 519–520; Cohen, *Sierra Club,* 260; Donald C. Swain, *Wilderness Defender: Horace M. Albright and Conservation* (Chicago: University of Chicago Press, 1970), 306–308.

148. Wirth, *Parks, Politics, and the People,* 260–261, 281; Carnes, "A Look Back to Look Ahead."

149. In his autobiography Wirth also claimed that the commission used about forty contrac-

tors to conduct different studies, including universities and bureaus of the federal government such as the Geological Survey and the Bureau of Sport Fisheries and Wildlife. Thus, he asked, "Why not the National Park Service?" In fact, the Service was accused of cooperating with the commission "only when it had to"—an indication of its basic resistance to the study. Wirth, *Parks, Politics, and the People,* 281–283; Foresta, *America's National Parks,* 62.

150. Foresta, *America's National Parks,* 63–65. Horace M. Albright to George B. Hartzog, Jr., January 15, 1971, Hartzog. The introduction and summary recommendations for the 1962 report reviewed the history of the nation's recreational planning and management, but did not even mention the National Park Service or the 1936 Park, Parkway, and Recreation Area Study Act. Outdoor Recreation Resources Review Commission, *Outdoor Recreation for America: A Report to the President and to the Congress* (Washington, D.C.: Government Printing Office, 1962), 1–10. Wirth's 1962 annual report noted matter-of-factly that the Park Service's survey work had become the responsibility of the Bureau of Outdoor Recreation. *Annual Report of the Secretary of the Interior for the Fiscal Year Ending June 30, 1962* (Washington, D.C.: Government Printing Office, 1962), 103–104.

151. Alston Chase, *Playing God in Yellowstone: The Destruction of America's First National Park* (Boston: Atlantic Monthly Press, 1986), 27–28.

152. Righter, *Crucible for Conservation,* 8–9, 148–149; Robert H. Bendt, "The Jackson Hole Elk Herd in Yellowstone and Grand Teton National Parks," in *Transactions of the Twenty-seventh North American Wildlife and Natural Resources Conference, March 1962* (Washington, D.C.: Wildlife Management Institute, 1962), 191–193.

153. Bendt, "The Jackson Hole Elk Herd," 193, 197–198.

154. With Grand Teton's dam and artificially enlarged lake, and with hunting and grazing allowed in the park, Superintendent Bill wondered if the national park should instead be a national recreation area. Harthon L. Bill to the Regional Director, January 31, 1961, NPS-HC. See also Wright, *Wildlife Research and Management,* 46–52, for a discussion of public sporthunting in the national parks.

155. Conrad L. Wirth to Anthony Wayne Smith, February 20, 1962, Advisory Board on Wildlife and Game Management Files, MVZ-UC. Earlier, in 1953, Sumner had written a strong and well-articulated indictment, entitled "Why Public Hunting Cannot Be Permitted in the National Park System," stating his belief that public hunting in the parks would "undermine the sciences of ecology and game management." Sumner based much of his argument on the rights of the "non-shooting" public (which he claimed made up ninety-two percent of the nation's population). "Shooters" were a "minority group," who had obtained the "special privilege" of hunting on public lands, and only in the national parks did the right of the nonhunting public to "protect its property in ways of its own choosing" exist. In reality, Sumner asserted, the sportsmen's associations were seeking to obtain "outside control of park wildlife for their own special form of recreation through local game departments which by their own admission are the agents of the shooters." Lowell Sumner, "Why Public Hunting Cannot Be Permitted in the National Park System," January 19, 1953, Records of Conrad L. Wirth, RG79.

156. Wirth to Smith, February 20, 1962.

157. Anthony Wayne Smith to Conrad L. Wirth, March 21, 1961, letter published in *National Parks Magazine* 35 (May 1961), 15, 19. Wirth's February 20 letter is found on pp. 14, 19.

In February 1963, as the debates on hunting continued, Ira N. Gabrielson, president of the Wildlife Management Institute and former head of the Fish and Wildlife Service, expressed his belief that the Park Service's recreational tendencies directly influenced attitudes toward public hunting in national parks. Writing to biologist A. Starker Leopold, Gabrielson claimed that the Service's promotion of "mass recreation" had caused the public to have "wrong impressions of the parks' functions." The result was a greater acceptance of public hunting in the parks. He added, "If a national park is, in fact, to be a mass recreation area then barring hunting is somewhat illogical." Ira N. Gabrielson to A. Starker Leopold, February 18, 1963, Advisory Board on Wildlife and Game Management File, MVZ-UC.

158. Howard Zahniser to Olaus Murie, March 22, 1961, O. Murie; Carl W. Buchheister to Stewart L. Udall, November 9, 1961, Records of Conrad Wirth, RG79. Buchheister, the Audubon Society president, stated in his letter to Udall that reductions of "excessive and injurious herds of big game are in order in the National Parks," but that recreational hunting was "irreconcilable" with national park purposes.

159. Lemuel A. Garrison to Regional Director, Region Two, March 24, 1961, YELL.

160. Lemuel A. Garrison to Regional Director, Region Two, May 12, 1961, Records of Conrad L. Wirth, RG79; Regional Director, Region Two, to the Director, May 22, 1961, Records of Conrad L. Wirth, RG79.

161. Conrad L. Wirth, "Wildlife Conservation and Management in the National Parks and Monuments," September 14, 1961, Records of Conrad L. Wirth, RG79.161.

162. Wirth, *Parks, Politics, and the People,* 310–311; John A. Carver, paper presented at the National Park Service Conference of Challenges, Yosemite National Park, October 13–19, 1963, typescript, NPS-HC. Wirth's formal submission of his policy statement to Secretary Udall came on October 25, 1961. Given the many informal communication channels within the Interior Department, it is almost certain that the secretary's office would have known of the statement by the time it was issued, if not before. Conrad L. Wirth to the Secretary of the Interior, October 25, 1961, Advisory Board on Wildlife and Game Management Files, MVZ-UC.

163. *Annual Report of the Secretary of the Interior* (1962), 86; Houston, *The Northern Yellowstone Elk,* 17; Wright, *Wildlife Research and Management,* 48–49.

164. National Academy of Sciences, "A Report by the Advisory Committee"; U.S. Department of the Interior, news release, April 25, 1962, NPS-HC; Stewart L. Udall to A. Starker Leopold, April 27, 1962, Advisory Board on Wildlife and Game Management Files, MVZ-UC; F. Fraser Darling and Noel D. Eichhorn, *Man and Nature in the National Parks,* 2nd ed. (Washington, D.C.: Conservation Foundation, 1969) 14–15; Wright, *Wildlife Research and Management,* 32. A third investigation got under way when the secretary requested that the Conservation Foundation examine the effects of human activity on national parks.

165. Lowell Sumner to A. Starker Leopold, May 16, 1962, Advisory Board on Wildlife and Game Management Files, MVZ-UC.

166. Lowell Sumner, "A History of the Office of Natural Science Studies," in "Proceedings of the Meeting of Research Scientists and Management Biologists of the National Park Service," Horace M. Albright Training Center, April 6–8, 1968, typescript, 4, Dennis.

167. Cahalane, interview with the author, February 25, 1992. See also Foresta, *America's National Parks,* 47–50.

168. Adams, "Yosemite—1958," 172. Wirth's views reflected the general consensus at the 1961 superintendents conference and were clearly expressed in his conference address. In part, belief that Mission 66 was inadequate resulted from the growth of the national park system. The establishment of new parks created additional construction and development demands, and helped raise the total number of park visitors in 1966 to 133 million, rather than the 80 million anticipated at the beginning of Mission 66. Conrad L. Wirth, "Catching Sight of the New Frontier," paper presented at the Mission 66 Frontiers Conference of the National Park Service, Grand Canyon National Park, April 24–28, 1961, NPS-HC; Wirth, *Parks, Politics, and the People,* 260–262, 280. For visitation figures from World War II to the mid-1950s, see Foresta, *America's National Parks,* 50.

169. Wirth to Crowe, January 7, 1958.

Chapter 6. Science and the Struggle for Bureaucratic Power

1. Horace M. Albright, interview by [] Erskine, Washington, D.C., January 28, 1959, typescript, 33, NPS-HC. Mission 66 did not have a definite ending date; some of its projects were not completed until well after 1966.

2. Conrad L. Wirth, *Parks, Politics, and the People* (Norman: University of Oklahoma Press, 1980), 199, 260–261.

3. Additions to the national park system during this period are listed in Barry Mackintosh, *The National Parks: Shaping the System* (Washington, D.C.: National Park Service, 1991), 80–84. See also Wirth, *Parks, Politics, and the People,* 261.

4. Wirth, *Parks, Politics, and the People,* 301. George B. Hartzog, Jr., Remarks Presented at the Annual Meeting of the American Society of Landscape Architects, Yosemite National Park, May 9, 1966, typescript, NPS-HC; George B. Hartzog, Jr., Memorandum to Each National Park Service Employee, January 3, 1967, NPS-HC. The quote is from George B. Hartzog, Jr., to Secretary of the Interior, October 12, 1965, NPS-HC. See also George B. Hartzog, Jr., "Parkscape U.S.A.: Tomorrow in Our National Parks," *National Geographic* 130 (July 1966), 48–92. Lacking the highly visible construction and development that distinguished Mission 66, Parkscape U.S.A. did not catch the public attention that its predecessor program had.

5. Greater backcountry use is one example of the pressure placed on the parks during this period. For instance, by the mid-1960s annual use of Yosemite backcountry was on the rise; it would increase 250 percent between 1968 and 1975, when it peaked at 219,000 "visitor nights" per year, then began to decrease. Similarly, backcountry use in Shenandoah National Park rose from 34,000 in 1967 to a high of 121,000 in 1973. National Park Service, "Wilderness Management Plan," Yosemite National Park, 1989, 2, YOSE; Robert R. Jacobsen, "The Management of Wilderness in Shenandoah National Park," expanded portion of a talk given to a U.S. Forest Service Wilderness Workshop at Gorham, New Hampshire, October 19, 1982, typescript, 2, Dennis. The effects of vastly increased backcountry use in Great Smoky Mountains National Park, one of the most heavily visited large natural parks in the system, are discussed in Susan P. Bratton, Linda Stromberg, and Mark E. Harmon, "Firewood-Gathering Impacts in Backcountry Campsites in Great Smoky Mountains National Park," *Environmental Management* 6 (January 1982), 63–71; William E. Hammitt and Janet Loy Hughes, "Characteristics of Winter Backcountry Use in Great Smoky Mountains National Park, *Environmental Management* 8 (March 1984), 161–166. See also George Alderson, "Instant Roads in the National Parks," *Sierra Club Bulletin* 54 (January 1969), 14.

6. Hartzog, "Parkscape U.S.A.," 52, 57. See also Arthur R. Gómez, *Quest for the Golden Circle: The Four Corners and the Metropolitan West, 1945–1970* (Albuquerque: University of New Mexico Press, 1994), 130–148, for a discussion of reservoirs, highway improvements, and new park proposals relative to local and regional tourism interests and to energy development in the Four Corners area. Interior secretary Udall envisioned a "Golden Circle" of parks and other recreation areas, predominantly in Arizona and Utah.

7. Hartzog, "Parkscape U.S.A.," 52, 57. Russell Martin, *A Story That Stands like a Dam: Glen Canyon and the Struggle for the Soul of the West* (New York: Henry Holt and Company, 1989), 45–47, 228–229; Mark W. T. Harvey, *A Symbol of Wilderness: Echo Park and the American Conservation Movement* (Albuquerque: University of New Mexico Press, 1994), 280–282, 298–301.

8. Anthony Wayne Smith, "Campaign for the Grand Canyon," *National Parks Magazine* 36 (April 1962), 12–15. Smith, executive secretary of the National Parks Association, asserted (p. 13) that the Bureau of Reclamation had become so powerful within the Department of the Interior that it was able to prevent the Park Service from making public its plans for expanding Grand Canyon National Park to the north, in an area targeted by the bureau for water control. Smith stated that "not until Director Wirth took the stand in [litigation against the Marble Canyon dam proposal], under subpoena of the [National Parks] Association, was it possible for the Service to make these plans public." The fight to protect Grand Canyon National Park ultimately centered on the proposed Marble Canyon and Bridge Canyon dams, which failed to gain approval. See Marc Reisner, *Cadillac Desert* (London: Stecker and Warburg, 1990), 283, 293–301. Ronald A. Foresta, in *America's National Parks and Their Keepers* (Washington, D.C.: Resources for the Future, 1984), 59–74, discusses the Park Service's declining status with the conservationists and environmentalists of the 1960s and 1970s.

9. Macintosh, *Shaping the System,* 102–103.

10. Department of the Interior, "Law Enforcement in Areas Administered by the National Park Service," Issue Support Paper no. 9, in "Recreation Use and Preservation," Fiscal Year 1970, typescript, 1, Hartzog; National Park Service, "1970 Summary Reports, Law Enforcement and Traffic Safety," Hartzog.

Director Hartzog's anticipation of park law-enforcement problems was evidenced in his 1965 statement that "Yosemite Valley is a great metropolitan area in the summertime . . . [so the Park Service has] all of the problems that you have in a metropolitan environment anywhere else in the United States in Yosemite Valley in the summertime." George B. Hartzog, Jr., "The National Parks, 1965," interview by Amelia R. Fry, Regional Oral History Office, University of California, Berkeley, April 4, 1965, typescript, NPS-HC.

Hartzog had instructed the Yosemite staff to take care of the increasingly difficult situation; nevertheless, the Park Service remained inadequately prepared. Russ Olsen, interview with the author, February 26, 1990. The riot in Yosemite's Stoneman Meadow is discussed in Alfred Runte, *Yosemite: The Embattled Wilderness* (Lincoln: University of Nebraska Press, 1990), 202. Jack Hope, in "Hassles in the Park," *Natural History* 80 (May 1971), 20–23, 80–91, gives an account of the riot that is critical of the Park Service.

11. Department of the Interior news release, "New Park Law Enforcement Division Readied for 1971 Vacation Season," March 6, 1971, Hartzog. Details of the planned law-enforcement training program were stated earlier in Deputy Assistant Director, Park Management, to Director, November 13, 1970, Hartzog; George B. Hartzog, Jr., to Hon. Julia Butler Hansen, February 4, 1971, Hartzog. Russ Olsen also described Hartzog's ability to seize opportunities: if he could "make people and dollar mileage" he would "take a program and run," as with law enforcement. Russ Olsen, interview with the author, February 26, 1990. See also Russ Olsen, *Administrative History: Organizational Structures of the National Park Service, 1917 to 1985* (Washington, D.C.: National Park Service, 1985), 93.

12. William R. Supernaugh, interview with the author, March 10, 1989. Longtime park superintendent Robert Barbee recalled that at Yosemite in the early 1970s an "empathy team" was established, which "wore beads and tried to explain the role of parks to hippies"—an example of the Service's effort to bridge the cultural gap. By contrast, biologist Susan Bratton, who worked in Great Smoky Mountains National Park in the 1970s, remembered the tougher image of the ranger that emerged during this period of increased law enforcement. "The old, gentle rangers," she said, were "replaced by SWAT teams," sometimes drawn from Vietnam War veterans. Robert D. Barbee, interview with the author, July 24, 1989; Susan P. Bratton, interview with the author, March 20–21, 1989. M. Peter Philley and Stephen F. McCool, in "Law Enforcement in the National Parks: Perceptions and Practices," *Leisure Sciences* 4 (1981), 355–371, discuss the relationship between the rangers and law enforcement in the parks. See especially 369.

13. The Service's increased interest in safety resulted in a number of measures taken in Yellowstone itself, as well as the evaluation and updating of safety programs throughout the national park system. The lawsuit was settled for $20,000, but the Hechts continued to give constructive criticism of national park safety standards. Statement prepared by J. H. Hast on *James L. Hecht, Admr., etc. v. United States,* Civil No. 344-71R, U.S.D.C., E.D. Virginia, typescript, January 14, 1974, Hartzog; George B. Hartzog, Jr., to James L. Hecht, October 21, 1970, Hartzog; George B. Hartzog, Jr., to Martin A. Cohen, January 6, 1971, Hartzog; Richard Halloran, "Boy's Death Spurs a Safety Campaign in National Parks," clipping from the *New York Times,* November 12, 1970, Hartzog; James L. Hecht to Ronald H. Walker, December 10, 1973, Walker; Richard W. Marks, interview with the author, November 14, 1989.

14. Director Hartzog shared President Johnson's deep dedication to the values of the Great Society, and he got to know the ex-President well when the Park Service was establishing the Lyndon B. Johnson National Historical Park in Texas. Both dynamic leaders, the two men developed a strong affinity for each other. Dining at the Lawyers Club in Washington in January 1973 when the announcement came of Johnson's sudden death, Hartzog, shaken by the news, recalled to his dinner guest his friendship with Johnson and their common sense of purpose. Among other

incidents, he told of his last trip to the Johnson Ranch, when he and the former President enjoyed a visit on the front porch of the Texas White House. At one point in their conversation, Johnson leaned over and slapped Hartzog on the knee, saying "George, I wish to hell I had known you when I was President because between the two of us we could have remade the fucking world." This story is related by Robert M. Utley, a former highly placed Park Service official and one of the dinner guests at the Lawyers Club, in Richard W. Sellars and Melody Webb, "An Interview with Robert M. Utley on the History of Historic Preservation in the National Park Service, 1947–1980," National Park Service, Southwest Cultural Resources Center, Professional Papers no. 16, 1988, typescript, 41. See also George B. Hartzog, Jr., *Battling for the National Parks* (Mt. Kisco, New York: Moyer Bell, 1988), 191–195.

15. Barry Mackintosh, *Interpretation in the National Park Service: A Historical Perspective* (Washington, D.C.: National Park Service, 1986), 67–71; Special Assistant to the Director, National Park Service, to Assistant to the Secretary, January 26, 1970, Hartzog. A discussion of the origins of "Summer in the Parks" is found in Hartzog, *Battling for the National Parks,* 141–142.

16. Mackintosh, *Interpretation in the National Park Service,* 52–53; Wirth, *Parks, Politics, and the People,* 348; William C. Everhart, *The National Park Service* (New York: Praeger Publishers, 1972), 50–51; Olsen, *Organizational Structures of the National Park Service,* 90–91. In late 1963 Wirth had opened another training facility—the Horace M. Albright Training Center, in Grand Canyon National Park—with a primary focus on training national park rangers (Wirth, p. 268).

17. Vernon L. Hammons, "A Brief Organizational History of the Office of Design and Construction, National Park Service, 1917–1962," ca. 1962, typescript, 6, NPS-HC; National Park Service, Denver Service Center, "Annual Report," 1981, 33, TIC; Denver Service Center, "On Board Count, 1981 to 1993;" National Park Service, "Composition of DSC Workforce," March 1983, Falb; John Luzader, "Some Chapters in the History of NPS Professions," manuscript, V-1, TIC; Denver Service Center, "Annual Report," 1990, 35 TIC.

18. G. Frank Williss, *"Do Things Right the First Time: The National Park Service and the Alaska National Interest Lands Conservation Act of 1980"* (Washington, D.C.: Government Printing Office, 1985), discusses the Park Service's deep involvement in the Alaska legislation and the early efforts to manage the new parks. See especially 69–156, 159–171, 237–296. See also Foresta, *America's National Parks,* 84–87, for a discussion of the decline of the Park Service's political strength, resulting from the Alaska effort and other developments of this period.

19. Joel V. Kussman, "National Park Service Wilderness Program," paper presented at the Society of American Foresters National Convention, October 1983, typescript, n.p., Dennis; National Park Service, Wilderness Task Force, "Report on Improving Wilderness Management in the National Park Service," September 3, 1994, typescript, 14–16, TIC; Michael McCloskey to George Hartzog, January 4, 1967, Hartzog; Director to Directors, Midwest, Northeast, Pacific Northwest, Southeast, Southwest and Western Regions, October 12, 1972, Hartzog; John W. Henneberger, interview with the author, June 17, 1989; "Park Wilderness in Danger," *National Parks Magazine* 38 (October 1964), 2. The commission had recommended six classifications for public-use lands: three for recreational lands; two for natural areas (specifically, "primitive areas" and "unique natural areas"); and one for historical and archeological sites.

Faced with probable passage of the Wilderness Act (and likely in an effort to ensure continued, unrestricted authority over national park backcountry), the Service by early 1963 had begun to identify potential wilderness areas. Nevertheless, revealing the long-standing ambivalence within the Service, an internal task force report prepared in 1994—thirty years after passage of the Wilderness Act—admonished the Service that it "should be proud" of wilderness areas in the parks and that it "should view the term 'wilderness' in a positive light." Director to All Regional Directors and Chief, Field Design Offices, January 11, 1963, Hartzog; National Park Service, Wilderness Task Force, "Report on Improving Wilderness Management in the National Park Service," 16.

20. The Great Smokies' transmountain road was intended as an alternative to a 1943 commit-

ment to build a highway along the shore of Fontana Lake. Forming part of the south boundary of the park, this Tennessee Valley Authority reservoir had inundated an existing road used by anglers and other locals. Neither the transmountain nor the lakeshore road proposal came to fruition. Luther J. Carter, "Wilderness Act: Great Smoky Plan Debated," *Science* 153 (July 1, 1966), 39–42; "Park Wilderness Planning: An Editorial," *National Park Magazine* 41 (February 1967), 2; National Park Service, Wilderness Task Force, "Report on Improving Wilderness Management in the National Park Service," 15; National Park Service, "General Management Plan, Cumberland Island National Seashore, Georgia," January 1984, typescript, 1–2, TIC. This analysis of the Service's response to the Wilderness Act benefited from a discussion with Jim Walters, National Park Service, Santa Fe, June 13, 1995.

The complex histories of wilderness designation in Fire Island National Seashore and Shenandoah National Park are discussed in Foresta, *America's National Parks,* 123–127; and Jacobsen, "The Management of Wilderness in Shenandoah National Park." Efforts in both parks were affected by the 1975 Eastern Wilderness Act, which eased the stipulations of size and primitive condition for proposals in the eastern United States. In Haleakala National Park, wilderness studies led to designation of the Kipahulu Valley as both a wilderness and a "scientific reserve," similar to the research reserves of the 1930s. Prized for its rare flora and fauna, having no trails, and closed to public access, it remains one of the most restricted natural areas in the park system. Henneberger, interview with the author, June 18, 1989; Donald Reeser, discussion with the author, June 21, 1995; National Park Service, "General Management Plan/Environmental Impact Statement, Haleakala National Park," January 1995, typescript, 15–16, Reeser.

21. Stanley Hulett, et al., "The National Park Service—A Prospectus for the Next Ten Years: A Special Report to the Director," February 2, 1973, typescript, Walker. Just two years before this prospectus was prepared, Horace Albright had written to Hartzog of his concern that the Service might take on too many responsibilities. As an old man who long before had pushed aggressively to build Park Service programs, Albright now feared that the Service was "already too big for one man and his few associates to plan for, direct and administer, do the job well, and keep morale high throughout the huge organization, without killing themselves." Unless it exercised caution, the Service would become a "size approaching that of a large industrial organization." Horace M. Albright to George B. Hartzog, Jr., January 15, 1971, Hartzog.

National park analyst Ronald Foresta viewed the situation during the era of expansion as one in which parks had become a "highly prized distributive good." In a time of public concern for outdoor recreation, members of Congress could gain favor with their constituencies by creating new parks. Foresta, *America's National Parks,* 78–79, 93. It took the administration of President Ronald Reagan to bring expansion of the national park system to a virtual halt. James M. Ridenour, Park Service director during the administration of President George Bush, believed that the system had come to include a number of unworthy parks. In 1993, studies for reduction of the system were proposed in a bill introduced in the House of Representatives. See James M. Ridenour, *The National Parks Compromised: Pork Barrel Politics and America's Treasures* (Merrillville, Indiana: ICS Books, 1994), 16–19; and H.R. 1508, A Bill to Provide for the Reformation of the National Park System, 103d Cong., 1st sess., March 29, 1993. Similar efforts continued through mid-decade. Dwight F. Rettie, *Our National Park System* (Urbana: University of Illinois Press, 1995), 163–170, discusses the funding and staffing problems generated by an expanding park system.

22. National Park Service, *Part Two of the National Park System Plan: Natural History* (Washington, D.C.: Department of the Interior, 1972), foreword, 1, 37.

23. A. Starker Leopold et al., "Wildlife Management in the National Parks," in *Transactions of the Twenty-eighth North American Wildlife and Natural Resources Conference,* ed. James B. Trerethen (Washington, D.C.: Wildlife Management Institute, 1963), 32, 34, 43; National Academy of Sciences, National Research Council, "A Report by the Advisory Committee to the National Park Service on Research," typescript, August 1, 1963, 1; Bruce M. Kilgore, "Above All . . . Naturalness: An Inspired Report on Parks," *Sierra Club Bulletin,* 48 (March 1963), 3. "Wildlife

Management in the National Parks" is the formal title of the Leopold Report. Because the National Academy Committee was chaired by William Robbins, the academy's report is sometimes referred to as the Robbins Report.

24. Conrad L. Wirth to the Secretary of the Interior, August 9, 1963, NPS-HC; A. Starker Leopold to Stewart Udall, March 4, 1963, NPS-HC. Similar views were expressed by former chief biologist Victor H. Cahalane, in Cahalane to A. Starker Leopold, March 26, 1963, Victor H. Cahalane Files, MVZ-UC.

25. Leopold et al., "Wildlife Management in the National Parks," 34–35.

26. National Academy of Sciences, "A Report by the Advisory Committee," 1.

27. National Academy of Sciences, "A Report by the Advisory Committee," x, xi, 31, 43.

28. National Academy of Sciences, "A Report by the Adivisory Committee," 3, x, 21, 58.

29. National Academy of Sciences, "A Report by the Advisory Committee," 44–48. At the time the report was issued, the assistant directors reported to the director, because no deputy or associate director positions existed. Olsen, *Organizational Structures of the National Park Service,* 83.

30. National Academy of Sciences, "A Report by the Advisory Committee," 53, 66–67, 71, 74.

31. Conrad L. Wirth to All Field Offices, July 26, 1963, Advisory Board on Wildlife and Game Management files, MVZ-UC.

32. Howard R. Stagner, interview with the author, April 15, 1989; Conrad L. Wirth to the Secretary of the Interior, September 16, 1963, NPS-HC. Park Service superintendent Robert Barbee recalled that the National Academy's report was "critical [of the Park Service] and forgotten." Barbee, interview with the author, July 24, 1989.

The National Academy Report was issued about two months before Conrad Wirth announced his retirement; to some this suggested that he was leaving in reaction to the criticism. Furthermore, surprisingly strong disapproval of the Service was expressed by Assistant Secretary of the Interior John Carver at the October 1963 superintendents conference in Yosemite (the conference at which Wirth advised the Park Service of his upcoming retirement in early 1964). The assistant secretary scolded the Service for its stubborn resistance to turning over programs to the newly created Bureau of Outdoor Recreation, and for acting too independently in matters of importance to Secretary Udall. In a hurtful remark, Carver attacked the Service for what he saw as its "mystic, quasi-religious" esprit de corps, which he compared to that of the Hitler Youth Movement (he was referring to the Service's "ranger mystique"). This affront to the Park Service and to Wirth stimulated further speculation that such disapproval had prompted the director's resignation. However, Wirth states in his autobiography that in early 1963, long before the criticisms surfaced, he had told Secretary Udall of his intention to resign at the end of the year. There appear to have been professional differences and differences of style between Wirth and Udall that may have brought Wirth to his decision. A holdover from the Truman-Eisenhower years, Wirth was not one of the "New Frontiersmen." John A. Carver, paper presented at the National Park Service Conference of Challenges, Yosemite National Park, October 13–19, 1963, typescript, NPS-HC; Wirth, *Parks, Politics, and the People,* 297–314.

33. National Academy of Sciences, "A Report by the Advisory Committee," x, 31; National Park Service, "Get the Facts, and Put Them to Work," October 1961, typescript, 4, NPS-HC; published in *George Wright Forum* 3 (Autumn 1983), 28–38. Howard Stagner's efforts to get prestigious scientific committees to review Park Service science programs were recalled by Lowell Sumner in a 1968 address to Service biologists. Lowell Sumner, "A History of the Office of Natural Science Studies," in "Proceedings of the Meeting of Research Scientists and Management Biologists of the National Park Service," Horace M. Albright Training Center, April 6–8, 1968, typescript, 4, Dennis. Starker Leopold was himself influenced by the reports of Stagner and the National Academy, and in his address to the October 1963 superintendents conference he reiterated their blunt criticism, at times quoting directly from them. A. Starker Leopold, "Wildlife Management in the Future," 11–12, address given at the National Park Service Conference of

Challenges, Yosemite National Park, October 1963, appended to Acting Director to All Field Offices, December 6, 1963, NPS-HC.

34. Wirth's comment is in Director, National Park Service, to Secretary of the Interior, August 9, 1963.

35. For the organizational chart effective in 1963, see Olsen, *Organizational Structures of the National Park Service,* 83.

36. Leopold et al., "Wildlife Management in the National Parks," 32, 34, 43.

37. For example, Assistant Director Howard W. Baker acknowledged to Starker Leopold in December 1965 that the Service was aware that resource management would present "complexities well beyond our present sophistication of ecological interactions." Baker added that it may take "many years, or perhaps decades" to arrive at an understanding of the parks' ecology and its relation to human activity. Howard W. Baker to A. Starker Leopold, December 29, 1965, Advisory Board on Wildlife and Game Management Files, MVZ-UC.

38. George B. Hartzog, Jr., to All Field Offices, March 29, 1965, NPS-HC.

39. Robert M. Linn, "The Natural Resources Committee—A Functional Concept," attached to Deputy Director to Natural Resources Committee Members, June 16, 1967, NPS-HC. In a somewhat similar way (and with somewhat similar results), U.S. Forest Service management contended with the question of independent internal scientific research. See Ashley L. Schiff, *Fire and Water: Scientific Heresy in the Forest Service* (Cambridge, Massachusetts: Harvard University Press, 1962), 169–181.

40. Director, National Park Service to Secretary of the Interior, September 16, 1963, with attachments, NPS-HC; National Academy of Sciences, "A Report by the Advisory Committee," 66; Sumner, "Biological Research and Management in the National Park Service: A History," *George Wright Forum* 3 (Autumn 1983), 20; Olsen, *Organizational Structures of the National Park Service,* 85.

41. George Sprugel, Howard Stagner, and Robert M. Linn, "National Parks as Natural Science Research Areas," *Trends in Parks and Recreation* 1 (July 1964), n.p.; Marietta Sumner et al., "Remembering Lowell Sumner," *George Wright Forum* 6, no. 4 (1990), 37; comments by Robert M. Linn, in "Proceedings of the Meeting of Research Scientists and Management Biologists of the National Park Service," April 6–8, 1968, 19; Garrett A. Smathers, "Historical Overview of Resources Management Planning in the National Park Service," 1975, typescript, 10–11, NPS-HC; and National Park Service, "Natural Resources Management Handbook," July 1968, part 1, chapter 2, 1; Roland H. Wauer, "Natural Resource Management—Trend or Fad?" *George Wright Forum* 4, no. 1 (1984), 27.

42. National Academy of Sciences, "A Report by the Advisory Committee," 66, 48–49.

43. Robert M. Linn, "The Science Program in the National Park Service," typescript, April 11, 1973, Dennis. George Sprugel, Jr., interview with the author, November 5, 1992. William Supernaugh also noted that many superintendents resented scientists being in their parks without being under their control. Supernaugh, interview with the author, November 4, 1993.

44. Sumner, "A History of the Office of Natural Science Studies," 3.

45. Olsen, interview with the author, February 26, 1990; Supernaugh, interview with the author, November 4, 1993; Roland H. Wauer, interview with the author, November 8, 1993.

46. Director to Secretary of the Interior, September 16, 1963, 5–6; Linn, "The Natural Resources Committee—A Functional Concept."

47. Acting Assistant Director, Administration, to Washington Office and All Field Offices, February 12, 1964, 3, Dennis. The Division of Resources Management and Visitor Protection received official sanction in the Service's revised organizational chart of December 1965. See Olsen, *Organizational Structures of the National Park Service,* 85.

48. Harthon L. Bill to Director, July 7, 1965, with attachment, Garrison; Supernaugh, interview with the author, November 4, 1993.

49. The quote is from Supernaugh, interview with the author, November 4, 1993.

50. Linn, "The Natural Resources Committee—A Functional Concept"; Linn, "The Science Program in the National Park Service"; Acting Assistant Director to S. Herbert Evison, September 13, 1966, NPS-HC (the acting assistant director's name is not indicated); Sumner, "A History of the Office of Natural Science Studies," 1.

51. Linn, "The Natural Resources Committee—A Functional Concept." Linn's statement had been written in late 1966.

52. Sprugel, Stagner, and Linn, "National Parks as Natural Science Research Areas," n.p.; Sumner, "Biological Research and Management," 21; Hartzog to All Field Offices, March 29, 1965. Hartzog took a pragmatic view of some research proposals, as evidenced by his reaction to Sprugel's request to study wild boars—a destructive, nonnative species in Great Smoky Mountains National Park, where Hartzog had once been stationed. When Sprugel told the director he needed to research the boars to know what to do with them, Hartzog roared: "Do with them? Do with them? I can tell you what to do with them. Shoot the goddamn beasts." This incident was witnessed by Robert M. Utley, who related it in correspondence with the author, January 17, 1994.

53. Sprugel, interview with the author, November 5, 1992.

54. Hartzog, *Battling for the National Parks,* 103. Linn, "The Science Program in the National Park Service," 8.

55. Linn, "The Science Program in the National Park Service," 8.

56. Sumner, "A History of the Office of Natural Science Studies," 2, 6.

57. Sprugel, interview with the author, November 5, 1992; Linn, "The Science Program in the National Park Service," 8, 9.

58. Sprugel, interview with the author, November 5, 1992; Smathers, "Historical Overview of Resources Management," 10. Sprugel's situation as chief scientist is further discussed in Alston Chase, *Playing God in Yellowstone: The Destruction of America's First National Park* (Boston: Atlantic Monthly Press, 1986), 247–248.

59. Harold Simons, "Science: Sense and Nonsense," *BioScience* (September 1966) 607–608. Sprugel, interview with the author, November 5, 1992.

60. Olsen, *Organizational Structures of the National Park Service,* 87. In his initial rejection of the job offer, Leopold wrote that it would be "impossible" to assume another large responsibility, and added that he was aware that "you can't accomplish everything in your lunch hour and I had better practice that bit of knowledge." A. Starker Leopold to George Hartzog, Jr., April 12, 1967, Hartzog.

61. Leopold, "Wildlife Management in the Future," 11–12.

62. Robert M. Linn to George Sprugel, Jr., May 16, 1967, Sprugel. Linn observed to Sprugel that the lack of representation was "one of the problems in the way the job was set up" when Sprugel was in office.

63. Clifford P. Hansen to Frank Dunkle, May 3, 1967, Hartzog; Résumé of Proposed National Park Service Natural Science Research Act, typescript, n.d., Hartzog.

64. Linn to Sprugel, May 16, 1967; George Sprugel, Jr., to Robert M. Linn, May 31, 1967, Sprugel; Harthon L. Bill to Frank H. Dunkle, June 15, 1967, Hartzog. Linn also met with Hartzog and Hansen on the science proposal. He later recalled that at the time he suspected Hansen's real concern was control of the Yellowstone elk management issue. Robert M. Linn, interview with the author, December 10, 1992.

65. A. Starker Leopold to George Hartzog, March 27, 1968, Leopold; A. Starker Leopold to George Hartzog, March 26, 1968, Leopold; Olsen, *Organizational Structures of the National Park Service,* 89.

66. Linn, "The Science Program in the National Park Service," 9. The regionalization is also discussed in Chief Scientist to NPS Scientists and Hydraulic Engineers, December 1, 1971, Dennis. When Hartzog regionalized the scientists, he made a similar rearrangement of the Service's history programs.

67. Robert M. Linn, correspondence with the author, October 25, 1992.

68. In reflecting on the scientists' dilemma, Lee Purkerson, veteran natural resource man-

ager at Redwood National Park, remarked that when Park Service scientists develop expertise in certain areas there are sometimes too many demands made on their time. They become what Purkerson called biopoliticians. Lee Purkerson, interview with the author, June 14, 1989. For parallel circumstances within the U.S. Forest Service, see Schiff, *Fire and Water,* 169–173. See also John G. Dennis, "Building a Science Program for the National Park System," *George Wright Forum* 4, no. 3 (1986), 17–18. Writing to Park Service Director Gary Everhardt in 1975, Purdue University biologist Durward L. Allen gave his opinion on the problems that occurred when scientists were not independent of superintendents. Allen commented that "in the field of natural sciences it is a mistake for management offices to be doing their own research. When done in this way, fact-finding tends to lose credibility with both scientists and the public." Durward L. Allen to Gary Everhardt, December 12, 1975, Linn. Park Service veteran Robert Utley, in correspondence with the author, January 17, 1994, recalled that those with high academic credentials who came into the Service during this era "did not harmonize with the old-line management, or even with Hartzog himself." He added that Hartzog developed a "growing disenchantment" with them and "an apprehension that they might gain more power than they should."

69. Linn, "The Natural Resources Committee—A Functional Concept."

70. Linn, interview with the author, October 26, 1992; Chief Scientist to Scientific Function Personnel, February 8, 1972, Dennis; Linn, correspondence with the author, October 25, 1992.

71. Linn, "The Science Program in the National Park Service," 9.

72. Ken Baker to Lowell Sumner, January 24, 1972, Leopold; Lowell Sumner to A. Starker Leopold, February 11, 1972, Leopold; A. Starker Leopold to Lowell Sumner, February 25, 1972, Leopold. Baker expressed similar concerns to the newly appointed regional chief scientist in San Francisco, in Ken Baker to Wally Wallis, April 6, 1972, Leopold. It is likely that Hawaii Volcanoes management was using at least part of the science money to address a resource management program of great urgency—the intensified (and ultimately successful) effort to eradicate exotic goats from the park.

73. Wauer, "Natural Resource Management—Trend or Fad?" 24–25.

74. Associate Director to All Regional Directors and Director, Denver Service Center, August 27, 1973, Dennis. The consequences of Hartzog's reorganization of the science programs are discussed in R. Gerald Wright, *Wildlife Research and Management in the National Parks* (Urbana: University of Illinois Press, 1992), 29–31. Citing an interview with longtime Park Service biologist Glen Cole, Wright states (p. 30) that the reorganization "diffused the power that the science program had accumulated over the previous decade . . . diminished the ability of scientists to react in a unified manner to service-wide problems . . . [and] created a staff of 'management scientists' who were required to work on those projects management considered to be important and not to work on those projects management did not want done or want to know anything about."

75. Submitted as a memorandum, the Leopold and Allen review was shorter and much less comprehensive than the 1963 Leopold Report. Durward L. Allen and A. Starker Leopold to the Director of the National Park Service, July 12, 1977, Dennis. The quote is from National Park Service, Briefing for Regional Director, Subject: Science Program Reorganization, to Be Given on January 16–17, 1978, Waggoner.

76. Resistance of the Park Service directorate to improving the bureaucratic status of science in accord with Assistant Secretary Herbst's wishes is discussed in Robert M. Linn, correspondence with the author, October 22, 1993, and by Wauer, interview with the author, November 8, 1993. See Olsen, *Organizational Structures of the National Park Service,* 99–101, for the organizational changes during this period.

77. Robert M. Linn to Gary Everhardt, November 29, 1975, Linn. See also Olsen, *Organizational Structures of the National Park Service,* 85–95.

78. Linn, interview with the author, December 10, 1992. Hartzog's resignation came in late 1972, after decisions he made at Biscayne National Monument (now a national park) were seen by President Nixon as an affront to his friend C. G. (Bebe) Rebozo, a frequent user of the park.

Apparently, the Nixon White House already wanted to remove Hartzog and put in one of its own people, such as Ronald Walker, who had been on Nixon's staff. See Hartzog, *Battling for the National Parks,* 233–247.

79. Richard H. Briceland, interview with the author, February 14, 1989. Ronald Foresta, in *America's National Parks,* 89–90, discusses the shifts in power during the post-Hartzog era.

80. Briceland, interview with the author, February 14, 1989. Yellowstone Superintendent Robert Barbee later recalled being ambivalent about this proposal, not unalterably opposed to it. He believed that its effect on park management would have been "as much symbolic as real." Barbee, interview with the author, July 24, 1989.

81. Jacob Hoogland, "The National Environmental Policy Act," in Michael A. Mantell, ed., *Managing National Park System Resources: A Handbook on Legal Duties, Opportunities, and Tools* (Washington, D.C.: Conservation Foundation, 1990), 41; John W. Henneberger, interview with the author, June 18, 1989.

82. William R. Supernaugh, "The Evolution of the Natural Resource Specialist: A National Park Phenomenon," paper prepared for the Department of Parks, Recreation and Environmental Education, Slippery Rock State University, August 25, 1987, typescript, 8–10, copy courtesy of the author; Supernaugh, interview with the author, March 10, 1989; Wauer, "Natural Resource Management—Trend or Fad?" 27; and Roland H. Wauer, "The Role of the National Park Service Natural Resources Manager," NPS Cooperative Park Studies Unit, College of Forest Resources, University of Washington, February 1980, typescript, 1–15.

83. Bruce M. Kilgore, "Views on Natural Science and Resource Management in the Western Region," keynote address at the NPS Pacific Northwest Region, Science/Resources Management Workshop, April 18–20, 1978, NPS Cooperative Park Studies Unit, College of Forest Resources, University of Washington, 1979, 7.

84. Separation had occurred in a number of parks by about the late 1970s, for example at Glen Canyon National Recreation Area, to which Supernaugh transferred in 1974 as a ranger with resource management duties. In a park with demanding law-enforcement needs, the resource management responsibilities grew rapidly during Supernaugh's tenure—a specialization that, he recalled, was "driven by the new environmental laws." In time these duties were separated from the ranger division, emerging as a new division that utilized persons trained in natural resource management.

Similarly, professionalization occurred at Channel Islands National Park, established in 1980, where natural resource management was initially "a collateral duty for island rangers." With an unusually clear mandate for scientific resource management, this park quickly converted a ranger position to resource management. Soon afterward, it hired a professional resource manager and established a separate resource management division. However, many other parks, such as Yellowstone, kept resource management under the control of the rangers, as did most small parks. Wauer, "Natural Resources Management, Trend or Fad?" 27; Supernaugh, interviews with the author, March 10, 1989, November 17, 1992, and November 4, 1993; Supernaugh, "Evolution of the Natural Resource Management Specialists," 8–10; Gary E. Davis, correspondence with the author, August 16, 1993; Bob Krumenaker, "Resource Management and Research in the NPS: An Uneasy Relationship," *Ranger* 7 (Spring 1991), 10–13; Olsen, *Organizational Structures of the National Park Service,* 97.

85. Memorandum of Understanding between University of Washington and National Park Service, United States Department of the Interior, April 14, 1970; Michael Soukup et al., "Cooperative Park Studies Units in the National Park Service: An Analysis by the Regional Chief Scientists," April 1988, typescript, 2, 7–8; Associate Director, Natural Resources, to Deputy Assistant Secretary, August 4, 1983, Dennis; Linn, "The Science Program in the National Park Service," 11; Napier Shelton and Marie Zack, "Scientific Research in the National Parks," August 1980, typescript, n.p., copy courtesy of Napier Shelton.

86. Officially established in 1973, the University of Massachusetts studies unit formalized an existing working relationship between the Park Service and biologist Paul J. Godfrey, who had

undertaken research for the Service at Cape Lookout National Seashore, near Cape Hatteras, before taking a position with the university's botany department. Director, National Park Service, to Assistant Secretary for Fish and Wildlife and Parks, July 26, 1973, CAHA. Robert D. Behn and Martha A. Clark, "The Termination of Beach Erosion Control at Cape Hatteras," *Public Policy* 27 (Winter 1979), 99–127, provide an overview of the history of the stabilization concerns. See also Paul J. Godfrey, "Management Guidelines for Parks on Barrier Beaches," *Parks* 2, no. 4 (1978), 5–10; Robert Dolan and Harry Lins, *The Outer Banks of North Carolina,* U.S. Geological Survey Professional Paper 1177-B (Washington, D.C.: Government Printing Office, 1986), 38; and National Park Service, *Management Policies* (Washington, D.C.: National Park Service, 1978), chapter 4, 22.

87. Napier Shelton, ed., "The National Park Service and Environmental Quality—An Overview," *Park Science* 6 (Summer 1986), 2, 3; Keith Yarborough, interview with the author, January 9, 1989; Samuel H. Kunkle, interview with the author, October 29, 1993.

88. National Academy of Sciences, "A Report by the Advisory Committee," 71; Director, National Park Service, to Secretary of the Interior, September 16, 1963. The use of the Jackson Hole Biological Research Station as a model is discussed in Natural Sciences Consulting Committee to Director, October 17, 1966, NPS-HC.

89. Gary Y. Hendrix, interview with the author, February 28, 1989; Susan Murphy, "Everglades Research Center Opens," *South Dade News Leader,* October 22, 1977, EVER.

90. Hendrix, interview with the author, February 28, 1989; Everglades National Park, "South Florida Research Center Base Funding History," typescript, 1989, EVER; Michael Soukup and Robert F. Doren, "Reorganization of the South Florida Research Center," *Park Science* 13 (Summer 1993), 1, 4. In the same issue of *Park Science* is an example of Nathaniel Reed's continuing concerns about science and resource management at Everglades: "Dare to Save the Everglades," p. 3.

91. Bratton, interview with the author, March 20–21, 1989. In a handwritten letter, Bratton described to veteran Great Smoky Mountains naturalist Arthur Stupka her concern about the "lack of management in the Park," adding that she hoped to "'advertise' some of the problems among the scientific community." Susan Bratton to Arthur Stupka, April 15, 1974, Bratton.

92. Bratton, interview with the author, March 20–21, 1989. As an example of the park's disinterest in natural history research during the period before Bratton and Evison arrived, Bratton stated that the park had "given away" a large assortment of birds, mammals, and salamanders that naturalist Arthur Stupka had collected in the park over many years. A brief history of the Uplands Laboratory is found in John McCrone et al., "Uplands Field Research Laboratory, Regional Review Team Package," June 21, 1982, typescript, GRSM.

93. National Park Service Science Center, "Annual Report," May 1976, appendix B: Captain, NPS Science Sub-Task Force to National Park Service Task Force Chairman, May 23, 1975, attachment, 12–13, TIC; National Park Service Science Center, "Annual Report," May 1975, 2–4, TIC; Henneberger, interview with the author, June 18, 1989.

94. National Park Service "Proposal to Accomplish Ecological and Environmental Management of Coastal Zones and Major Upland Parks Utilizing NASA's Mississippi Test Facility/Slidell Computer Center," September 12, 1972, draft, Dennis; National Park Service, "History of Science in NPS," typescript, 4, Waggoner; Gary S. Waggoner, interview with the author, August 10, 1993. Waggoner, who was stationed at the center, believed the office also suffered from the regional chief scientists' desire to maintain close control of research in their regions; thus they viewed the center as a competitor for a "limited pot of money to do research."

95. National Park Service Science Center, "Annual Report," May 1976, 2, and appendix B, 2, 3, 16. As Susan Bratton recalled it, the center often had to seek projects, such as a mimosa control proposal at Great Smoky Mountains—a project that was considered unessential and tended to undermine the center's credibility. Bratton, interview with the author, March 20–21, 1989. About the time the science center was terminated, the Park Service set up the Coastal Field Research Laboratory, using facilities near Bay Saint Louis. This effort was designed to coordinate research on the southeastern barrier islands and beaches. Stephen V. Shabica, "Southeast Regional Office

Coastal Research Laboratory, NSTL Station, Mississippi—Research Perspectives," 1979, October 16, 1978, CAHA.

96. Robert M. Linn to George Sprugel, Jr., April 7, 1967, Sprugel.

97. Lary M. Dilsaver and William C. Tweed, *Challenge of the Big Trees* (Three Rivers, California: Sequoia Natural History Association, 1990), 316; William R. Supernaugh, correspondence with the author, October 28, 1993; "Channel Islands National Park, 1992 Annual Report—Natural and Cultural Resources Management," copy courtesy of Gary E. Davis; Gary E. Davis, correspondence with the author, August 16, 1993.

98. Purkerson, interview with the author, June 14, 1989; National Park Service, "Redwood Renaissance," brochure, n.d. (ca. 1980s); James K. Agee, "Issues and Impacts of Redwood National Park Expansion," *Environmental Management* 4 (September 1980), 409–419. Douglas Warnock, Redwood superintendent during the 1980s, recalled being told by representatives of timber companies clear-cutting lands adjacent to the park that the redwoods were "just like tomatoes—when they're ripe you pick 'em." But the public paid a price for this. In addition to the money needed to restore the forest, the government paid the companies that clear-cut the lands hundreds of millions of dollars for the expansion acreage, much of it stripped and barren because of the cutting. Most of this acreage was upslope from the original national park lands, and, if not restored, would—through erosion—devastate the downslope, old-growth park redwoods and the streams. Moreover, additional federal funds had to be expended to restore the streams and their aquatic life, including commercially important salmon populations, which were seriously affected by the silting that followed the clear-cutting. Douglas G. Warnock, interview with the author, June 13, 1989; National Park Service, "The Redwood National Park Watershed Rehabilitation Program: A Progress Report and Plan for the Future," June 1984, typescript, 3–5, REDW; U.S. Secretary of the Interior, "Report to Congress in Compliance with Section 104(a), P.L. 95–250, on the Status of Implementation of the Redwood National Park Expansion Act of March 27, 1978," Sixth Annual Report, typescript, 3–5, REDW.

99. Linn, comments in "Proceedings of the Meeting of Research Scientists and Management Biologists of the National Park Service," April 6–8, 1968, 7; Assistant Director, Service Center Operations, to Director, September 1, 1972, Hartzog.

100. Assistant Director, Service Center Operations, to Director, September 1, 1972.

101. William P. Gregg, interview with the author, July 14, 1993.

102. Gregg, interview with the author, July 14, 1993; Henneberger, interview with the author, June 18, 1989.

103. Gregg, interview with the author, July 14, 1993; R. Gerald Wright, interview with the author, August 11, 1993, and October 28, 1993. To gain a broader understanding of resources, the Service hired from diverse disciplines, such as wildlife, geology, and aquatic science.

104. Wright, interview with the author, August 11, 1993. Wright worked at the service center from 1972 to 1975.

105. Wright, interview with the author, August 11, 1993.

106. National Park Service, *State of the Parks—1980, A Report to the Congress* (Washington, D.C.: National Park Service, 1980), 36; Supernaugh, interview with the author, November 4, 1993; National Academy of Sciences, "A Report by the Advisory Committee," x, xi, 31, 43.

107. Leopold et al., "Wildlife Management in the National Parks," 39–40, 42.

108. Leopold et al., "Wildlife Management in the National Parks," 34, 35–37.

109. Leopold et al., "Wildlife Management in the National Parks," 29, 32, 43. The committee may have been influenced by Fauna No. 1's goal of maintaining a stable scene—that there was "one point in time which satisfies wild-life survey requirements as regards a particular [park]. This is the period between the arrival of the first whites and the entrenchment of civilization in that vicinity." George M. Wright, Joseph S. Dixon, and Ben H. Thompson, *Fauna of the National Parks of the United States: A Preliminary Survey of Faunal Relations in National Parks*, Contributions of Wild Life Survey, Fauna Series no. 1 (Washington, D.C.: Government Printing Office, 1933), 10.

110. A. Starker Leopold, "Wildlife Management in the Future," address presented at the Yo-

semite National Park Conference of Challenges, October 1963, typescript, appended to George B. Hartzog, Jr., to All Field Offices, December 6, 1963, 4–5, NPS-HC.

111. Udall's May 1963 declaration is cited in the management policy book, *Compilation of the Administrative Policies for the National Parks and National Monuments of Scientific Significance (Natural Area Category)* (Washington, D.C.: Government Printing Office, 1970), 23. (The administrative policies of the Service were commonly referred to as management policies.) See also Hartzog, *Battling for the National Parks*, 102; Mackintosh, *Shaping the System*, 63–64; National Park Service, "National Park Wilderness Planning Procedures," August 8, 1966, typescript, 3, NPS-HC; and National Park Service, Wilderness Task Force, "Report on Improving Wilderness Management in the National Park Service," 15–16. The National Park Service's Denver Service Center library houses an extensive collection of the handbooks.

112. National Park Service, *Administrative Policies* (1970), 77. For the full text of the Lane Letter and Leopold Report, see 68–71 and 97–112.

113. Stagner, interview with the author, April 15, 1989.

114. See Michael A. Mantell, ed., *Managing National Park System Resources: A Handbook on Legal Duties, Opportunities, and Tools* (Washington, D.C.: Conservation Foundation, 1990), 12–15.

115. National Park Service, *Administrative Policies* (1970), 17.

116. National Park Service, *Administrative Policies* (1970), 22–26; Hartzog's September 1967 memorandum is on pp. 113–116.

117. Hartzog's announcement was dated before the meeting with Udall and McGee took place, and its release may have been delayed until after the meeting—perhaps an indication that the Service had already decided to change its elk policy and was gathering support from the two officials. George B. Hartzog, "Management Program, Northern Yellowstone Elk Herd, Yellowstone National Park," March 1, 1967; attached to National Park Service news release, "National Park Service Director Hartzog Initiates Elk Management Program for Yellowstone National Park," March 1, 1967, Hartzog; *Hearings before a Subcommittee of the Committee on Appropriations, United States Senate, Ninetieth Congress, First Session, on Control of Elk Populations, Yellowstone National Park* (Washington, D.C.: Government Printing Office, 1967), 1, 6–7, 33–34, 89–90; Hartzog, *Battling for the National Parks*, 104, 252–253; Wright, *Wildlife Research and Management*, 42–44, 78–79, 89; Karl Hess, Jr., *Rocky Times in Rocky Mountain National Park: An Unnatural History* (Niwot, Colorado: University Press of Colorado, 1993), 22–23. See also Michael B. Coughenour and Francis J. Singer, "The Concept of Overgrazing and Its Application to Yellowstone's Northern Range," 1989, typescript, 8, YELL.

118. *Hearings before a Subcommittee of the Committee on Appropriations, on Control of Elk Populations*, 89–90; National Park Service, "Management Objectives for Northern Yellowstone Elk," September 19, 1967, typescript, Hartzog; National Park Service, "Natural Control of Elk," December 5, 1967, typescript, Hartzog. A paper prepared in 1971 by Yellowstone biologist Glen Cole declared predators to be "non-essential" to natural regulation. Glen F. Cole, "An Ecological Rationale for the Natural or Artificial Regulation of Native Ungulates in Parks," draft paper prepared for the Thirty-sixth North American Wildlife and Natural Resources Conference, Portland, Oregon, March 7–10, 1971, Hartzog. As recently as 1995, a park document acknowledged that termination of the elk reduction was "due largely to public controversy." Yellowstone National Park, Resource Management Plan, approved 1995, typescript, PS page 0001, YELL.

119. Wright, Thompson, and Dixon, *Fauna of the National Parks* (1933), 147. William Barmore to Glen Cole, cited in Charles Edward Kay, "Yellowstone's Northern Elk Herd: A Critical Evaluation of the 'Natural Regulation' Paradigm," Ph.D. diss., Utah State University, 1990, 8–9. William J. Barmore, "Conflicts in Recreation—Elk versus Aspen in Yellowstone National Park," paper presented at the Twentieth Annual Meeting of the American Society of Range Management, Seattle, February 13–17, 1967, typescript, 1, YELL. Kay (pp. 7–10) repeatedly charges that the park acted with little or no scientific information. Indicative of this, no significant northern range research was cited by Cole in his 1971 paper, "An Ecological Rationale for the Natural or

Artificial Regulation of Native Ungulates in Parks," 15–18. Much later, in 1980, Barmore would modify his position, reducing the emphasis on the effects of grazing and attributing changes in aspen to a variety of causes. William Barmore, "Population Characteristics, Distribution and Habitat Relationships of Six Ungulates in Northern Yellowstone Park," final report, 1980, n.p., YELL.

120. Leopold et al., "Wildlife Management in the National Parks," 38–41, 43. Leopold stated to the Senate committee that "we recommended that direct control continue. And I have not changed my mind on this." *Hearings . . . on Control of Elk Populations, Yellowstone National Park,* 20; National Park Service, "National Park Service Director Hartzog Initiates Elk Management Program for Yellowstone National Park." See also Jack K. Anderson to Horace M. Albright, December 22, 1970, Hartzog. Hartzog states flatly in his autobiography that direct reduction was "consistent with the Leopold Report." Hartzog, *Battling for the National Parks,* 104.

121. A. Starker Leopold, interview by Carol Holleuffer, June 14, 1983, Sierra Club Oral History Project, typescript, 19; A. Starker Leopold to Boyd Evison, June 9, 1983, Leopold-FRM.

122. Paul Schullery, *The Bears of Yellowstone* (1986; 3rd ed., Worland, Wyoming: High Plains Publishing Company, 1992), 109.

123. John J. Craighead and Frank C. Craighead, Jr., "Management of Bears in Yellowstone National Park," July 1967, typescript, YELL; Schullery, *Bears of Yellowstone,* 118–120; Chase, *Playing God in Yellowstone,* 149–152.

124. Frank C. Craighead, Jr., *Track of the Grizzly* (San Francisco: Sierra Club Books, 1979), 194–195; Schullery, *Bears of Yellowstone,* 113, 120. The Service's concern that a gradual closing of the dumps should not result in an incident such as those that occurred at Glacier is expressed in Deputy Chief Scientist Robert M. Linn to Associate Director Joseph P. Linduska, Bureau of Sport Fisheries and Wildlife, May 21, 1968, Leopold.

125. John Craighead to Jack K. Anderson, July 24, 1970, YELL; Craighead and Craighead, "Management of Bears in Yellowstone National Park," iii–iv. Later, Frank Craighead, *Track of the Grizzly* (p. 194), wrote that the project was not "initiated or significantly funded" by the Park Service.

126. In the late 1970s, John Craighead wrote to Yellowstone historian Paul Schullery, charging that at the time of the controversy "very little" of the park's grizzly bear management "could be backed by scientific evidence" and that the park's position on bear management was "taken without the benefit of *any* field research. I emphasize the *any,* since the decisions that were made and later had to be defended were not based on *any* NPS research." Schullery later wrote that certainly "nobody in the Park Service knew [the key grizzly bear data]." He added that "in fact there was no one in the world with [the Craigheads'] scientific credentials as grizzly bear authorities." As a former park archivist at Yellowstone, Schullery also commented that the park had done a "lousy job" of keeping records on bears. In 1974 a National Academy of Sciences review panel stated that the park's research program was still "inadequate to provide the data essential for devising sound management policies for the grizzly bears of the Yellowstone ecosystem." John Craighead to Paul Schullery, April 6, 1978, Leopold-FRM; Schullery, *Bears of Yellowstone,* 124, 120–121. The academy's quote is in Schullery (p. 144).

127. Craighead, *Track of the Grizzly,* 195 (see also 196–199); Bill Gilbert, "The Great Grizzly Controversy," *Audubon* 78 (January 1976), 69. Park Service objection to the Craigheads' use of the media is expressed in, for example, Jack K. Anderson to Dr. John J. Craighead, February 9, 1971, Leopold.

128. Jack K. Anderson to John J. Craighead, April 7, 1969, YELL.

129. Jack K. Anderson to John J. Craighead, July 9, 1970, YELL; Anderson to John Craighead, April 7, 1969; National Academy of Sciences, "A Report by the Advisory Committee," 62; Natural Sciences Advisory Committee, draft report, attached to A. Starker Leopold to George Hartzog, October 6, 1969, Leopold. The committee did not take a position on the how fast the dumps should be closed. Schullery comments in *The Bears of Yellowstone* (p. 135) that termination of the Craigheads' research at this point was a "great misfortune." The research would not

only have been valuable to the park, but would have been "priceless for managers of other areas." See also Craighead, *Track of the Grizzly,* 201–202.

130. Nathaniel P. Reed to Robert Barbee, February 14, 1983, Leopold—FRM. For a conspiratorial view of the controversy, see Chase, *Playing God in Yellowstone,* 142–194. Chase indicates possible Park Service malfeasance in attempting to justify its grizzly bear management actions (see for example pp. 155–160). He traces the continuing controversy into the 1980s (pp. 157–194). See also Frederic H. Wagner et al., *Wildlife Policies in the U.S. National Parks* (Washington, D.C.: Island Press, 1995), 102–103; and Craighead, *Track of the Grizzly* (for example, pp. 196–197), which angrily accuses the Park Service of creating difficulties and obstructions for the Craigheads. Schullery's *Bears of Yellowstone,* 114–146, is more supportive of the Park Service and at times critical of the Craigheads.

Stressing the point of Park Service responsibility (which had long been explicit in legislation), Glen Cole told a 1970 meeting of Park Service scientists and resource managers that the "situation has reached a stage where the Park Service must assert that it is responsible both for the preservation of the grizzly and for the protection of visitors within Yellowstone National Park. This responsibility cannot be transferred to other individuals or agencies. While basic studies of park fauna are encouraged, these cannot take precedence over operational management which is necessary for human safety and, ultimately, for the preservation of the grizzly population itself." Glen F. Cole, "Grizzly Bear Management in Yellowstone National Park," in National Park Service, "Proceedings of the Meeting of Research Scientists and Resource Managers of the National Park Service," April 18–20, 1970, typescript, 34, NPS-HC.

131. Chase, *Playing God in Yellowstone,* 148, 155; Wright, *Wildlife Research and Management,* 114–115; Craighead, *Track of the Grizzly,* 226; Schullery, *Bears of Yellowstone,* 118, 135–136, 143–144. On the safety issue, biologist Cole had reported to Anderson in August 1969 that, statistically, "the grizzly bear represents a low risk to Yellowstone Park visitors." Supervisory Research Biologist to Superintendent, August 18, 1969, Leopold. Yet the initial bear mortality count after closure was unusually high because the Service was making an extra effort to protect park visitors at a time when, with the dumps closed, the bears were believed to be more likely to seek other sources of human foods, as in campgrounds.

132. Yellowstone National Park, "Interagency Grizzly Bear Study Team," September 1979, typescript, Wauer; Interagency Grizzly Bear Study Team, Briefing Statement, July 9, 1981, typescript, YELL; Schullery, *Bears of Yellowstone,* 142–161; Wright, *Wildlife Research and Management,* 114–118.

133. Edward C. Stone, in "Preserving Vegetation in Parks and Wilderness," *Science* 150 (December 3, 1965), 1261–67, discusses the Service's failure to develop an understanding of plant ecology.

134. Leopold et al., "Wildlife Management in the National Parks," 37. Following Secretary Udall's request for a report, Lowell Sumner had written to his friend Starker Leopold to express alarm that insect control was being "expanded and extended." Sumner urged the committee to "present an opinion on this subject." E. Lowell Sumner to A. Starker Leopold, May 16, 1962, Advisory Board on Wildlife and Game Management files, MVZ-UC.

135. Leopold et al., "Wildlife Management in the National Parks," 33. Like the Service's biologists, Starker Leopold had long believed that fire policies needed revision. As early as 1952 he had predicted that the Service would "eventually come around to controlled burning," as biologist Bruce Kilgore later recalled it. In the mid-1950s Leopold told a wilderness conference that fire was a "dominant molding element" for national park flora. In Kilgore's recollection, Leopold had stated that he was "convinced that ground fires some day will be reinstated in the regimen of natural factors permitted to maintain the parks in something resembling a virgin state and that conditions in the parks would force this issue sooner or later." Recalled in Bruce M. Kilgore to A. Starker Leopold, November 29, 1957, Bruce M. Kilgore files, MVZ-UC.

136. Director, National Park Service, to Secretary of the Interior, August 9, 1963, attachment, 8–9.

137. Rachel Carson, *Silent Spring* (Boston: Houghton Mifflin, 1962). Murie stated that the spraying program in Grand Teton had affected 318,604 trees, with a total of 328,809 gallons of fuel oil mixed with 101, 071 pounds of the chemical ethylene dibromide. The 1966 program would, he predicted, treat more than seventy thousand lodgepole pines. Adolph Murie, "Pesticide Program in Grand Teton National Park, *National Parks Magazine* 40 (June 1966), 17–19.

138. Stanley A. Cain to Paul M. Tilden, July 21, 1966, A. Murie; Pete Hayden, discussion with the author, April 22, 1993; National Park Service, *Administrative Policies* (1970), 20–21; National Park Service, "Pest Control, National Park System," May 1977, 1–4, TIC.

139. Director, National Park Service, to Secretary of the Interior, August 9, 1963, attachment, 3; Bruce M. Kilgore, "Research Needed for an Action Program of Restoring Fire to Giant Sequoias," in "The Role of Fire in the Intermountain West," Intermountain Fire Research Council Symposium, 1970, typescript, 177–178. Cook's opposition to releasing a study in Sequoia by Richard J. Hartesveldt was recalled by former chief scientist Robert M. Linn in a discussion with the author, March 22, 1993. Cook had once been the park forester in Sequoia. Dilsaver and Tweed, *Challenge of the Big Trees,* 170.

140. William B. Robertson, Jr., "A Survey of the Effects of Fire in Everglades National Park," February 15, 1953, typescript, 7, 11, EVER. See also Kilgore, "Restoring Fire," 17.

141. Kilgore, "Research Needed for an Action Program," 175–176; Robertson, "A Survey of the Effects of Fire in Everglades National Park," 14; David M. Graber, "Coevolution of National Park Service Fire Policy and the Role of National Parks," in *Proceedings—Symposium and Workshop on Wilderness Fire,* Missoula, Montana, November 15–18, 1983 (Ogden, Utah: Intermountain Forest and Range Experiment Station, 1985), 346–347. On Indians and fire, Graber, for instance, writing to Starker Leopold (his former professor), stated that in Sequoia the "limited evidence at our disposal suggests that Indians were a major factor in the 'natural' fire cycle." David M. Graber to A. Starker Leopold, April 5, 1983, Leopold-FRM. See also the chapter on fire and the American Indian in Stephen J. Pyne, *Fire in America: A Cultural History of Wildland and Rural Fire* (Princeton: Princeton University Press, 1982), 71–83.

142. Kilgore, "Research Needed for an Action Program," 174, 177–178; Dilsaver and Tweed, *Challenge of the Big Trees,* 263–265. In 1975 Kilgore stated that there were "still individual and organizational hangups that sometimes get in the way of trying new concepts in forest management." These hangups were, he believed, "gradually being overcome." Bruce M. Kilgore, "Integrated Fire Management on National Parks," in "Proceedings of the 1975 National Convention of the Society of American Foresters," typescript, 2, Hartzog. Additional comments on reluctance to accept changes in fire policies are found in Bruce Kilgore, "From Fire Control to Fire Management: An Ecological Basis for Policies," in Kenneth Sabol, ed., *Transactions: Forty-first North American Wildlife and Natural Resources Conference* (Washington, D.C.: Wildlife Management Institute, 1976), 478. See also Kilgore, "Restoring Fire to National Park Wilderness," 17; and Stephen J. Botti and Tom Nichols, "The Yosemite and Sequoia–Kings Canyon Prescribed Natural Fire Programs, 1968–1978," n.d., typescript, 4, YOSE.

143. National Park Service, *Administrative Policies* (1970), 17.

144. Kilgore, "Research Needed for an Action Program," 173. Many critics believed that in certain situations prescribed burning would not be effective; others such as former director Horace Albright simply opposed altogether the change from traditional fire suppression policies. In July 1972 Albright had written to Director Hartzog to express opposition to proposed new fire policies "before something terrible happens." He warned that Yellowstone superintendent Jack Anderson would need "only one 'experiment' to burn up Yellowstone Park." To the park's claim that fire has a natural ecological role, Albright replied, "I don't think so at all," and stated flatly that Yellowstone "was not created to preserve an 'ecosystem.'" Horace M. Albright to George B. Hartzog, July 12, 1972, Hartzog; Yellowstone National Park, "Information Paper No. 16," March 1972, YELL. See also Horace Marden Albright, speech given at the Eleventh Cosmos Club Award, April 15, 1974, 14–15, Hartzog; and Horace Albright, "Former Directors Speak Out," *American Forests* 82 (June 1976), 51.

145. Kilgore, "Integrated Fire Management," n.p.; Kilgore, "From Fire Control to Fire Management," 483; Pyne, *Fire in America,* 303. See Schiff, *Fire and Water,* 51–115, on the U.S. Forest Service management's longtime resistance to controlled burning.

146. Department of the Interior, news release, December 12, 1974, Garrison; Department of the Interior, news release, February 18, 1976, Garrison; John F. Chapman, "The Teton Wilderness Fire Plan," *Western Wildlands* (Summer 1977), 15; National Park Service, *Management Policies* (1978), chapter 4, 13.

147. National Park Service, *Administrative Policies* (1970), 56; National Park Service, "Actions Recommended in the Leopold Report to Advance the Ability of the Public to View Wildlife," July 3, 1967, 3–5, Dennis. An early 1980s report stated that most parks had exotic species. John G. Dennis, "National Park Service Research on Exotic Species and the Policy behind That Research: An Introduction to the Special Session on Exotic Species," 1980, prepared for "Proceedings of the Second Conference on Scientific Research in National Parks," November 1979, San Francisco, typescript, 241–243, 245, Dennis; Wright, *Wildlife Research and Management,* 92; Wright, Dixon, and Thompson, *Fauna of the National Parks* (1933), 148. The Leopold Report's brief mention of exotics in national parks is found in Leopold et al., "Wildlife Management in the National Parks," 32, 34.

148. Wright, *Wildlife Research and Management,* 95–101; Milford R. Fletcher, discussion with the author, August 22, 1995; National Park Service, "European Wild Hogs in Great Smoky Mountains National Park," May 1, 1985, typescript, 32, 36–41, GRSM.

149. James K. Baker and Donald W. Reeser, *Goat Management Problems in Hawaii Volcanoes National Park: A History Analysis and Management Plan* (Washington, D.C.: National Park Service, 1972), 2–5.

150. Donald W. Reeser, "Establishment of the Resources Management Division, Hawaii Volcanoes National Park," paper presented at the George Wright Society meeting, Jacksonville, Florida, November 1992, typescript, n.p., Reeser.

151. Reeser, "Establishment of the Resources Management Division," n.p.; George B. Hartzog, Jr., to Earl Pacheco, October 20, 1970; Robert L. Barrell to Michio Takata, November 23, 1970, Reeser; National Park Service, *Administrative Policies* (1970), 56. In August 1963, responding to the Leopold Report, the Park Service had specifically acknowledged the "most extreme examples of severe ecological dislocation" caused by exotics in Hawaiian parks. Director, National Park Service, to Secretary of the Interior, August 9, 1963.

152. Anthony Wayne Smith to George B. Hartzog, Jr., May 12, 1971. The Hawaii Botanical Society denounced Hartzog's decision merely to control rather than to eradicate the goats, arguing that the Park Service's true mandate was "protection of native plant and animal species under pristine conditions." Hawaii Botanical Society, "Resolution Regarding Goats in National Parks in Hawaii," April 5, 1971, typescript, Reeser.

153. George B. Hartzog, Jr., to Anthony Wayne Smith, June 17, 1971.

154. A view similar to that of Balaz is expressed in Edward A. Hummel to Hon. Patsy T. Mink, June 22, 1971, Reeser. Ken Baker to Bob Linn, July 2, 1971, Reeser; Gene J. Balaz to William L. Canine, May 4, 1971, Reeser; Reeser, "Establishment of the Resource Management Division," n.p.

155. Reeser, "Establishment of the Resources Management Division." Through hunting, trapping, and fencing, rangers at Hawaii Volcanoes also waged war on feral pigs. Following the lead of Hawaii Volcanoes, Haleakala National Park resolved its exotic goat problem by fencing and killing, and was rid of most goats by the end of the 1980s. Wright, *Wildlife Research and Management,* 105–107; Russell W. Cahill to Ruth Gay, March 30, 1972, Reeser.

156. Douglas B. Houston, Edward G. Schreiner, Bruce Moorhead, *Mountain Goats in Olympic National Park: Biology and Management of an Introduced Species* (Denver: National Park Service, 1994), 10–12, 190–197; Wagner et al., *Wildlife Policies,* 104–105; Rolf O. Peterson and Robert J. Krumenaker, "Wolves Approach Extinction on Isle Royale: A Biological and Policy Conundrum," *George Wright Forum* 6, no. 1 (1989), 14–15; Stephen Nash, "The Wolves of Isle

Royale," *National Parks* 63 (January/February 1989), 21–26, 42; Wright, *Wildlife Research and Management,* 101–105, 142–143.

157. "Science and the NPS Work Session," attached to Boyd Evison to Participants, April 11, 1978, typescript, DEVA.

158. National Parks and Conservation Association, "NPCA Adjacent Lands Survey: No Park Is an Island," *National Parks and Conservation Magazine* 53 (March 1979), 4–9, and (April 1979), 4–7. The National Parks and Conservation Association was formerly the National Parks Association.

159. The Redwood National Park Expansion Act has been viewed as somewhat ambiguous in its declaration on protecting park resources. See United States General Accounting Office, *Parks and Recreation: Limited Progress Made in Documenting and Mitigating Threats to the Parks* (Washington, D.C.: General Accounting Office, 1987), 51–57; Robert B. Keiter, "National Park Protection: Putting the Organic Act to Work," in David J. Simon, ed., *Our Common Lands: Defending the National Parks* (Washington, D.C.: Island Press, 1988), 76–78; and Mantell, *Managing National Park System Resources,* 14–15. The Park Service itself, in National Park Service, *National Parks for the Twenty-first Century: The Vail Agenda* (Washington, D.C.: National Park Service [1993]), 126, stated that the act "*appears* to authorize the Service to take reasonable measures to protect park resources from degradation" (emphasis added), noting also that "the Service has been reluctant to use this authority, and courts have not vigorously enforced this provision." For an extended discussion of the "Emergence of External Threats as a Policy and Management Issue," see John C. Freemuth, *Islands under Siege: National Parks and the Politics of External Threats* (Lawrence: University of Kansas Press, 1991), 9–36.

160. Wauer, interview with the author, November 8, 1993; and National Park Service, *State of the Parks—1980.* An account of the evolution of State of the Parks is found in Roland H. Wauer, "The Greening of Natural Resources Management," *Trends* 19, no. 1 (1982), 2–6; Wauer, "Natural Resources Management—Trend or Fad?" 24–28; William R. Supernaugh, "Threats to Parks: Five Years Later," paper prepared at Slippery Rock University, December 6, 1985, typescript, 1–4, copy courtesy of the author; William R. Supernaugh, "An Assessment of Progress Made between 1980 and 1992 in Responding to Threats to the National Park System," thesis, Slippery Rock University, January 1994, 1–4, copy courtesy of the author.

The congressmen emphasized that they did "*not* have in mind" traditional park administrative and facility concerns, such as "personnel and equipment; local concessions, operational problems, [or] maintenance inadequacies." Rather, they were expressly interested in natural resource problems such as "air or water pollution, encroaching development, troublesome visitor use pressures, . . . adverse adjacent resource uses, exotic plant and/or animal intrusion, . . . rights to exercise incompatible uses within the park[s], and the like." Phillip Burton and Keith G. Sebelius to William Whalen, July 10, 1979, NPS-HC.

161. *State of the Parks—1980,* ix, 3–5.

162. *State of the Parks—1980,* viii, 34–36. The inadequacy of documentation on the parks' resources and their threats is discussed on 5–7.

163. Wauer, interview with the author, November 8, 1993.

164. *State of the Parks—1980,* ix–x; Wauer, interview with the author, November 8, 1993.

165. National Park Service, *State of the Parks: A Report to the Congress on a Servicewide Strategy for Prevention and Mitigation of Natural and Cultural Resources Management Problems* (Washington, D.C.: National Park Service, 1981), 2–13. See also Supernaugh, "An Assessment of Progress Made between 1980 and 1992," 16, 25–27, 34–35, 69–70, 75–78; and Wauer, "The Greening of Natural Resource Management, 2–7. During the 1970s, while head of natural resource management activities in the Service's southwest region, Wauer had developed a prototype natural resource management training program, first at Bandelier National Monument, then at other parks in the region. At the same time he had urged completion of the region's resource management plans. Later, seeking Servicewide implementation, he put these two elements into

the State of the Parks reports. Wauer, interview with the author, November 8, 1993; and the author's personal recollections.

166. National Park Service, *State of the Parks: A Report to Congress,* 4, 9, 35; National Park Service, "Summary of 'State of the Parks' in the Southwest Region," January 25, 1980, author's files.

Chapter 7. A House Divided

1. National Park Service, draft, *Our National Parks: Challenges and Strategies for the Twenty-first Century* (Washington, D.C.: National Park Service, 1991), 112–113.

2. F. Fraser Darling and Noel D. Eichhorn, *Man and Nature in the National Parks: Reflections on Policy* (1967; 2nd ed., Washington, D.C.: Conservation Foundation, 1969); Conservation Foundation, *National Parks for the Future: Task Force Reports* (Washington, D.C.: Conservation Foundation, 1972); Durward L. Allen and A. Starker Leopold, to the Director of the National Park Service, July 12, 1977, Dennis; National Park Service, *State of the Parks—1980, A Report to the Congress* (Washington, D.C.: National Park Service, 1980); National Park Service, *State of the Parks: A Report to the Congress on a Servicewide Strategy for Prevention and Mitigation of Natural and Cultural Resources Management Problems* (Washington, D.C.: National Park Service, January 1981). The National Parks and Conservation Association reports include *Investing in Park Futures: A Blueprint for Tomorrow,* vol. 2, *Research in the Parks: An Assessment of Needs* (Washington, D.C.: National Parks and Conservation Association, 1988); *National Parks: From Vignettes to a Global View* (Washington, D.C.: National Parks and Conservation Association 1989); and *A Race against Time: Five Threats Endangering America's National Parks and the Solutions to Avert Them* (Washington, D.C.: National Parks and Conservation Association, 1991).

3. National Park Service, draft, *Our National Parks,* 105; National Academy of Sciences, National Research Council, *Science and the National Parks* (Washington, D.C.: National Academy Press, 1992), 10–11 (see p. 42 for a partial listing of reports since the Leopold Report); National Park Service, *National Parks for the Twenty-first Century: The Vail Agenda* (Washington, D.C.: National Park Service, [1993]), 128 (hereafter referred to as the Vail Agenda). A more recent analytical study was conducted by a committee headed by Utah State University ecologist Frederic H. Wagner. See Frederic H. Wagner et al., *Wildlife Policies in the U.S. National Parks* (Washington, D.C.: Island Press, 1995).

4. George M. Wright, Joseph S. Dixon, and Ben H. Thompson, *Fauna of the National Parks of the United States: A Preliminary Survey of Faunal Relations in National Parks,* Contributions of Wild Life Survey, Fauna Series no. 1 (Washington, D.C.: Government Printing Office, 1933), 148; National Park Service, "Research in the National Park System, and Its Relation to Private Research and the Work of Research Foundations," February 10, 1945, typescript, 4, NPS-HC; National Park Service, "Get the Facts, and Put Them to Work," October 1961, typescript, 1–2, NPS-HC. Director, National Park Service, to Secretary of the Interior, September 16, 1963, attachment, 4, NPS-HC; Assistant Director, Operations, to All Field Offices, October 14, 1965, NPS-HC.

5. National Park Service, *State of the Parks—1980,* viii–ix, 5–7, 34–35; R. Gerald Wright, *Wildlife Research and Management in the National Parks* (Urbana: University of Illinois Press, 1992), 184–185; Shenandoah National Park, "Natural Resource Inventory and Long-Term Ecological Monitoring System Plan for Shenandoah National Park," August 1991, 6–25, copy courtesy of Robert J. Krumenaker; Gary E. Davis, "Design of a Long-Term Ecological Monitoring Program for Channel Islands National Park, California," *Natural Areas Journal* 9 (2) (1989), 80–82; Channel Islands National Park, "1992 Annual Report—Natural and Cultural Resources Management," copy courtesy of Gary E. Davis; Robert Cahn, "Inventory and Monitoring in the National Park System," 1988, draft, NPS-HC; National Parks and Conservation Association, *From Vignettes to a Global View,* i, 5–7, 10; National Park Service, draft, *Our National Parks,* 48; National Park Service, Vail Agenda, 129, 131–132. See also National Park Service, *Twelve-Point Plan: The Challenge, The Actions* (Denver: National Park Service, 1986), 2–3.

6. William R. Supernaugh, "An Assessment of Progress Made between 1980 and 1992 in Responding to Threats to the National Park System," thesis, Slippery Rock University, January 1994, copy courtesy of the author, 158–160, 163, 165–166. See also United States General Accounting Office, *Parks and Recreation: Limited Progress Made in Documenting and Mitigating Threats to the Parks* (Washington, D.C.: General Accounting Office, 1987), 12, 24–25, 36.

7. Supernaugh, "An Assessment of Progress Made between 1980 and 1992," 75–89, 96–110. William H. Walker, Jr., interview with the author, October 17, 1995; Roland H. Wauer, interview with the author, November 8, 1993. See also William H. Walker, Jr., "The Natural Resource Specialist Trainee Program," *Trends* 23, no. 2 (1986), 39–42; and United States General Accounting Office, *Parks and Recreation,* 32.

8. Abigail Miller, correspondence with the author, December 12, 1995, provided the principal data for the budget and staffing comparisons and helped sort out much contradictory information. The difficulties in tracking funding and staffing during this period are apparent in the diverse conclusions reached in different reports. See National Park Service, *State of the Parks—1980,* 36; National Academy of Sciences, *Science and the National Parks,* 6–7, 77–79; Wagner et al., *Wildlife Policies,* 94–96; and National Parks and Conservation Association, *Research in the Parks,* 33–35. Regarding fully professional resource managers, a 1995 internal document stated that the Service had "just under 500 permanent and temporary full-time" such positions located in the parks. Bob Krumenaker and Abby Miller, "The Natural Resource Management Challenge: The NR-MAP Report," March 3, 1995, typescript, 4, copy courtesy of Robert J. Krumenaker.

9. National Academy of Sciences, *Science and the National Parks,* 9; National Academy of Sciences, National Research Council, "A Report by the Advisory Committee to the National Park Service on Research," August 1, 1963, typescript, x; David A. Haskell, "Is the U.S. National Park Service Ready for Science?" *George Wright Forum* 10, no. 4 (1993), 102.

10. Shenandoah National Park, "Natural Resource Inventory and Long-Term Ecological Monitoring System Plan," 9–15; Krumenaker and Miller, "The Natural Resource Management Challenge," 3–5.

11. Carlsbad Caverns National Park, "Cave and Karst Management Plan," 1995, typescript, 34–37, copy courtesy of Dale Pate; discussions with Ronal C. Kerbo and Dale Pate, January 29, 1996. Lechuguilla's numerous vertical inclines and other conditions that require advanced technical caving skills form a barrier to extensive public use.

12. Rolf O. Peterson, *The Wolves of Isle Royale: A Broken Balance* (Minocqua, Wisconsin: Willow Creek Press, 1995), 165–188; Rolf O. Peterson and Robert J. Krumenaker, "Wolves Approach Extinction on Isle Royale: A Biological and Policy Conundrum," *George Wright Forum* 6, no. 1 (1989), 10–15; Gary E. Davis and William L. Halvorson, *Science and Ecosystem Management in the National Parks* (Tucson: University of Arizona Press, 1996), 74–95; discussion with Robert J. Krumenaker, February 9, 1996. Possible causes for the drop in the wolf population include diminished food supply, decline in genetic variability, and a recently introduced canine disease.

13. National Academy of Sciences, *Science and the National Parks,* 27. The most thorough Park Service study of the northern range is Douglas B. Houston, *The Northern Yellowstone Elk: Ecology and Management* (New York: Macmillan Publishing Co., 1982). A more general account is found in Don Despain et al., *Wildlife in Transition: Man and Nature on Yellowstone's Northern Range* (Boulder, Colorado: Roberts Rinehart, 1986). For accounts of recent research, see Francis J. Singer, "Yellowstone's Northern Range Revisited," *Park Science* 9 (Fall 1989), 18–19; and Research Division, Yellowstone National Park, "Interim Report, Yellowstone National Park Northern Range Research," April 1992, typescript, YELL. Forthcoming is Douglas B. Houston and Margaret Mary Meagher, *Yellowstone and the Biology of Time: Photographs Across the Century,* to be published by the University of Oklahoma Press in 1998. In addition to Alston Chase, *Playing God in Yellowstone: The Destruction of America's First National Park* (Boston: Atlantic Monthly Press, 1986), critical analysis of the natural regulation policy includes Fred-

eric H. Wagner et al., *Wildlife Policies* (Washington, D.C.: Island Press, 1995), see for example 48–58, 127–134; and Charles Edward Kay, "Yellowstone's Northern Elk Herd: A Critical Evaluation of the 'Natural Regulation' Paradigm," Ph.D. diss., Utah State University, 1990.

14. Chase, *Playing God in Yellowstone*, 247–260; Karl Hess, Jr., *Rocky Times in Rocky Mountain National Park: An Unnatural History* (Niwot: University Press of Colorado, 1993), 77–88, 98–100.

15. A discussion of the definition of ecosystem management and the concept's potential is found in National Park Service, "Ecosystem Management in the National Park Service," September 1994, typescript, 1–8, copy in author's files.

16. The Greater Yellowstone Ecosystem (sometimes referred to as the Greater Yellowstone Area) has been scrutinized and administered by a complex of coordinating associations, among them the Greater Yellowstone Coordinating Committee and the Interagency Grizzly Bear Study Team. The latter was established by the Interior Department in 1973 and directed first by the Grizzly Bear Steering Committee, then by the Interagency Grizzly Bear Committee. The Greater Yellowstone Coalition was formed in 1983 to represent major environmental organizations. See Chase, *Playing God in Yellowstone*, 159, 180, 363–367. Gray wolf recovery is being coordinated by the Northern Rocky Mountain Wolf Recovery Team, in cooperation with the U.S. Fish and Wildlife Service. Analysis of many aspects of the Greater Yellowstone Ecosystem is found in Robert B. Keiter and Mark S. Boyce, eds., *The Greater Yellowstone Ecosystem: Redefining America's Wilderness Heritage* (New Haven: Yale University Press, 1991).

17. The Greater Yellowstone Coordinating Committee, "The Greater Yellowstone Postfire Assessment," March 1989, typescript, vii–ix, 1–4, YELL; Special Directive 89–7, Acting Director to Directorate, Field Directorate, WASO Division Chief and Park Superintendents, with attachments, July 12, 1989, YELL; Yellowstone National Park, "Yellowstone National Park Wildland Fire Management Plan," March 1992, 16–17, 47–61, YELL. A critical analysis of Yellowstone's fire policies is found in Stephen J. Pyne, "The Summer We Let Wild Fire Loose," *Natural History* [98] (August 1989), 45–49. A perspective from within the Service is Paul Schullery and Don G. Despain, "Prescribed Burning in Yellowstone National Park: A Doubtful Proposition," *Western Wildlands* 15 (Summer 1989), 30–34.

18. U.S. Fish and Wildlife Service and the Northern Rocky Mountain Wolf Recovery Team, "Northern Rocky Mountain Wolf Recovery Plan," August 3, 1987, iv–vi, 1–11; Northern Rocky Mountain Wolf Recovery Team, "Yellowstone Wolf Tracker: A Monthly Bulletin on Wolf Recovery in Yellowstone" (April 1995), YELL.

19. Partnership and external programs are discussed in Dwight F. Rettie, *Our National Park System: Caring for America's Greatest Natural and Historic Treasures* (Urbana: University of Illinois Press, 1995), 7, 39, 172–173.

20. National Park Service, Vail Agenda, 1, 4, 128, 137.

21. National Park Service, Vail Agenda, 11, 29, 104, 106, 111.

22. National Park Service, Vail Agenda, 18.

23. National Park Service, Vail Agenda, 105–121, 123, 124–142. For comparison see National Park Service, *State of the Parks—1980*, viii–ix, 20–23; National Parks and Conservation Association, *From Vignettes to a Global View*, 4–12; National Academy of Sciences, *Science and the National Parks*, 8–13, 87–111. The academy (pp. 41–57) notes the similarity of recommendations made over time.

24. National Park Service, Vail Agenda, 95–97. Even though the Agenda vacillated in its commitment to full compliance by the Service, it recommended (p. 126) "more effective and positive use" of environmental laws to deal with problems *outside* park boundaries—apparently seeking full compliance by other land-managing bureaus.

25. National Park Service, Vail Agenda, 113; Jake Hoogland, "Defending the Pristine Canyons," and "NEPA Compliance—What Have We Done?" in *Courier* 35 (June 1990), 16, 9. Two years after official publication of the Vail Agenda, national park authorities Robert and Patricia

Cahn commented on the Service's continuing reluctance to give full-faith compliance to the National Environmental Policy Act. See "Policing the Policy," *National Parks* 69 (September–October 1995), 37–41.

26. Even as the Vail conference was urging environmental leadership, the Park Service was cooperating with a major chemical company in a recycling program for a number of parks. This resulted in promotional advertisements on national television, featuring the company's involvement with Everglades National Park and suggesting a kind of innocence by association—although chemical pesticides have been a major cause of species extinction nationwide and around the world, and portions of Everglades itself have been devastated by use of toxic, environmentally harmful pesticides and chemical fertilizers on nearby lands. Everglades National Park, "Recycling Proposal for Everglades National Park," 1993, EVER.

The damaging effects of agricultural chemicals draining from adjacent lands into Everglades had even been recognized in the park's own literature. A 1989 park report stated that an estimated thirty thousand acres of "native Everglades wetlands have already been destroyed due to the introduction of nutrient-rich water [saturated with fertilizer ingredients such as nitrogen, phosphorus, and potassium] from upstream agricultural lands." Statement by hydrologist Daniel Scheidt, in National Park Service, "Everglades National Park: Status of Major Issues," February 1, 1989, EVER. Among many other examples, see National Park Service, "An Assessment of Fishery Management Options," January 1979, 9, EVER. See also Rettie, *Our National Park System,* 188–193, for a discussion of the benefits and problems resulting from corporate support of park programs. Although not specifically addressing pesticide concerns, Rettie comments (p. 192) that in such commercial arrangements the Service should be "sensitive to . . . constructive association." He adds as examples that commercials for cigarettes and liquor "should not be associated with the national park system."

27. Samuel P. Hays, in collaboration with Barbara D. Hays, *Beauty, Health, and Permanence: Environmental Politics in the United States, 1955–1985* (Cambridge: Cambridge University Press, 1987), details broadening environmental interests in the post–World War II era. (A succinct statement is found on pp. 21–39.)

28. Such caution can influence the outcome of the ecosystem management efforts. Park Service reluctance to speak out even on issues directly threatening parks is discussed in Joseph L. Sax and Robert B. Keiter, "Glacier National Park and Its Neighbors: A Study of Federal Inter-Agency Cooperation," in David J. Simon, ed., *Our Common Lands: Defending the National Parks* (Washington, D.C.: Island Press, 1988), 175–240. Soon after the Vail conference, veteran Park Service manager Richard B. (Rick) Smith chided the Service for not having taken a "strong stand on major environmental issues such as overgrazing of public lands, irresponsible mineral development, or the failure to add to the nation's Wilderness Preservation System." Rick Smith, letter to the editor, *Park Science* 12, (Spring 1992), 14.

29. The Park Restoration and Improvement Program did not provide much funding beyond what had been regularly appropriated. A discussion of park construction budgets is found in Rettie, *Our National Park System,* 180–183.

30. Rettie, *Our National Park System,* 7, 172–173. Under President Jimmy Carter, the Bureau of Outdoor Recreation had been abolished. Its functions were taken over by the newly created Heritage Conservation and Recreation Service, which also assumed direction of the Park Service's National Register and related external cultural resource programs. All of these activities were returned to the Park Service in 1981 by Secretary Watt. See William R. Lowry, *The Capacity for Wonder: Preserving National Parks* (Washington, D.C.: Brookings Institution, 1994), 71; Napier Shelton and Lissa Fox, *An Introduction to Selected Laws Important for Resources Management in the National Park Service* (Washington, D.C.: National Park Service, 1994), 23–24.

31. Denver Service Center, "Professional/Technical Staffing of Denver Service Center, 1986–1992," Falb. The employee figures represent numbers of individuals on board; some were in less-than-full-time and some in temporary positions.

32. National Academy of Sciences, *Science and the National Parks*, 54, 60, 105, 111; National Park Service, Vail Agenda, 42, 104.

33. Figures for Carlsbad Caverns National Park provide an example of the scope of tourism-related management and economics in just one park. In 1991 the 46,766-acre park contained 76 known caves and 33,125 acres of wilderness. Visits to the park that year totaled 679,450, with daily figures as high as 5,000—people mostly there to see the vast cavern for which the park had been established in 1923. To accommodate the public, the park had eleven miles of paved roads, with pullouts and roadside exhibits; ten miles of gravel roads; a picnic area; thirty miles of backcountry trails; three miles of cavern trails; four elevators descending 750 feet into the caverns; a restaurant in the caverns; a large visitor center with exhibits, theater, restaurant, souvenir and book sales areas, a nursery, and a kennel; and water, electrical, and sewage systems.

In 1991 the park had the equivalent of ninety-five full-time employees; the concessionaire employed about thirty-five people year-round, plus fifty-five for the summer season. The park's support organization (or "cooperating association") employed eighteen people in permanent or seasonal positions. Total payroll for all employees amounted to about $3.3 million. (Three of the full-time Park Service employees were natural resource managers.)

Much of the payroll went for living expenses in local communities. In addition, in 1991 tourism to Carlsbad Caverns generated approximately $50 million in benefits to the local economy (the local area defined as within about a hundred-mile radius), $5 million in "secondary" benefits, and $3.5 million in increased tax revenues. Tourism-related jobs in the area were estimated at 1,636. Glen Kaye, "New Mexico Parks and Their Economic Impact," 1992, typescript, copy courtesy of the author; discussions with Ronal C. Kerbo and Dale Pate, January 29, 1996.

34. Before the social revolutions of the 1960s and 1970s, all Park Service leaders were white males. In the 1970s, women and minorities began to attain leadership roles; however, their ascendancy seems not to have had any effect on overall natural resource policy and practice—mainly, it diversified the composition of the decisionmaking cadres. Through early 1997, with women and minorities serving as superintendents, regional directors, and associate and deputy directors, the Park Service directorship remained the last bastion of the white male. A history of women's involvement with the national parks is found in Polly Welts Kaufman, *National Parks and the Woman's Voice* (Albuquerque: University of New Mexico Press, 1996).

35. In a 1993 article on resource management, park superintendent Jonathan B. Jarvis remarked critically that "there are *cultural* barriers within the NPS that prevent research and resource information from playing a significant role in management decision making." Jarvis provided his own list of cultural attributes, among them "old school, scenery management," superintendents' lack of understanding of research, and the assignment of resource managers to ranger divisions. Jonathan B. Jarvis, "Action vs. Rhetoric: Resource Management at the Crossroads," *Park Science* 13 (Summer 1993), 10. See also Jonathan B. Jarvis, "Principles and Practices of a Research and Resource Management Program," *George Wright Forum* 8, no. 3 (1991), 2–11. In recent years, for example, Everglades Superintendent Michael V. Finley was recognized by the Florida Audubon Society as government Conservationist of the Year; and the National Parks and Conservation Association has honored superintendents Bill Wade (Shenandoah), Robert Barbee (Yellowstone), and Regional Director Boyd Evison with its Stephen Tyng Mather Award for achievements in resource protection and preservation. Mather Award information from Laura Loomis, National Parks and Conservation Association.

36. Joseph Sax and Robert Keiter, in "Glacier National Park and Its Neighbors," 175–240, reveal the persistence of tradition in park management. The powerful allegiance a culture can have to its fundamental assumptions has been analyzed by sociologist Edgar Schein, who states that "to understand a group's culture, one must attempt to get at its shared basic assumptions. . . . Once a set of shared basic assumptions is formed . . . it can function as a cognitive defense mechanism. . . . [In defense of a group's cultural values], it is easier to distort new data by denial, projection, rationalization, or various other defense mechanisms than to change the basic assump-

tion. . . . Culture change, in the sense of changing basic assumptions is, therefore, difficult, time consuming, and highly anxiety provoking." Edgar H. Schein, *Organizational Culture and Leadership* (San Francisco: Jossey-Bass Publishers, 1992), 26.

37. A list of annual numbers of visits throughout the system is found in Rettie, *Our National Park System,* 252–254. On the Service's continual popularity with the public, see 125, 147n.

38. National Parks and Conservation Association, *From Vignettes to a Global View,* 6; National Academy of Sciences, *Science and the National Parks,* 10, 88; National Park Service, Vail Agenda, 134. Perhaps revealing an uncertainty about the role of science in park management, the Agenda mentioned (p. 29) that the Park Service was the "*apparent* natural home of scientific resource management and research" (emphasis added).

39. National Park Service, *Science and the National Parks II: Adapting to Change* (Washington, D.C.: National Park Service, [1993]), viii, 10. Among others, the committee included former Alaska regional director Boyd Evison, Sequoia superintendent Thomas Ritter, and Big Bend superintendent Robert Arnberger. In a somewhat similar vein, Park Service natural resource manager Robert J. Krumenaker observed earlier that "without a specific mandate for research we have backed into it through a need for information." He asserted that an "*internal* mandate clearly exists . . . even if many managers choose not to see it." Bob Krumenaker, "Resource Management and Research in the NPS: An Uneasy Relationship," *Ranger* 7 (Spring 1991), 11.

40. National Parks and Conservation Association, *A Race against Time,* 17, 25.

41. As the international programs evolved in the 1970s, the majority of requests from other nations were for planners and landscape architects to advise on the development and management of parks and protected areas, many newly authorized. Once these countries developed their own planning expertise, such requests tapered off and the new focus was assistance in interpretation. At the peak of the programs in the early 1980s, an average of about 175 requests were received annually. Then, by mid-decade, shortages of funds and staffing (both foreign and domestic—the State Department funded most Park Service efforts) brought a decline in international programs. Limited assistance to foreign countries continues to be provided in fields such as park management, ranger and protection work, and interpretive activities. Robert C. Milne, interview with the author, September 25, 1995. The National Park Service, in the Vail Agenda, 87, asserts that the "evolution, growth, and development of NPS interpretation has been one of the most significant contributions that the agency has made in the world park movement." See also 77–78.

42. Department of the Interior, news release, "Scientific Research to Be Reorganized under National Biological Survey at Interior Department," April 26, 1993, Office of the Secretary, Department of the Interior. See also Wagner et al., *Wildlife Policies,* 107–110, for speculation on "potential gains and losses" from the new research arrangement; and Rettie, *Our National Park System,* 219–220.

43. National Park Service, *Restructuring Plan for the National Park Service* (Washington, D.C.: National Park Service, 1994), 1–3; National Park Service press release, "National Park Service Reorganization Marks Most Significant Organizational Change in Agency's 79-Year History," May 15, 1995, National Park Service, Office of Public Affairs, Washington, D.C.

44. National Park Service, Vail Agenda, 34.

45. National Park Service, Vail Agenda, 106.

Acknowledgments

When the opportunity arose to write the history of natural resource management in the national park system, my wife, Judy, urged me to seek the assignment. And when I discussed the idea with John E. Cook, then director of the National Park Service's Southwest Regional Office in Santa Fe, he readily and enthusiastically agreed. Through funding allocations and other administrative decisions, John cleared the way for this project. When politics within the Service several times threatened early termination, he made sure the project was given the time and thought necessary, asking only that I proceed at all deliberate speed. John is a third-generation Park Service employee, who loves the Service more than anyone I know; yet his devotion to its many traditions never led him to interfere with my study of the Service's history or to place constraints on my writing. I shall always be grateful for his generous support and trust.

Serving as a volunteer for the National Park Service while taking a leave of absence from her position as librarian with the Museum of International Folk Art in Santa Fe, Judy Sellars accompanied me on numerous research trips, becoming closely familiar with the topics being studied and giving invaluable assistance with the research. Throughout the project, during countless conversations on the road and at home, she provided important insights that helped me revise and refine this analysis of Park Service history. She also commented on each chapter of the manuscript. I am thankful for her continued support and encouragement.

Although I am responsible for the research and writing, this study was in many ways a broad cooperative effort. Dozens of generous people gave their time and support to nurture the project to conclusion. They critiqued all or most of the manuscript, or gave me advice and support at crucial times during my work. Special thanks go to Edwin C. Bearss, William E. Brown, Susan L. Flader, Jane N. Harvey, F. Eugene Hester, Robert J. Krumenaker, Barry Mackintosh, Charles H. McCurdy, Mary Meagher, George J. Minnucci, Jr., Jerry L. Rogers, Jay Shuler, William R. Supernaugh, and Robert M. Utley. I am indebted to JoAnn Y. Ortiz and Rebecca M. Post, who undertook a wide variety of secretarial tasks to move this work along. The library staff of the National Park Service's Santa Fe office was always exceedingly helpful. I am grateful to Amalin Ferguson, Peg Johnson, and Michael Gonzales for all their efforts on my behalf. And for their support in the metamorphosis of my manuscript into publication, I thank especially Charles Grench, editor in chief of Yale University Press, and Margaret Otzel, production editor. I also thank copyeditor Vivian B. Wheeler and indexer Laura Moss Gottlieb for their excellent work.

I am grateful for the special support I received from Roland T. Bowers, Susan P. Bratton, Victor H. Cahalane, Judy Chetwin, Vanessa Christopher, John Conoboy, Gary E. Davis, John G. Dennis, Tom DuRant, Linda Eade, S. Norman Ewanciw, Nancy Potts Flanagan, Milford Fletcher, Vivian M. Garber, John W. Henneberger, Robert J. Howard, Glen Kaye, Ronal C. Kerbo, Gerald D. McCrea, David Nathanson, Dwight T. Pitcaithley, James B. Snyder, and Roger Young.

Many others made contributions to this effort: Craig D. Allen, Leslie P. Arnberger, Robert L. Arnberger, Lisa Backman, Robert D. Barbee, Barbara Baroza, Reginald H. Barrett, Robert Beauchamp, James M. Benedict, Mark S. Boyce, Jim Brady, Jeffrey Bradybaugh, Willys E. Bramhall, Edgar B. Brannon, Jr., Richard H. Briceland, Harry A. Butowsky, Clara Cassidy, Howard H. Chapman, Stuart E. Coleman, Wayne B. Cone, Richard (Rick) Cook, Tim Coonan, Bruce Craig, Richard Crawford, Mary Shivers Culpin, Richard Curry, John P. Debo, Jr., Don Despain, Michael J. Devine, Rolf Diamant, Russell E. Dickenson, Lary M. Dilsaver, James J. Donoghue, David J. Donohue, F. Dominic Dottavio, Tom Ela, Boyd Evison, Rick Ewig, Richard D. Falb, Michael V. Finley, Ronald A. Foresta, Dawn McGilvrey Foy, Denis P. Galvin, Marshall Gingery, Art Gómez, Jerome A. Greene, Linda W. Greene, William P. Gregg, Neal G. Guse, Jr., Aubrey Haines, William L. Halvorson, Wayne Hamilton, Shirley Harding, David Harmon, Annette Hartigan, Thomas L. Hartman, Peter Hayden, Gary Y. Hendrix, Ruthanne Heriot, Darlene S. Herrera, Marjorie Hoffman, David

Humphrey, Jonathan B. Jarvis, Einar Johnson, Renee Jussaud, Bruce Kaye, Richard Keigley, Bruce M. Kilgore, Zack Kirkland, Samuel H. Kunkle, Keith R. Langdon, Gary L. Larson, Robert M. Linn, Laura Loomis, Lourdes Lujan, John Mack, Manuel Mandell, Kitty Manscill, Stephen Mark, Richard W. Marks, Robert Martin, Clifford Martinka, Mary Maruka, Henry E. McCutchen, Brian McHugh, Abigail Miller, Sue Mills, Robert C. Milne, Rene C. Minnick, Ben Moffett, Kenneth Morgan, Ernest Morrison, Jody Morrison, Duncan Morrow, Louise Murie-MacLeod, Susan Myers, Dan Nealand, O. V. Olsen, Russ Olsen, Gordon C. Olson, Mary Padilla, Dale Pate, Jim Peaco, John Peine, Grant A. Petersen, Randall R. Pope, L. Lee Purkerson, Edie Ramey, Charles E. Rankin, Donald W. Reeser, Linda Ries, John Thomas Ritter, William B. Robertson, Molly N. Ross, Edwin L. Rothfuss, Jimmy Rush, Joseph L. Sax, Paul Schullery, Susan Schultz, Marvel Schumacher, Ellen Seeley, C. Mack Shaver, Napier Shelton, Steven Slack, Richard B. (Rick) Smith, Michael Soukup, Laura Soulliere, George Sprugel, Jr., Howard R. Stagner, Barbara R. Stein, Robert T. Steinholtz, Tom Tankersley, Julie Thomas, Ben H. Thompson, Vivie Thue, Ed Trout, Kent Turner, John D. Varley, Stephen D. Veirs, Jr., Gary S. Waggoner, Frederic H. Wagner, William H. Walker, Jr., Glennie Wall, Louis S. Wall, Jim Walters, Douglas Warnock, Janelle Warren-Findley, Thomas H. Watkins, Michael D. Watson, Betty Wauer, Roland H. Wauer, Robert A. Webb, David J. Weber, Donald H. Weir, Jannette Wesley, Beverly Whitman, Lee Whittlesey, Arthur Wilcox, Robin W. Winks, Kenna Wood, Terry Wood, Bebe Woody, David G. Wright, R. Gerald Wright, Keith A. Yarborough, and Barbara Zafft. Beyond those listed above, I thank the many staff members of the National Park Service's Santa Fe office who gave me support.

In addition to hundreds of informal discussions, I interviewed the following: Leslie P. Arnberger, Robert L. Arnberger, Robert D. Barbee, James M. Benedict, Mark S. Boyce, Jeffrey Bradybaugh, Susan P. Bratton, Richard H. Briceland, Victor H. Cahalane, Stuart E. Coleman, John E. Cook, Tim Coonan, Richard Curry, John P. Debo, Jr., John G. Dennis, F. Dominic Dottavio, Tom Ela, Michael V. Finley, Milford Fletcher, William P. Gregg, Neal G. Guse, Jr., Aubrey Haines, Thomas L. Hartman, Peter Hayden, Gary Y. Hendrix, John W. Henneberger, F. Eugene Hester, David Humphrey, Einar Johnson, Bruce M. Kilgore, Zack Kirkland, Robert J. Krumenaker, Samuel H. Kunkle, Keith R. Langdon, Robert M. Linn, Richard W. Marks, Robert Martin, Charles H. McCurdy, Brian McHugh, Mary Meagher, Kenneth Morgan, Russ Olsen, John Peine, Randall R.

Pope, L. Lee Purkerson, William B. Robertson, Molly N. Ross, Edwin L. Rothfuss, Richard B. (Rick) Smith, James B. Snyder, Laura Soulliere, George Sprugel, Jr., Howard R. Stagner, William R. Supernaugh, Ben H. Thompson, Ed Trout, Kent Turner, Stephen D. Viers, Jr., William H. Walker, Jr., Douglas Warnock, Roland H. Wauer, Arthur Wilcox, Garree Williamson, David G. Wright, R. Gerald Wright, and Keith A. Yarborough.

Index